ADVANCES IN BIOLOGICAL TREATMENT OF LIGNOCELLULOSIC MATERIALS

Proceedings of a Workshop on Advances in Biological Treatment of Lignocellulosic Materials, held in Lisbon, Portugal, from 25 to 27 October, 1989, under the auspices of COST (European Cooperation in Scientific and Technical Research)—COST 84-bis, organized with the support of the Commission of the European Communities by Departamento de Tecnologia de Indústrias Alimentares, Laboratório Nacional de Engenharia e Tecnologia Industrial (DTIA-LNETI), Ministerio da Indústria e Energia, Lisbon, Portugal.

ADVANCES IN BIOLOGICAL TREATMENT OF LIGNOCELLULOSIC MATERIALS

Edited by

M.P. COUGHLAN

Department of Biochemistry, University College, Galway, Ireland

and

M.T. AMARAL COLLAÇO

DTIA-LNETI, Lisbon, Portugal

ELSEVIER APPLIED SCIENCE
LONDON and NEW YORK

ELSEVIER SCIENCE PUBLISHERS LTD
Crown House, Linton Road, Barking, Essex IG11 8JU, England

Sole Distributor in the USA and Canada
ELSEVIER SCIENCE PUBLISHING CO., INC.
655 Avenue of the Americas, New York, NY 10010, USA

WITH 75 TABLES AND 93 ILLUSTRATIONS

British Library Cataloguing in Publication Data

Advances in biological treatment of lignocellulosic
materials.
1. Lignocellulosics. Microbiological treatment
I. Coughlan, M. P. (Michael P.) II. Collaço, M. T. Amaral
661.802

ISBN 1-85166-542-0

Library of Congress CIP data applied for

Publication arrangements by Commission of the European Communities, Directorate-General Telecommunications, Information Industries and Innovation, Scientific and Technical Communication Unit, Luxembourg

EUR 12671

Printed in Northern Ireland by The Universities Press (Belfast) Ltd.

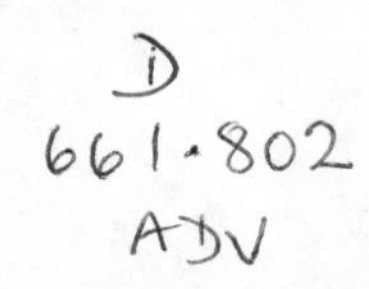

Contents

Session II: Changes in Lignocellulosic Materials during Biological Treatment

Session III: Bioconversion of Lignocellulosic Materials in Submerged and Solid-State Cultivation and in Reactors

Session IV: Bioconversion of Lignocellulosic Materials *in vivo*

Session V: Use of White-Rot Fungi for Food, Feed and Industrial Purposes

INTRODUCTION TO THE LISBON WORKSHOP

In the late 1960s the European Community launched the idea of facilitating the scientific endeavours of the countries of Europe by promoting a flexible set of arrangements for scientific cooperation. This initiative led to the creation of COST (Cooperation in Scientific and Technical Research) with the participation of the 12 Member Countries, the EFTA Countries, Turkey and Yugoslavia. In 1971 the first 7 COST concerted actions were implemented. This number has now increased to more than 60. COST 84 bis, one such activity, coordinates ongoing multidisciplinary research, within the Community and other contributing countries, on the use of lignocellulose-containing byproducts and other plant residues for animal feeding and industrial purposes. This it does by the holding of regular committee meetings at which appropriate representatives participate; by providing funding for the exchange of personnel between laboratories engaged in relevant research; by assisting the setting up and operation of centres of excellence in specific analytical techniques to which investigators may send/bring samples for analysis; and by the provision of funding for the holding of Workshops at regular intervals.

Several Workshops, dealing with a range of relevant topics, have been held since the foundation of COST 84 bis. The theme of the sixth such Workshop, held in Lisbon (October 25-27, 1989), was "Advances in Biological Treatment of Lignocellulosic Materials." Twenty three papers dealing with a variety of topics within the general theme were presented - and, as in previous Workshops, led, as indeed they should, to vigorous discussion. Each of the papers presented is included in this proceedings as are the summary reports by the relevant chairmen of each session.

Huge amounts of lignocellulosic wastes and residues, of agricultural, forest, industrial and domestic origin are generated annually. Such materials are comprised for the most part of cellulose, hemicellulose and lignin. Clearly, the successful exploitation of the potential of lignocellulosic substances, as sources of animal or human feedstuffs or chemical feedstocks, requires that each of these polymers be utilized to the fullest extent possible. For various reasons, including environmental considerations, biological rather than chemical conversion is the preferred route. This, in turn means that an understanding of the organisms involved, and their relevant enzyme systems, is *sine qua non.* Thus, the theme of the Lisbon Workshop was timely.

The opening lecture of the Workshop reported on the promising results obtained on using carbohydrases to increase the nutritional value of straw as fodder or as a fodder supplement. Session I dealt with solid-state fermentation of straw with white-rot fungi. Emphasis was placed on new developments in composting procedures, the production of animal feeds and chemical feedstocks, the problems attendant on scale-up of solid-state processes, the possible utility of the soluble lignocellulose produced during fermentation, and on the modelling of the physical process parameters during lignin degradation.

The ultrastructural changes accompanying the biodegradation of plant materials are, as yet, poorly understood. This was essentially the topic of Session II. Papers presented included, the use of various chemical and spectroscopic techniques for studying the effects of chemical and biological treatment on the composition and ultrastructure of lignocelluloses, and procedures for the rapid determination of substrate quality during solid-state fermentation. Session III dealt with the production of enzymes during liquid- and solid-state fermentation of straws and pulps by fungi, the enrichment of seed shells with protein, the use of mixed cultures in bioconversion, the hydrolysis of cellulose in enzyme reactors, and the effects of various parameters on the operational stability of the enzymes involved. Topics discussed in Session IV included, the modelling of rumen processes in vitro and in vivo, the role of anaerobic fungi in the degradation of lignocellulosics, the use of white-rot fungi to increase forage digestibility, and the contribution of rumen fungi and bacteria to the degradation of straw. The last Session dealt specifically with the cultivation of edible fungi on plant residues, the use of white-rot fungi and their enzyme systems in biopulping and biobleaching, the decontamination of polluted air and soils using biofilters based on white-rot fungi growing on straw, and various aspects of the use of such fungi in upgrading lignocellulosic wastes to food, fodder and compost-based products.

As we have said before, the organization of a scientific meeting, the editing and retyping of manuscripts and the preparation of proceedings for publication, cannot be done without the generous assistance of others who are whole-heartedly committed to the project. For this reason we are pleased to acknowledge our gratitude to Dr. Peter Reiniger, Secretary of COST 84, CEC, to Instituto de Promoção Turistica, Lisbon, CFT do LNETI, Editorial do LNETI and all of the staff of DTIA-LNETI, to Sandy Lawson, for her typing excellence, to Rita Richardson, Yvonne Egan and Dorothy Fox who, with unfailing good humour, took care of mountains of FAX messages, and, to our families who saw little of us for several months.

M. Teresa Amaral Collaço and Michael P. Coughlan

Lisbon and Galway

OPENING LECTURE

ENZYME TREATMENT OF CROP RESIDUES

E.R. Ørskov and Y. Nakashima
The Rowett Research Institute,
Bucksburn, Aberdeen AB2 9SB, Scotland

The results of an investigation of the use of crude preparations of enzyme mixtures (see activities below) to preserve and/or upgrade straw is discussed. When moist straws (600 $g.kg^{-1}$) were stored at room temperature with the enzyme mixtures for 30 days the pH was reduced due to fermentation of sugars released by the enzyme action and subsequent production of lactic and volatile fatty acids. If fermentation of the released sugars was prevented by the addition of propionic acid both the solubility of straws and the potential extent of fermentation assessed by *in sacco* incubation was increased. Greater amounts of soluble sugars were released from leaves than from stems of temperate cereal straws. The opposite was found to be the case with rice straw. With combinations of alkali, oxidative and enzyme treatment it is possible to produce high quality feeds from straw.

INTRODUCTION

In this paper the discussion will centre on biological treatment of straw, not by use of micro-organisms, but by use of cell-free enzyme mixtures. Essentially, it is a summary of work carried out during the past 2 years. However, the use of crude enzymes to improve the nutitive value of forages is not a new concept. It has been applied mainly to preserves forages such as silage. In silage, enzymes sometimes have the advantage of increasing the acidity due to the release of sugar from ß-linked polysaccharides that are subsequently fermented to yield lactic and volatile fatty acids (see Henderson *et al.*, 1982 and Bertin *et al.*, 1985). Sometimes, enzyme treatments are also claimed to increase digestibility of silage. In our work we concentrated on the application of enzymes to straw and examination of the degradation characteristics.

The use of enzymes as a possible method of upgrading straw has some advantages. The most important of these is that there are no undesirable chemical end-products. To be successful the preparations used must contain a wide spectrum of enzymes. The commercial preparation (Meicelase) used in these studies contained the activities listed in Table 1.

Table 1. Activities exhibited by Meicelase (from Nakashima *et al.*, 1988).

Substrate	pH of test	Activity (units.g^{-1})
Xylan	5.0	106.3
ß-1,4-galactan	5.0	4.8
Carboxymethylcellulose	5.0	53.9
Avicel	5.0	37.9
Polygalacturonic acid	4.5	5.4
Pectin	4.5	20.9
Starch	4.5	0.4
Arabinan	4.5	2.4
Lichenan	5.0	55.0
Barley mixed link glucan	4.5	3619.5

Table 2. Effects of concentration of polysaccharidase enzymes, moisture contents and particle size on ensiling characteristics of rice straw.

Ensiling concentrations	Final pH	Organic acid (g.kg^{-1} fresh) Lactic	Acetic	Butyric
Cellulase concentrations (g.kg^{-1} DM)				
0	5.21	1.98	3.90	1.24
5	4.87	3.90	2.71	0.57
10	4.82	3.67	3.09	0.21
Significance of linear trend	**	**	*	**
Moisture contents (g/kg)				
500	5.44	2.34	2.06	0.14
600	5.00	3.04	3.33	0.58
700	4.46	4.17	4.31	1.30
Significance of linear trend	**	**	**	**
Particle size (mm)				
20	5.33	1.67	3.38	1.12
5	4.74	4.49	2.56	0.30
2	4.84	3.39	3.75	0.60
Significance of linear trend	**	**	*	**

* = $P<0.01$; ** = $P<0.001$

In the first trials rice straw was used. The effects of moisture, enzyme concentrations and particle size were examined. The results in terms of final pH and concentrations of lactic and volatile fatty acids after 30 days of incubation at 20°C are given in Table 2. The concentrations of acids were increased by increasing the amount of enzyme used, by increasing the moisture content and by reducing the particle size of the straw. Measurement of degradation characteristics using the nylon bag technique showed that the effects of moisture and particle size were relatively small. The effects of enzyme concentrations are given in Table 3. Solubility as well as the 48 h dry matter loss increased with the addition of enzyme. However, the total digestive potential of the straw, i.e. the asymptote (a + b) from the exponential equation, $p = a + b(1 - e^{-ct})$, where (p) is degradation at time (t) (Ørskov and McDonald, 1979) did not increase. By contrast, the rate constant c was increased. In other words the solubility and the rate of degradation were increased but the extent of degradation was not increased. While intake studies were not carried out on the samples, Ørskov *et al.* (1988) had shown clearly that, as the rate constant increased, food intake increased even if digestibility did not.

Table 3. Effects of the concentration of polysaccharidase enzymes on the solubility, dry matter loss (DML), maximum potential degradability (a + b) and the rate constant (c) of enzyme-treated rice straw in nylon bags in the rumen of sheep using the equation $p = a + b(1 - e^{-ct})$.

	Solubility ($g.kg^{-1}$ DM)	48 h DML ($g.100\ g^{-1}$)	(a + b) ($g.kg^{-1}$ DM)	c (fraction per h)
	Mean values			
Enzyme concentration ($g.kg^{-1}$ DM)				
0	152	477	624	0.0498
5	196	533	621	0.0677
10	212	565	628	0.0817
Significance of linear trend	**	**	*	**

*, $P<0.01$; **, $P<0.001$

Table 4. Effect of cellulase enzyme mixtures on degradation characteristics of stem leaf sheath and leaf blade from barley straw.

Botanical fraction	Enzyme addition ($g.kg^{-1}$)	pH	Solubility ($g.kg^{-1}$)	48-h loss ($g.kg^{-1}$)	Potential ($g.kg^{-1}$)	Rate constant ($fraction.h^{-1}$)
Stem	0	5.6	135	299	403	0.0244
Stem	5	4.7	174	293	383	0.0212
Leaf sheath	0	5.0	185	680	812	0.0347
Leaf sheath	5	4.8	290	677	832	0.0276
Leaf blade	0	5.2	227	775	851	0.0454
Leaf blade	5	4.8	461	789	834	0.0484

Nakashima and Ørskov (1990) also examined the effects of chemical and enzymic pretreatment on different botanical fractions of barley straw. The results for stems, leaf sheath and leaf blade are summarized in Table 4. The degradation with or without enzymes was compared. The pH of the straw silage after 30 d was consistently reduced and solubility, particularly of leaf, was increased. On the other hand, the potential degradability, the 48 h dry matter losses and the rate constants were not changed.

Table 5. Effect of chemical pretreatment followed by enzyme treatment on degradation characteristics of internode of barley straw in polyester bags in the rumen of sheep according to the equation $p = a + b(1 - e^{-ct})$.

Chemical pretreatment	Enzyme addition ($g.kg^{-1}$)	Solubility ($g.kg^{-1}$ DM)	48-h loss ($g.kg^{-1}$ DM)	Potential ($g.kg^{-1}$ DM)	c (fraction per h)
Untreated	0	136	299	403	0.0244
	5	174	293	383	0.0212
NaOH	0	117	283	380	0.0238
	5	179	332	423	0.0238
NaOH + H_2O_2	0	113	380	690	0.0139
	5	195	430	695	0.0145

The effects of treatment of stems (Table 5) and leaves (Table 6) with NaOH, with or without enzyme treatment, are shown above and below, respectively.

The effect of H_2O_2 here is very apparent. The application of enzymes increased solubility particularly for leaves, but, as before, while the potential was increased by chemical treatment it was not consistently increased using enzymes. In fact, in the case of leaves the rate constants were actually decreased.

We examined the actual loss of substrate during treatment with the enzyme preparation and the extent to which the loss could be reduced by inhibiting fermentation of the released sugar. Accordingly propionic acid was used in different concentrations to inhibit fermentation. Concentrations higher than 30 $g.kg^{-1}$ are not included in Table 7 since there was little effect on adding more. It is clear that dry matter losses were associated with fermentation of the solubilized sugars.

Table 6. Effect of chemical pretreatment followed by enzyme treatment on degradation characteristics of leaf and leaf sheath of barley straw in polyester bags in the rumen of sheep according to the equation $p = a + b(1 - e^{-ct})$.

Chemical pretreatment	Enzyme addition ($g.kg^{-1}$)	Solubility ($g.kg^{-1}$ DM)	48-h loss ($g.kg^{-1}$ DM)	Potential ($g.kg^{-1}$ DM)	c (fraction per h)
Untreated	0	185	680	810	0.0347
	5	290	677	832	0.0276
NaOH	0	181	735	875	0.0354
	5	376	733	888	0.0272
$NaOH + H_2O_2$	0	155	739	903	0.0334
	5	382	773	937	0.0275

It is also clear that propionic acid prevented most of such losses. As a result of preventing fermentation during enziling, it could be demonstrated that both the 48 h dry matter loss and total potential degradability were increased as a result of enzyme treatment.

A comparison between the effects of ammonia and enzyme treatments of leaf of rice straw is given in Table 8. The solubilized sugars were here preserved with propionic acid as for Table 7. While the solubility was greater with enzyme treatment, the 24 h dry matter loss was similar for both treatments although the potential was greatest with ammonia treatment. In Table 9, the same information is given for stems of rice straw which is generally more nutritious than leaf. Solubility was again greater for the enzyme-treated material. The 24 h loss was only little affected. Potential digestiblity was increased to almost the same extent by the two treatments.

The effectiveness of using enzymes as a practical method of improving value of lignocellulosic residues still remains to be tested. By separating straw into its botanical fractions, it was found that enzyme treatment is very effective for leaves. The

Table 7. Effects of addition of propionic acid on the dry-matter loss (DML) during fermentation.

Propionic acid addition ($g.kg^{-1}$)	Enzyme addition ($g.kg^{-1}$)	DML during fermentation ($g.kg^{-1}$)	Solubility ($g.kg^{-1}DM$)	48-h loss ($g.kg^{-1}DM$)	Potential ($g.kg^{-1}DM$)	c (fraction per h)
0	0	20.3	190	569	640	0.0471
	5	57.7	234	533	614	0.0418
10	0	44.8	185	556	635	0.0424
	5	15.3	257	543	692	0.0250
30	0	23.7	195	549	634	0.0399
	5	9.7	255	578	673	0.0347

improvement in the digestibility of straw leaf following enzyme treatment is so great that the product could be considered as a feed for monogastric animals, possibly even as a source of fibre in diets for humans. The use of cell free enzymes has the advantage over live microbes in that losses during ensiling are very small and can be almost totally prevented.

Table 8. Effects of ammonia and enzyme treatment on the washing loss and degradation characteristics of leaf from rice straw according to the equation $p = a + b(1 - e^{-ct})$ when incubated in polyester bags in the rumen of sheep.

	Solubility ($g.kg^{-1}$ DM)	24-DML ($g.kg^{-1}$ DM)	Potential ($g.kg^{-1}$ DM)	c (fraction per h)
Treatments				
Untreated	151	325	525	0.0342
Ammonia	169	375	586	0.0372
Cellulase	207	358	572	0.0294
s.e.d.		3.1	9.2	0.0019
Significance		**	**	**

Table 9. Effect of ammonia and cellulase treatment on the washing loss and degradation characteristics of stems of rice straw according to the equation $p = a + b(1 - e^{-ct})$ when incubated in polyester bags in the rumen of sheep.

	Solubility ($g.kg^{-1}$ DM)	24-DML ($g.kg^{-1}$ DM)	Potential ($g.kg^{-1}$ DM)	c (fraction per h)
Treatments				
Untreated	300	502	638	0.0487
Ammonia	317	531	683	0.0424
Cellulase	364	527	681	0.0342
s.e.d.		4.3	11.1	0.0400
Significance		**	**	**

Enzyme treatments are environment-friendly and it is probable that developments in the next few years will lead to further substantial progress. It is also possible that, with the help of biotechnology, some fungal organism can be engineered to produce appropriate enzyme mixtures at cost-effective prices. The disadvantage is that straw has to be treated in the wet form. On the other hand, a lot of wet paddy straw is wasted that could perhaps be adequately ensiled and preserved with the use of enzymes because the released sugars can produce sufficient acidity for preservation.

REFERENCES

Bertin, G., Hellings, Ph. and Vanbelle, M. (1985). The effect of cellulolytic enzyme preparations as *in vitro* improvement for forage digestibility. Processing and conservation of forages including leaf protein in research. In *Proceedings of the 15th International Grassland Congress,* Kyoto, Japan, pp. 9-20.

Henderson, A.R., McDonald, P. and Anderson, D. (1982). The effect of a cellulase preparation derived from *Trichoderma viride* on the chemical changes during the ensilage of grass, lucerne and clover. J. Sci. Fd. Agric. **33**, 16-20.

Nakashima, Y., Ørskov, E.R., Hotten, P.M., Ambo, K. and Takase, Y. (1988). Rumen degradation of straw. 6. Effect of polysaccharidase enzymes on the degradation characteristics of ensiled rice straw. Anim. Prod. **47**, 421-427.

Nakashima, Y. and Ørskov, E.R. (1989). Rumen degradation of straw. 7. Effects of chemical pretreatment and addition of propionic acid on the characteristics of botanical fractions of barley straw treated with a cellulase preparation. Anim. Prod. **48**, 543-551.

Nakashima, Y. and Ørskov, E.R. (1990). Rumen degradation of straw. 9. Effect of cellulase and ammonia treatment on different varieties of rice straw and their botanical fractions. Anim. Prod. in press.

Ørskov, E.R. and McDonald, I. (1979). The estimation of protein degradability in the rumen from incubation measurements weighted according to rate of passage. J. Agric. Sci. (Camb.), **92**, 499-503.

Ørskov, E.R., Reid, G.W. and Kay, M. (1988). Prediction of intake by cattle from degradation characteristics of roughages. Anim. Prod. **46**, 29-34.

SESSION I

Solid-state fermentation of plant residues with white-rot fungi

SESSION I

Solid state fermentation of plant residues with white-rot fungi

NEW DEVELOPMENTS IN INDOOR COMPOSTING: THE TUNNEL PROCESS

J.P.G. Gerrits and L.J.L.D. van Griensven
Mushroom Experimental Station
P.O. Box 6042, 5960 AA Horst, The Netherlands.

The emission of NH_3 and an offensive odour during phase I of mushroom compost preparation is increasingly considered as a nuisance. This paper presents results of three series of trials concerning mushroom substrate production in specially designed climatized rooms called "tunnels". A normal productive substrate (or compost) can be obtained from a homogeneous mixture of raw organic materials such as horse or chicken manure by maintaining the temperature of the air supply in a tunnel at 40-45°C. After one week the substrate is free from NH_3 and is selective for mushroom mycelium.

Additional pasteurization is detrimental and so the substrate has to be transported hygienically. The release of NH_3 depends on the total N level and must be kept to a minimum. The remaining NH_3 can be removed from the air by washing or biofiltration. The production of an offensive odour during this process is expected to be far less than that during the existing outdoor process.

INTRODUCTION

Mushroom compost is prepared in two phases. Phase 1 is an outdoor process in windrows (or stacks) based on spontaneous heating and natural aeration (Gerrits, 1988a). The NH_3 and odour produced during this process are increasingly being considered as a nuisance. Phase II is carried out in a mushroom house (in trays or shelves) or in bulk (in a tunnel). The latter process has been discussed previously (Gerrits, 1988b).

Mushrooms can be produced on a sterile substrate. A composting process is not a prerequisite for production (Huhnke *et al.*, 1965). Because sterilization is expensive, Huhnke (1972) tried to make the substrate selective by submitting it to a phase II process. Laborde *et al.* (1979) tried with variable success to prepare a horse manure compost, and Smith (1983) a straw compost using a strongly reducing phase I. Excessive temperatures in the compost layer occurred because activity could not be controlled by the ambient air. This problem had already been observed by Lambert (1941) who suggested that "the outdoor composting process may be considered merely as a means of mixing and moistening the manure while carrying it through the initial explosive fermentation to reduce the tendency to overheat and lose moisture during the subsequent prolonged sweating out".

This problem is largely overcome in a tunnel where temperature differences are much smaller than in trays or shelves. The temperature in a tunnel is controlled by forcing air through the compost, whereas in trays or shelves the temperature is controlled by the ambient air that penetrates slowly to the compost centre. Derks (1973) described a 3-phase-1 process involving phase I, phase II and spawn run in tunnels. However, phase I in a tunnel was not successful. Bech (1979) combined phase I and phase II in a tunnel during a period of only 5 days. Perrin and Gaze (1987) did much the same with a mixture of straw and chicken manure. Laborde *et al.* (1986) used horse manure plus a high temperature treatment in their process.

In the Netherlands indoor composting is being developed because two large-scale plants have environmental problems with phase I compost production. Small quatities of compost (up to 500 tonnes per week) can be produced in closed sheds. The air can be extracted and treated by washing or biofiltration. During large-scale production this is impossible because too many technical problems are involved. On the other hand, reduction of odour and NH_3 emission by altering the composting process itself is hardly possible. There are too many anaerobic spots (Randle and Flegg, 1978) and even methane is produced (Derikx *et al.*, 1989). If phase I could be carried out in tunnels, the exhaust air could easily be washed or filtered. The environmental problem would then be solved. The question that remains is whether the productivity of such a compost equals that of compost from the existing process. To check this, three series of three experiments were carried out in the tunnels at the Mushroom Experimental Station. The four individually-controlled tunnels with four containers each have previously been used for studies on phase II and spawn running (Gerrits, 1981, 1985).

In the first series the effects of a wide range of compost temperatures (as occurring in a windrow) were compared with temperatures considered as being optimal for thermophilic microflora. Phase I in tunnels was followed by a normal phase II in a growing room.

At a temperature of 50°C the substrate was found to be freed rapidly of ammonia. Therefore, in a second series of trials the optimum temperature for the tunnel process was studied in detail. A mixture of raw materials was first pasteurized and then kept at 40°, 45°, 50° or 55°C. Spawn run took place in the tunnel.

To prevent the introduction of pathogens and pests, the handling and transport of spawned or spawn run compost to the growing room must be very hygienic. This is expensive. Transport of compost (ready for spawning) in a non-hygienic way is much cheaper. However, this procedure can only be successful if the compost is pasteurized after it has been filled into the growing room. To check how this affects compost productivity the third series of trials was performed.

The substrate was not pasteurized in the tunnel. Instead, it was submitted from the beginning to the same temperatures as in series 2 followed by a short phase II in the growing room.

The results from these three series of trials are reported in this paper.

MATERIALS AND METHODS

Experimental procedures have been described previously (Gerrits, 1981, 1985). Only some additional information is given here. Process parameters are temperature and air circulation. Data on NH_4-N, N and ash are expressed as a percentage of dry matter. A good indication for an optimal process is given by weight losses (fresh and dry matter), the decrease of NH_4-N and pH and the increase in N. The average composition of the horse and chicken manures used are given in Table 1.

Table 1. Composition of horse and chicken manures.

		$\%H_2O$	%N	%ash
Horse manure	series 1	61.2	1.30	16.7
	series 2	66.7	1.26	19.1
	series 3	61.1	1.39	16.9
Chicken manure	series 1	44.8	4.83	19.8
	series 2	43.3	4.41	23.1
	series 3	36.9	4.81	18.1

The percentage N in horse manure is the sum of NH_4-N determined in fresh samples and Kjeldahl-N determined in dried samples. The percentage N in chicken manure is Kjeldahl-N determined directly in fresh samples. In horse manure almost 20% of the N is NH_4-N, in chicken manure this is 30%.

In each trial 16 heaps of horse manure (usually 800 kg each) were watered for 4 days to a moisture content of approximately 75%. Recipes, weight and N losses in this paper have been converted to 1000 kg horse manure. In series 1, four different compost formulae were used:

A. No nitrogen + gypsum (25 kg)

B. Chicken manure (100 kg) + gypsum (25 kg)

C. Milli Champ 3000 (33 kg) + gypsum (25 kg)

D. $(NH_4)_2SO_4$ (11 kg) + $CaCO_3$ (33 kg)

Milli Champ 3000 is based on soya bean meal and its use in supplementing spawn run compost is standard practice (Gerrits, 1986).

The quantity of N added in B, C and D is the same. The ingredients (except $CaCO_3$) were mixed prior to filling the 16 containers. Each of the 4 were filled with the mixtures A, B, C and D each in one container. After 3 days the substrates in the containers were mixed again and $CaCO_3$ was added to D. The following conditions were maintained in the tunnel:

I Temperature supply air 50°C; normal circulation

II Temperature supply air 60°C; normal circulation

III Temperature supply air 25°C; reduced circulation

IV Door open; no circulation; aeration 1 min per day

After 7 days the substrate from each container was used to fill 5 experimental plots in the growing room (1.3 m^2, 100 kg per m^2); peakheating followed normally for 9 days (phase II). The mushroom strain used at spawning was Horst® U3.

Two substrates were used in series 2:

A. Horse manure + gypsum (25 kg)

B. Horse manure + gypsum (25 kg) + chicken manure (100 kg)

Each tunnel was filled with 2 containers of the mixtures A and B. After filling, the tunnels were pasteurized as usual for phase II (supply air temperature 8 h at 56°C) and then immediately dropped to the desired value of 40°, 45°, 50° or 55°C. After 3 days the substrate was mixed again and spawned after another 4 days. In one trial, strain Horst ® U3 was used, and in two trials Horst ® U1. The substrates were subjected to spawn run in the tunnels for about 2 weeks. From each container 5 experimental plots in a growing room were filled with 100 kg spawn run compost (i.e. 76.9 kg per m^2). Half of the containers had previously been supplemented with 1% Milli Champ 3000 (0.77 kg per m^2).

Series 3 was identical to series 2, except that no pasteurization took place at the beginning. After one week, the substrates were filled in a growing room (100 kg per m^2)

and submitted to a normal pasteurization process followed by a short conditioning. Phase II in the growing room took 3 or 4 days. The compost was then spawned (strain Horst ® U1) and spawn-running took place in the growing room. In two of the three experiments, half of the substrates were supplemented with Milli Champ 3000 (1 kg per m^2).

All 9 crops were harvested for 4 weeks. About 2/3 of the mushrooms were harvested when closed (as buttons) and 1/3 when open. Average yields of 5 replicate plots (representing the yield of one container) were used in the final analysis of variance. The 3 replicate experiments in each series were considered as blocks of one experiment.

RESULTS

Table 2 shows some data on temperature and air circulation in tunnels of experimental series 1.

Table 2. Temperature (°C) of air and compost in four tunnels under different conditions as measured at day 1 and 4.

Tunnel	I		II		III		IV	
Day	1	4	1	4	1	4	1	4
Air (return)	59	52	64	61	59	54	27	27
Compost (top)	65	55	69	63	72	60	60	60
Compost (bottom)	52	51	61	61	50	44	47	45
Air (supply)	50	50	57	59	27	30	20	22
Circulation (m^3/1000 kg/h)	158	238	141	176	30	50	0*	0*

* aerated for 1 min daily

Temperature in tunnels III and IV (representing those in a compost stack) varied from 25° to over 70° and the CO_2 concentration in the substrate was high. In tunnels I and II, narrower temperature ranges were created by increasing the air circulation.

Table 3 shows the composition of compost B (with chicken manure) after tunnel treatment. In tunnel I, pH and NH_4-N are low and N high, indicating that the microflora are active. Tunnel IV shows the least favourable conditions for the microflora. The data on the other composts are not given, but compost C behaved as did B; A has a lower N content. Compost D had a lower pH and higher NH_4-N as was expected because of the use of ammonium sulphate.

Table 3. Effect of tunnel conditions on some characteristic data of composts B (horse manure and chicken manure) in series 1.

	At emptying tunnels					At spawning			
Tunnel	kg*	%H_2O	pH	NH_4-N	N	%H_2O	pH	NH_4-N	N
I	1219	72.5	8.0	0.05	1.91	69.7	7.8	0.04	2.31
II	1209	72.3	8.2	0.11	1.67	68.1	8.1	0.01	2.02
III	1320	73.0	8.3	0.23	1.65	70.9	8.1	0.06	2.25
IV	1448	74.1	8.4	0.41	1.38	71.1	8.3	0.23	1.93

*kg per 1000 kg horse manure

Yields are given in Table 4. Compost A (the one without N addition) tends to yield less, although there are no statistically significant differences between the various composts. In the same growing room, conventionally-prepared compost yielded 19.1 kg per m^2.

In the first trial of series 1 the compost in the 50° tunnel proved to be free of NH_3 after one week and the compost looked "ready to spawn". Two trays were actually filled with this material and produced a normal crop. Therefore, experimental series 2 was designed to determine the optimum temperature for the process in a tunnel without an additional phase II as in series 1.

Table 4. Mushroom yield in kg per m^2 of four composts prepared in four tunnels in series 1.

	Tunnel			
Compost	I	II	III	IV
A	20.8	17.9	19.3	17.5
B	21.3	19.3	21.0	20.1
C	18.7	19.4	21.9	22.1
D	20.6	21.7	21.1	20.2

Lsd ($P_{0.05}$) = 4.1

Table 5 shows that the temperatures measured at the four different supply air temperatures are very similar to those in a previous study about phase II in tunnels (Gerrits, 1981). The compost with chicken manure has a higher temperature in the top. This could be the result of more microbial activity or of a denser structure of the substrate with chicken manure allowing less air to pass.

Table 5. Temperature (°C) of air and compost during pasteurization and conditioning (at day 5 after filling) in four tunnels (series 2).

	Past	------- Conditioning @ --------			
		40°	45°	50°	55°
Air (return)	61.0	43.5	49.2	52.8	58.2
Compost top (+ ch.m)	68.8	51.0	50.4	54.8	60.5
Compost top (- ch.m)	63.1	44.9	50.6	53.5	59.1
Compost bottom	59.0	41.6	46.7	51.6	58.0
Air (supply)	55.5	40.3	45.1	50.0	56.3

Table 6 shows that the addition of chicken manure leads to more substrate with higher NH_4-N and N at filling. Under the influence of temperature, differences arise. At 55° the substrate has a higher pH, higher NH_4-N and lower N and more weight remains.

Table 6. Effect of temperature during conditioning on some characteristic data of horse manure with (+) and without (-) added chicken manure (series 2).

	At filling tunnel				
	kg*	%H_2O	pH	NH_4-N	N
- ch.m	1157	74.8	8.1	0.11	1.33
+ ch.m	1377	75.4	8.1	0.24	1.42

Temp.		At spawning (tunnel) kg*	%H_2O	pH	NH_4-N	N	After spawn run kg*	%H_2O	pH
40°	-	868	72.1	7.8	0.01	1.83	781	69.4	6.9
	+	1010	71.9	7.7	0.02	1.94	894	69.2	7.0
45°	-	847	70.6	7.9	0.01	1.78	782	68.3	7.0
	+	967	72.0	7.8	0.02	1.96	875	69.1	7.0
50°	-	825	70.8	7.9	0.03	1.73	752	67.4	7.2
	+	972	70.6	8.1	0.08	1.78	889	69.0	7.3
55°	-	868	71.3	8.4	0.14	1.47	807	70.0	7.7
	+	1005	71.2	8.5	0.16	1.58	925	69.2	7.7

***kg per 1000 kg horse manure**

After spawn run this compost still has the highest pH, indicating poor mycelial growth. This also applies at 50° but to a lesser extent. Obviously microbial processes are less active at higher temperatures. There is no difference between 40° and 45°.

Table 7. Mushroom yield in kg per m^2 in two types of substrate (with and without chicken manure, ch. m.) prepared in four tunnels with and without supplementation (Milli Champ 3000) in series 2.

Substrate		Tunnel temperature			
ch.m.	MC 3000	40°	45°	50°	55°
0	0	21.4	21.7	21.1	2.6
0	1	27.2	27.1	25.8	3.4
100	0	23.7	23.0	18.6	2.7
100	1	27.0	26.5	24.6	3.9

Lsd (P 0.01) = 2.5

Table 7 shows the yield from the substrates with and without supplementation. Yields from substrates in the 40° and 45° tunnels were similar. Substrate with chicken manure produced about 2 kg per m^2 more. Supplementation stimulated yields. There was an interaction between chicken manure and supplementation. If compost was supplemented yield differences were no longer present. At 50° the addition of chicken manure decreased yield. This was only partly compensated by supplementation. At 55°, yield was strongly reduced and lots of ink caps (two species) were present. In the experiments, a conventionally-produced and spawn run compost yielded 24.0 kg per m^2 without and 27.0 kg per m^2 with supplementation. This equalled the yields in the 40° and 45° treatments.

Table 8 shows that the substrate temperatures in series 3 (without initial pasteurization) were similar to those in series 2 (Table 5). A higher supplied air temperature was accompanied by more CO_2 and less 0_2. Also, NH_3 (measured with Draeger tubes) was higher. A possible explanation is that at a higher temperature less fresh air was taken in for cooling.

Greater weight losses were found at lower temperatures. This indicates a more active microflora. The effects of the temperature values on the composition of the substrates were similar to series 2 (compare Tables 9 and 6). Note that the NH_4-N of the

45° compost in Table 9 was not a value between those of the 40° and 50° compost, but higher. This was the result of a technical problem in one of the three experiments causing too high a compost temperature. This high value has been omitted in calculating temperature averages in Table 8.

Table 8. Temperature, %O_2 and %CO_2 in air and compost during conditioning in 4 tunnels (series 3, average of day 1, 3, 4 and 7) and temperature in the growing room.

	Tunnel temperature			
Temperature	40°	45°	50°	55°
Air (return)	43.9	50.4	52.7	58.8
Compost top (+ch.m)	48.3	54.4	58.3	61.9
Compost top (-ch.m)	46.4	50.9	52.6	60.1
Compost bottom	42.8	49.5	52.5	57.9
Air (supply)	41.2	46.7	51.0	56.2
% O_2 (supply)	20.1	19.5	19.0	18.7
% CO_2 (supply)	1.0	1.6	2.0	2.2
	Temperature (in growing room)			
Pasteurisation (peak)	72.2	67.9	63.3	59.1
Conditioning (day 1)	50.1	50.7	51.5	48.2
Spawn run (week 1)	24.2	25.7	24.9	26.0

Unexpected events happened during pasteurization in the growing room (Table 8). The 40° composts, already completely free of NH_3, rose to a high temperature value. The temperature increase in the 45° and 50° composts was less and the 55° compost was the least active. NH_4-N at spawning was highly affected, increasing to a high level in the 40° and 45° compost. In the 50° compost, it increased only slightly and it decreased in the 55° compost. In the latter, the temperature rose to high values during spawn run. The average increase was moderate but, incidentally, values over 35° were measured. The pH after spawn run indicated poor growth in the 40° compost and the best growth in the 50° compost.

Table 10 shows that the yield correlated well with the pH of the spawn run compost. The 50° compost gave the best yield followed by the 45°, 55° and 40° composts. On average, chicken manure had a slight (non-significant) negative effect of 1 kg per m^2. However, the tendency was that the effect of chicken manure was positive at

50° (highest yield) and negative at 40° (lowest yield). Only two experiments were supplemented. This resulted in the percentage increase or decrease indicated. The lower yields were accompanied by the development of considerable numbers of ink caps.

Table 9. Effect of temperature during conditioning in tunnels (followed by a short phase II) on some characteristic data of horse manure with (+) and without (-) chicken manure (series 3).

At filling tunnel

	kg*	%H_2O	pH	NH_4-N	N
- ch. m	1380	75.8	7.6	0.18	1.39
+ ch. m	1592	75.8	7.6	0.29	1.53

At emptying tunnels

Temp		kg*	%H_2O	pH	NH_4-N	N
40°	-	964	72.9	7.8	0.02	2.10
	+	1094	72.8	8.0	0.02	2.15
45°	-	997	72.0	8.0	0.08	1.94
	+	1122	72.3	8.0	0.09	2.04
50°	-	1017	73.3	7.9	0.04	1.83
	+	1184	72.8	8.0	0.06	1.79
55°	-	1082	72.3	8.3	0.15	1.65
	+	1169	71.5	8.3	0.15	1.70

		At spawning				At casing	
Temp		%H_2O	pH	NH_4-N	N	%H_2O	pH
40°	-	70.5	8.1	0.15	1.93	70.2	7.4
	+	71.2	8.1	0.20	2.07	70.4	7.5
45°	-	71.1	8.0	0.15	2.02	69.1	7.2
	+	70.6	8.1	0.16	2.11	69.0	7.4
50°	-	72.1	7.8	0.08	2.04	70.3	6.8
	+	70.6	8.0	0.10	2.01	70.1	6.8
55°	-	72.3	8.0	0.09	1.91	70.2	7.1
	+	71.8	8.1	0.11	1.93	69.6	7.3

*kg per 1000 kg horse manure

Table 10. Mushroom yield in kg per m^2 in two types of substrate prepared in four tunnels (followed by a short phase II) without supplementation in series 3.

ch. m (kg)	MC3000 (kg)	Tunnel temperature 40°	45°	50°	55°
0	0	13.9	19.9	21.9	15.0
0	1	-40%	+1%	+5%	+41%
100	0	12.2	18.5	22.3	13.7
100	1	+4%	-8%	+10%	+22%

DISCUSSION

From experimental series 1 it is obvious that phase I can take place under various circumstances without affecting yield, if it is followed by a normal phase II process in a growing room. This agrees with commercial practice in which several phase I systems are used successfully. However, the most interesting observation was that the raw materials were already free of NH_3 after one week at 50°C. In series 2, optimum temperatures for this process, no longer involving phase II, were determined. Highly reproducible results with good yields were obtained in the three individual trials. Obviously, high temperatures occurring in a compost stack are not necessary for normal mushroom production.

Laborde *et al.* (1986) maintain high temperatures (75°-80°C) for some time in the tunnel to enable chemical reactions to take place. As this partly sterilizes the substrate, it is necessary to inoculate it again with some compost or a microflora suspension. In a phase I compost stack, however, a range of temperatures occurs that enables all kinds of microorganisms to survive. During turning, high temperature zones are reinoculated. In our trials the process was optimal between 40° and 45°C. At these temperatures substrates are soon free of NH_3 and beneficial thermophilic fungi develop spontaneously (Straatsma *et al.*, 1989). Obviously reinoculation is unnecessary if excessively high temperatures are avoided. Ross and Harris (1982, 1983) had already mentioned temperatures of 40° and 45° as being optimal for the disappearance of NH_3 and for obtaining a selective substrate in which thermophilic fungi play an important role.

Series 3 experiments show that it is undesirable to pasteurize a substrate that is already free of NH_3 and in which thermophilic fungi have developed. Therefore, it is necessary to handle such a substrate hygienically and convey it into mushroom growing rooms as spawned or preferably as spawn run compost. In series 3, the 50° compost gave better yields than the 55° compost. This is unlike series 1 in which 50° and 60° compost behaved in the same way. The difference could be a result of unequal phase II time. At 55° or 60°, more N is present as ammonium-N and, therefore, conditioning takes more time. If 55° in tunnels had been followed by a 9 day phase II the same result as in series 1 would be expected. The increase in temperature during spawn run in the 55° compost might result from the activity of thermophilic fungi that could not develop previously. The fungi concerned can grow at 25° although the growth rate is much less than at 45°C.

The overall results of these experiments could be summarized as follows. After a high temperature treatment in a tunnel (>50°), NH_3 is still present, thermophilic fungi cannot develop and the substrate will not produce mushrooms. The product resembles a usual phase I compost and has to be submitted to an additional phase II before it will be productive. At a slightly lower temperature (50°) less NH_3 remains and thermophilic fungi develop only weakly. A short phase II is sufficient to remove the NH_3 and pasteurization does little damage to the ill-developed fungi.

If an organic mixture is kept in a tunnel at a temperature of 40° or 45° thermophilic fungi develop well, ammonia disappears quickly and the remaining substrate is immediately suitable for mushroom production. If such a substrate is pasteurized, however, the temperature increases to high values. The reason for this increase is unknown. NH_3 is generated again and selectivity is lost, probably because thermophilic fungi disappear. This situation cannot be restored in a short time. This phenomenom has also been described by Ross and Harris (1982, 1983).

From the data in Tables 1 and 3, it follows that, until the tunnel is emptied, 1.2 kg N, i.e. 1.5 kg NH_3, disappears from a tonne of horse manure mixed with chicken manure (mixture B). From mixture C and D the loss of N is about the same, but from mixture A (without extra N) the loss is negligible. Series 2 (Table 1 and 6) agrees well with series 1; with chicken manure, 1.1 kg N disappears and without chicken manure practically nothing is lost. Series 3 (Tables 1 and 9) follows the same tendency but the losses are higher. In this series, 2.2 kg N disappears if chicken manure is added. This results mainly from the higher percentage N and dry matter in the basic materials. Thus, N-losses depend greatly on the quantity of N already present or added. It is important to determine more accurately the minimum N content of the substrate guaranteeing both a maximum yield as well as a minimum loss of NH_3 to the environment. Minimizing NH_3 losses at the same time minimizes the cost of air washing. The smell problem may already be solved to a great extent by the indoor composting process itself. Oxygen concentration during this process is much higher than in an outside stack. Therefore, the production of offensive smell will be considerably less. This expectation is already confirmed in commercial practice where phase II in tunnels is applied on a large-scale without odour problems.

ACKNOWLEDGEMENT

The authors sincerely thank Lynn Moore for correcting the English text.

REFERENCES

Bech, K. (1979). Preparing a productive commercial compost as a selective growing medium for *Agaricus bisporus* (Lange) Sing. Mushroom Sci. **10**, 77-83

Derikx, P.J.L., op den Camp, H.J.M. , Bosch, W.P.G.M., Vogels, G.D., Gerrits, J.P.G. and van Griensven, L.J.L.D. (1989). Production of methane during preparation of mushroom compost. Mushroom Sci.12(1), 353-360.

Derks, G. (1973). 3-phase-1. Mushroom J. 9, 396-403.

Gerrits, J.P.G. (1981). Factors in bulk pasteurization and spawn-running. Mushroom Sci. **11**(1), 351-365.

Gerrits, J.P.G. (1985). Further studies on factors in bulk pasteurization and spawn-running. The Mushroom J. **155**, 385-395.

Gerrits, J.P.G. (1986). Supplementation with formaldehyde treated soya bean meal. Mushroom J. **161**, 169-174.

Gerrits, J.P.G. (1988a). Nutrition and compost. In *The cultivation of mushrooms,* ed. L.J.L.D. van Griensven, Darlington Mushroom Laboratories Ltd., Rustington U.K, pp. 29-72.

Gerrits, J.P.G. (1988b). Compost treatment in bulk for mushroom growing, In *Treatment of Lignocellulosics with White-rot Fungi,* eds. F. Zadrazil and P. Reiniger, Elsevier Applied Science, London, pp. 99-104.

Huhnke, W., Lemke, G. and Sengbusch, R. von (1965). Die Weiterentwicklung des Tillschen Champignonkulturverfahrens auf nicht kompostiertem sterilem Nehrsubstrat (Zweite Phase). Die Gartenbauwissenschaft, **30**, 189-207.

Huhnke, W. (1972). Die Weiterentwicklung des Champignonanbauverfahrens auf nicht kompostiertem Nehrsubstrat. Mushroom Sci. **8**, 503-515.

Laborde, J., Delmas, J. and Delpech, P. (1979). Préparation rapide de substrats pour la culture du champignon de couche: questions posés et tentative de réponse. Mushroom Sci. **10**(2), 85-103.

Laborde, J., Olivier, J.M., Houdeau, G., Delpech, P. (1987). Indoor static composting for mushroom (*Agaricus bisporus* Lange Sing) cultivation, In *Cultivating Edible Fungi,* eds. P.J. Wuest *et al.*, Elsevier, Amsterdam, pp. 91-100.

Lambert, E.B. (1941). Indoor composting for mushroom culture. U.S. Dept. Agriculture Circ. No. 609, 1-15.

Perrin, P.S. and Gaze, R.H. (1987). Controlled environment composting. Mushroom J. **174**, 195-197.

Randle, P. and Flegg, P.B. (1978). Oxygen measurements in a mushroom compost stack. Scientia Horticulturae, **8**, 315-323.

Ross, R.C. and Harris, P.J. (1982). Some factors involved in phase II of mushroom compost preparation. Scientia Horticulturae, **17**, 223-229.

Ross, R.C. and Harris, P.J. (1983). An investigation into the selective nature of mushroom compost. Scientia Horticulturae, **19**, 55-64.

Smith, J.F. (1983). The formulation of mixtures suitable for economic, short-duration mushroom composts. Scientia Horticulturae, **19**, 65-78.

Straatsma, G., Gerrits, J.P.G., Augustijn, M.P.A.M., op den Camp, H.J.M., Vogels, G.D., van Griensven, L.J.L.D. (1989). Population dynamics of *Scytalidium thermophilum* in mushroom compost and stimulatory effects on growth rate and yield of *Agaricus bisporus*. J. Gen. Microbiol. **135**, 751-759.

PILOT-SCALE REACTOR FOR SOLID-STATE FERMENTATION OF LIGNOCELLULOSICS WITH HIGHER FUNGI: PRODUCTION OF FEED, CHEMICAL FEEDSTOCKS AND SUBSTRATES SUITABLE FOR BIOFILTERS

F. Zadrazil*, H. Janssen**, M. Diedrichs* and F. Schuchardt ***

*Institut für Bodenbiologie and ***Institut für Technologie
Bundesforschungsanstalt für Landwirtschaft
Bundesallee 50, 33 Braunschweig, FRG

**Technisch Handels en Adviesbureau,
Napoleonsbaan 82, 6086 AH Neer, Nederland

In this paper we describe the solid-state fermentation-pilot reactor that has been in use at the Institute of Soil Biology, FAL, Braunschweig since 1985. The pilot reactor may, with some technical modifications, also be used as a model for large-scale technology. However, additional theoretical and practical work must be done before such reactors may be used in commercial applications.

INTRODUCTION

In recent years many different reactors for solid-state fermentation (SSF) have been designed, developed and constructed. Some are used in Koji processes for the production of soya bean sauce, or in the production of substrates for the cultivation of edible fungi such as *Agaricus bisporus* (White mushroom; Francescutti, 1972; Gerrits, 1988), *Pleurotus* spp. (Schuchardt and Zadrazil, 1982). Another proposed use for these fungal substrates is the conversion of lignocellulosics into animal feed (Zadrazil, 1977, 1980, 1985; Bajracjarva and Mudget, 1980; Levonen-Munoz and Bone, 1980; Matteau and Derek, 1980; Zadrazil and Brunnert, 1980, 1981; Ulmer *et al.*, 1981; Reid and Seifert, 1982; Laukevics *et al.*, 1984; Abdullah *et al.*, 1985; Agosin and Odier, 1985; Agosin *et al.*, 1985; Kamra and Zadrazil, 1985, 1986; Weiland, 1988), chemical feedstocks (Hatakka and Pirhonen, 1985), and for biological pulping (Kirk *et al.*, 1980; Yang *et al.*, 1980; Ander and Eriksson, 1986; Eckstein, 1985). Prepared substrates can also be used in environmental control, e.g. as process biofilters (Hüttermann *et al.*, 1988) and in the decontamination of xenobiotics in soils and waste materials (Hüttermann *et al.*, 1988, 1989).

Definition of solid-state fermentation

SSF can be defined as a process, in which solid substrates are decomposed by known mono- or mixed cultures of microorganisms, mainly fungi, that can grow on and through the substrate under controlled environmental conditions, with the aim of producing a high quality standardized product. The substrate (mixtures of different particles) is characterized by a relatively low water content. Since much of the water is chemically or physically bound, the physical properties, e.g. porosity and density, of the substrate are uniform. The substrate is neither mixed nor moved during the process.

The principle of the SSF process

The pretreated substrate is filled into the reactor in 1.5-2.0 m deep layers and incubated by percolation of the gas phase through the substrate. The temperature in the substrate is indirectly controlled by conditioned, percolated gas. In the cultivation of *Agaricus* spp., fresh air is added in order to cool the substrate during cultivation. By contrast, high concentrations of CO_2 and low concentrations of O_2 are required during the period of colonization by wood-decaying fungi (Zadrazil, 1975).

The technique for substrate production for growth of the white mushroom (*Agaricus bisporus*) is now well developed. Two factories in the Netherlands produce about 10,000 tonnes of substrate per week. Examples of this technology are shown in Figs. 1 and 2. Similar technology and equipment can, as mentioned earlier, be used for the utilization of lignocellulosics.

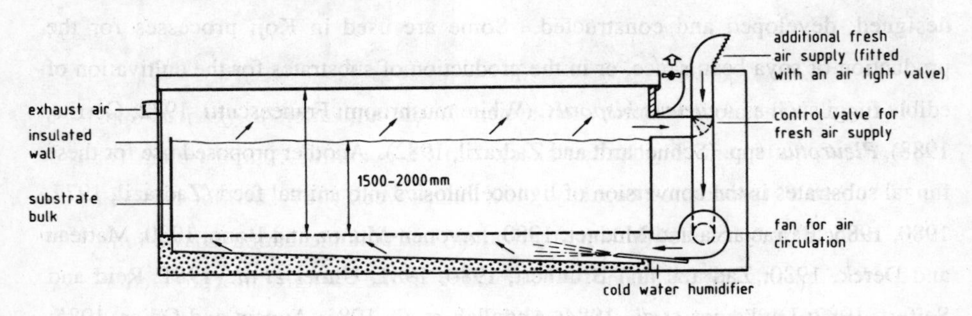

Fig. 1 Scheme of a reactor for production of substrate for *Agaricus bisporus*.

The equipment and the technology used in the production of *Agaricus bisporus* cannot be used for the incubation of wood-decaying fungi on cereal straw or other plant residues.

The substrates used for cultivation of such fungi are not as selective as that used for the for cultivation of *A. bisporus* and need additional pretreatment and process control.

Fig. 2. The modern Dutch shelf-system for the cultivation of *Agaricus* spp. (CNC, Ottersum, The Netherlands). (a) Co-operative compost plant for "phase I" (production capacity 7400 tonnes/week); (b) Corridor between tunnels for "phase II" (bulk pasteurization) and bulk incubation (spawn run). Note machinery for filling and spawning in the background (production capacity 1000 tonnes/week of fully grown compost); (c) Tunnel-system machines for filling, spawning and emptying the substrate; (d) Compost plant, Blanchaud, France; (e) Machinery for filling shelves; (f) Cropping room equipped with aluminium shelves. Photographs (b, c, f) by Vorkamp Agroprojecten, The Netherlands.

WORK AT INSTITUT FÜR BODENBIOLOGIE, BRAUNSCHWEIG

Description of the SSF pilot reactor

The principle of the design of a small-scale reactor and its control were described by Schuchardt and Zadrazil (1982). On the basis of prior experience, the reactor was constructed of polyurethane foam sandwich panels covered on both sides with polyester board. The filling height is approximately 2.0 m, the internal width is 2.3 m and the depth is 2.0 m. This results in a net filling volume of 6.9 m^3, i.e. equivalent to 1.5 tonnes of straw or 3.0 tonnes of wood chip substrate.

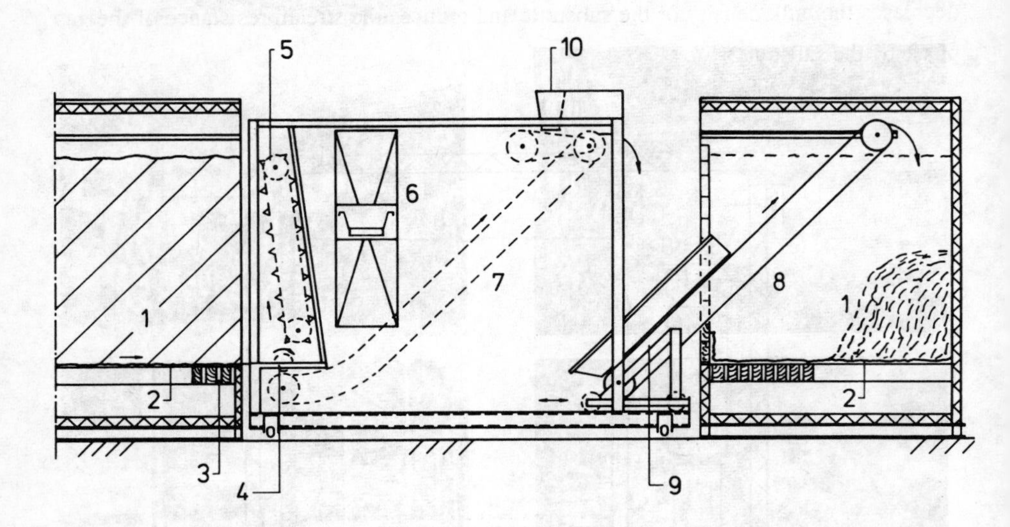

Fig. 3. Horizontal cross-section of reactors for SSF and spawning-emptying machine. Numbers refer to the following: 1, SSF reactor; 2, Net for emptying the substrate; 3, Panels for aerating the substrate; 4, Equipment for emptying the substrate; 5. Equipment for fragmenting the substrate; 6, Equipment for filling the substrate; 7, Conveyor to bring substrate to spawning machinery; 8, Conveyor for filling the reactor with substrate; 9, Carrier for moving the conveyor; 10, Spawning machine.

Two similarly insulated swing doors, situated at the front, occupy the full width of the reactor. Inside the reactor there is a raised slatted floor covered firstly with a gliding net and then with a drag net. At the Institute of Soil Biology two reactors of the same shape and construction (Figs. 3, 4, 5) are used, one for substrate pretreatment (pasteurization or microbial pretreatment) and the other for substrate colonization. The substrate is inoculated on being moved from one reactor to the other.

The reactor is filled with substrate using a specially-designed machine that deposits it on the drag net. A removable front panel keeps the substrate from falling out of the container during filling. The substrate, which is removed from the container by attaching the drag net to a winch, is loosened as it is pulled through a set of toothed bars before falling onto the elevator. It is then fed into another reactor either for incubation or for treatment of the fully-colonized substrate. This translocation from one reactor to the other decreases the bulk density of the substrate and reduces the stream resistance of the gas phase by the substrate.

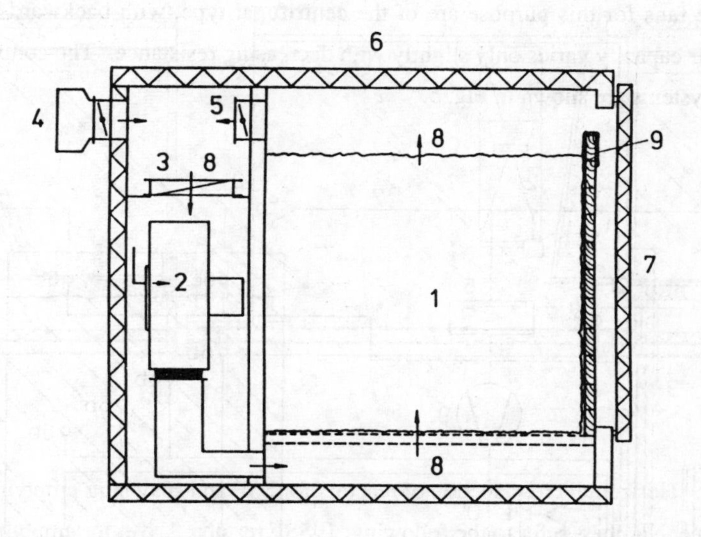

Fig. 4. Vertical cross-section of the reactor for SSF. Numbers refer to the following: 1, Substrate; 2, Fan for gas recycling; 3, Cooling and heating system; 4, Fresh air supply with anemometer; 5, Regulation of air movement; 6, Body of reactor; 7, Reactor door; 8, Direction of gas movement; 9, Net for emptying the substrate.

Air-conditioning

Fungal growth and heat exchange from the substrate is controlled by re-circulating the air within the reactor. During the incubation of *Agaricus bisporus,* the reactor is supplied with largge quantities of fresh air. For the cultivation of *Pleurotus* spp. and other wood-decaying fungi, the gaseous phase is recycled and gas composition is controlled by monitoring CO_2 and O_2 concentrations (Schuchardt and Zadrazil, 1982). To control and reduce the growth of competitive microorganisms, CO_2 could be added at the commencement of the fermentation process.

The reactor has an air-conditioning system installed on the roof or in the gas-tubes (Figs. 4 and 5). A system of aluminium air ducts, with supply duct below, and return ducts in the roof, ventilates the substrate from below. The total quantity of circulated air can be varied by changing the speed of electronically-controlled fans. A centrifugal type fan was chosen to ensure that the various processes could be controlled adequately. It has a capacity of 800 $m^3.h^{-1}$ at a static pressure of 1200 Pa, or, with respect to substrate, approximately 200-500 $m^3.tonne^{-1}.h^{-1}$ (depending on its specific volume weight). The most suitable fans for this purpose are of the centrifugal type, with backward curved blades. Their capacity varies only slightly with decreasing resistance. The control and registration systems are shown in Fig. 5.

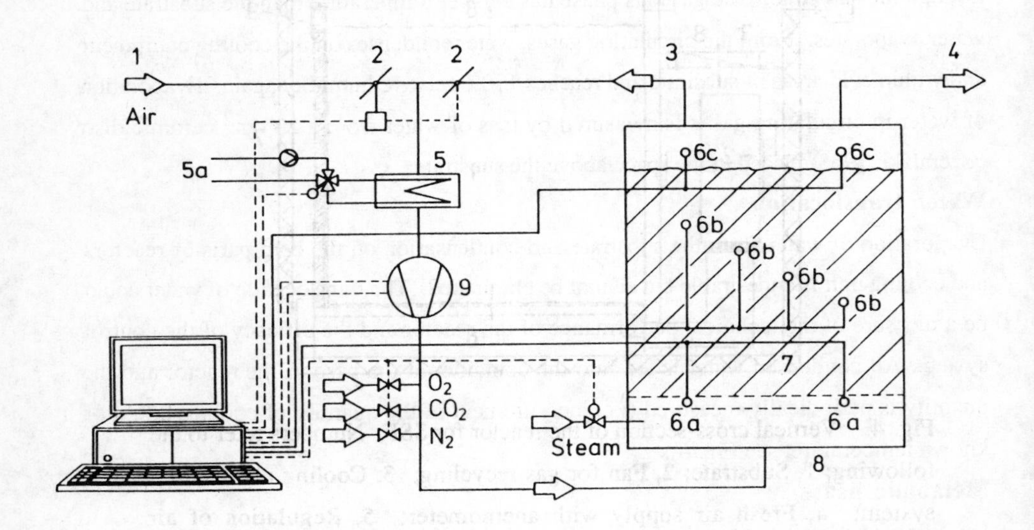

Fig. 5. Schematics of the control system for the solid-state reactor. 1, Fresh air supply; 2, Air damper; 3, Gas phase recirculation; 4, Emptying of gas; 5, Gas-cooling and heating system; 5a, Liquid-cooling system; 6, Temperature control sensors; 6a, Gas-input; 6b, Substrate temperature; 6c, Gas-output; 7. Substrate; 8. Body of SSF reactor; 9. Fan.

In order to humidify the air and raise the substrate temperature for pretreatment of substrate (e.g. pasteurization), a steam injection pipe is installed under the slatted floor.

Steam injection is controlled by a 2-way valve operated by a servomotor. Heat exchange takes place by cooled refrigeration in the air-circulation system controlled by a servomotor driven a 4-way valve. The lowest the temperature of the cooling liquid can be is -15°C. Fresh air, O_2 or CO_2 can be added by computer-controlled valves after gas analysis (Fig. 5). Exhaust gases leave the fermentor through an over-pressure valve located above the substrate.

Control of gas humidity

The humidity of the gaseous phase is controlled by hygrometers (Hygrotest 720) situated in different parts of the reactor. The air humidity fluctuates between 95 and 100%. At these values, commercially-available hygrometers are not sufficiently sensitive to give more precise control.

Control of water evaporation

Saturation deficit of water in the gaseous phase increases during the penetration of gas through the substrate. The gaseous phase has a lower temperature than the substrate and water evaporates. From the circulating gases, water condenses on the cooling equipment and within cold areas of substrate and reaches 100% relative humidity again. Evaporation of water in circulating gases is measured by loss of water from a 20 cm^2 ceramic disc (Czeratzki, 1968) placed in the space above the substrate.

Water translocation

Evaporation of water from the substrate and condensation on the cold parts of reactors and cooling unit is undesirable but cannot be eliminated. The translocation of water could be a measure of the technical performance of the reactor and the efficacy of the control system. All condensed water is led into the container at the base of the reactor and the quantity is periodically measured (i.e. measurement of the quantity of cooling liquid of known temperature; see Fig. 6).

Metabolic heat

Ducts must be insulated to eliminate uncontrolled heat loss from the reactor. The rate of heat liberation during the fermentation process is measured by a flow meter installed in the cooling system of the reactor.

Temperature control

The temperature of the gas phase is monitored by placing four PT100 resistance thermometer sensors at the inlet and outlet ducts (Fig. 5, Position nos. a, b, c, in different layers of substrate). The temperature of the substrate in 4 different layers is measured using 8 thermometer sensors and registered in 2 independent computers. The sensors themselves are installed in stainless steel tubes in order to protect against damage.

Minimum and maximum limits of temperature can be adjusted. If the temperature of the air rises above or falls below the set limit, the heating or cooling system is activated and an alarm could be calibrated for warning of fluctuations in areas where temperature control is critical.

The function and sensitivity of the steam and cooling valves in maintaining a desired environment is demonstrated in Fig. 6. By operating temperature controls, temperature differences as a function of substrate layer can be estimated. Mathematical and physical models, and practical possibilities, for temperature control are discussed elsewhere in this book [Teifke and Bohnet (1990) this volume].

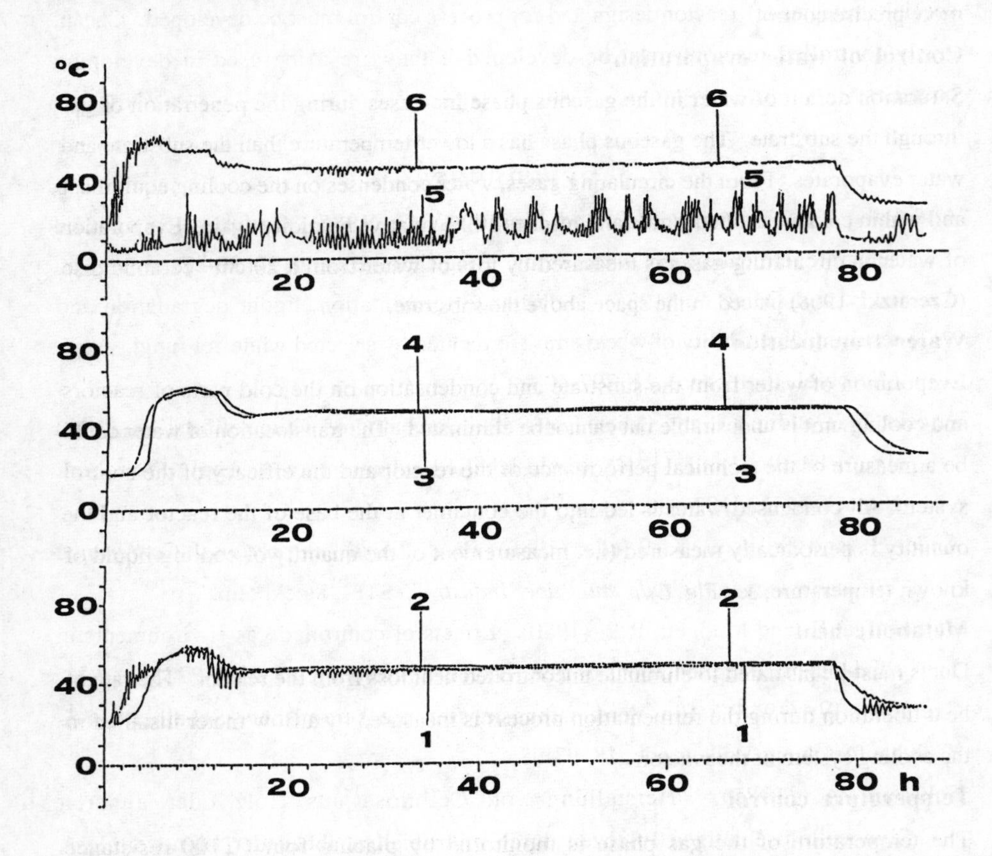

Fig. 6. Control of temperature in different parts of SSF-reactor during substrate pretreatment: 1, Air input; 2, Air output; 3, Substrate; 4, Substrate; 5, Cooling liquid temperature (frequency of opening and closing of valves); 6. Temperature fluctuation near mouth of steam pipe.

Air supply and control of circulated air

The amount of air circulated through the substrate in the reactor is measured on the supply side of the fan by measuring the difference in pressure across a gauge ring. The volume of circulated gas is indicated in m^3. The accuracy of this apparatus is very good and air resistance is very low. The required fresh air is added with an air pump and measured with a flowmeter.

Research needs

Knowledge of solid-state fermentation and of reactors suitable for this process is insufficient. There is a need for more basic and applied research before this "mushroom" technology can be transferred in other areas of biotechnology. For each organism, special strategies for reactor design and for process control must be developed. Cheap, easily-controlled reactors must be developed if they are to be used in developing countries.

REFERENCES

Abdullah, A.L., Tengerdy, R.P. and Murphy, V.G. (1985). Optimization of solid substrate fermentation of wheat straw. Biotechnol. Bioeng. **27**, 20-27.

Agosin, E. and Odier, E. (1985). Solid-state fermentation, lignin degradation and resulting digestibility of wheat straw fermented by selected white-rot fungi. Appl. Microbiol. Biotechnol. **21**, 397-403.

Agosin, E., Daudin, J.J. and Odier, E. (1985). Screening of white-rot fungi on (^{14}C) lignin-labelled and (^{14}C) whole-labelled wheat straw. Appl. Microbiol. Biotechnol. **22**, 132-138.

Ander, P. and Eriksson, K.-E. eds. (1986). *Proc. Third International Conference "Biotechnology in the Pulp and Paper Industry"*, STFI, Stockholm.

Bajracjarya, R. and Mudgett, R.E. (1980). Effects of controlled gas environments in solid-substrate fermentations of rice. Biotechnol. Bioeng. **22**, 2219-2235.

Czeratzki, W. (1968). Ein Verdunstungsmesser mit keramischer Scheibe. Landbauforschung Volkenrode, **18**, 93-98.

Eckstein, A. (1985). Herstellung von Cellulose aus Holz oder anderen lignocellulosehaltigen Pflanzen durch mikrobiellen abbau der Lignocellulose. Europaische Patenschrift No. 0 060 467.

Francescutti, B. (1972). Verfahren und Vorrichtung zum Anbau von Pilzen. Deutsches Patentamt, Offenlegungsschrift (1972).

Gerrits, J.P.G. (1988). Compost treatment in bulk for mushroom growing. In *Treatment of Lignocellulosics with White-rot Fungi,* eds. F. Zadrazil and P. Reiniger, Elsevier Applied Science, London, pp. 99-104.

Hatakka, A.J. and Pirhonen, T.J. (1985). Cultivation of wood-rotting fungi on agricultural lignocellulosic material for the production of crude protein. Agricultural Wastes, **12**, 81-97.

Hüttermann, A., Loske, D., Braun-Lüllemann, A. and Majcherczyk, A. (1988). Der einsatz von Weißfaulepilzen bei der Sanierung kontaminierter Boden und als Biofilter. Bio-Engineering, **4** (3), 156-160.

Hüttermann, A., Loske, A. Majcherczyk, A. and Zadrazil, F. (1989). Einsatz von Weißfaulepilzen bei der Sanierung kontaminierter Boden. *Der Bundesminister für Forschung und Technologie,* pp. 115-121.

Kamra, D.N. and Zadrazil, F. (1985). Influence of oxygen and carbon dioxide on lignin degradation in solid-state fermentation of wheat straw with *Stropharia rugosoannulata*. Biotechnol. Lett. **7**, 335-340.

Kamra, D.N. and Zadrazil, F. (1986). Influence of gaseous phase, light and substrate pretreatment on fruit-body formation, lignin degradation and *in vitro* digestibility of wheat straw fermented with *Pleurotus* spp. Agricultural Wastes, **18**, 1-17.

Kirk, T.K., Higuchi, T. and Chang, H., eds. (1980). *Lignin Biodegradation: Microbiology, Chemistry and Potential Applications,* Vols. I and II. CRC Press, Boca Raton, Florida.

Laukevics, J.J., Apsite, A.F., Viesturs, U.E. and Tengerdy, R.P. (1984). Solid substrate fermentation of wheat straw to fungal protein. Biotechnol. Bioeng. **26**, 1465-1474.

Levonen-Munoz, E. and Bone, D.H. (1985). Effect of different gas environments on bench-scale solid-state fermentation of oat straw by white-rot fungi. Biotechnol. Bioeng. **22**, 382-387.

Matteau, P.P. and Derek, D.H. (1980). Solid-state fermentation of Maple wood by *Polyporus anceps*. Biotechnol. Lett. **2**(3), 127-132.

Reid, J.D. and Seifert, K.A. (1982). Effect of an atmosphere of oxygen on growth, respiration and lignin degradation by white-rot fungi. Can. J. Bot. **60**, 252-260.

Schuchardt, F. and Zadrazil, F. (1982). Aufschluß von Lignocellulose durch hohere Pilze - Entwicklung eines Feststoff-Fermenters. In *Proc. Fifth Symp. Techn. Microbiologie "Energie durch Biotechnologie",* ed. H. Dellweg, Berlin, pp.

422-428.

Teifke, J. and Bohnet, M. (1990). Modelling of physical process parameters of technical lignin degradation by *Pleurotus* spp. In *Advances in Biological Treatment of Lignocellulosic Materials,* eds. M.P. Coughlan and M.T. Amaral Collaço, Elsevier Applied Science, London, pp. 71-84.

Ulmer, D.C., Tengerdy, R.P. and Murphy, V.G. (1981). Solid-state fermentation of steam-treated feedlot waste fibres with *Chaetomium cellulolyticum.* Biotechnol. Bioeng. Symp. **11**, 449-461.

Yang, H.H., Effland, J.J. and Kirk, T.K. (1980). Factors influencing fungal degradation of lignin in a representative lignocellulosic, thermomechanical pulp. Biotechnol. Bioeng. **22**, 65-77.

Weiland, P. (1988). Principles of solid-state fermentation. In *Treatment of Lignocellulosics with White-rot fungi,* eds. F. Zadrazil and P. Reiniger, Elsevier Applied Science, London, pp. 64-76.

Zadrazil, F. (1975). Influence of CO_2 concentration on the mycelium growth of three *Pleurotus* species. Eur. J. Appl. Microbiol. **1**, 327-335.

Zadrazil, F. (1977). The conversion of straw into feed by Basidiomycetes. Eur. J. Appl. Microbiol. **4**, 291-294.

Zadrazil, F. (1980). Conversion of different plant waste into feed by Basidiomycetes. Eur. J. Appl. Microbiol. Biotechnol. **9**, 243-248

Zadrazil, F. (1985). Screening of fungi for lignin decomposition and conversion of straw into feed. Angewandte Botanik, **59**, 433-452.

Zadrazil, F. and Brunnert, H. (1980). The influence of ammonium nitrate supplementation on degradation and *in vitro* digestibility of straw colonized by higher fungi. Eur. J. Appl. Microbiol. Biotechnol. **9**, 37-44.

Zadrazil, F. and Brunnert, H. (1981). Investigation of physical parameters on solid-state fermentation by (saprophytic) white-rot fungi. Eur. J. Appl. Microbiol. Biotechnol. **11**, 183-188.

LARGE SCALE SOLID-STATE FERMENTATION OF CEREAL STRAW WITH *PLEUROTUS* SPP.

F. Zadrazil *, M. Diedrichs *, H. Janssen **, F. Schuchardt *, and J.S. Park ***

*Institut fur Bodenbiologie, Bundesforschungsanstalt für Landwirtschaft
Bundesallee 50, 33 Braunschweig, FRG

**Technisch Handels en Adviesbureau
Napoleonsbaan 82, 6086 AH Neer, Nederland

***Agricultural Sciences Institute, Rural Development Administration,
Suweon 170, Korea.

At present, it is difficult to describe the best technology for substrate pretreatment. Temperature and the composition of gas phase and water content influence the development of thermophilic microorganisms in the substrate. More research is required for identification of the interactions between indigenous microflora, and the cultivated fungi. Solid-state fermentation is a promising technology for biological treatment of lignocellulosics and other solid substrates. From a technical viewpoint this system is still undeveloped. Acquisition of the relevant practical knowledge is necessary for large-scale, but low cost, industrial applications.

INTRODUCTION

Processes that have been proposed for biological upgrading (using white-rot fungi) of lignocellulosics into animal feed, substrates for growth of edible fungi, or other valuable products, e.g. raw material for industry, are complex (Zadrazil and Brunnert, 1980, 1981). Many factors influence the course of solid-state fermentation, but only some factors, such as temperature, humidity and the composition of gas phase, can be controlled and changed (Zadrazil, 1979; Schuchardt and Zadrazil, 1982, 1988).

Different types of reactors for solid-state fermentation have been described and tested on different scales. The most promising technology for solid residue management is that for the production, by way of a tunnel process, of white mushroom substrate (Francescutti, 1972). This technique could be modified for conversion of lignocellulosics into animal feed (Zadrazil, 1985), production of fungal substrates for environmental control (Loske *et al*., 1990; Majcherczyk *et al*., 1990), biological pulping (Eckstein, 1985) and other industrial processes. Lignocellulosic materials have low bulk density and, as such, must be used in large-scale, low-cost processes.

Condition of growth of higher fungi in solid particle substrates

Defining the ideal conditions for fungal growth is both difficult to describe and to determine. It is a combination of many factors, only some of which can be estimated. Knowledge of the relevant factors and cultural limits, the scale or size that ensures good mycelial growth and degradation of the substrate, can only be determined empirically for each size of reactor (Zadrazil, 1979). Temperature optima for fungal growth vary between 25° and 35°C depending on species (Zadrazil, 1985) and the water content of the substrate can vary between 70 and 80% of dry weight (Zadrazil and Brunnert, 1981).

Wood-decaying fungi are not as sensitive to high concentrations of gaseous CO_2 as are other microorganisms. They can withstand relatively high CO_2 concentrations. Indeed, growth of *Pleurotus* spp. can be stimulated by CO_2 concentrations up to 30% (Zadrazil, 1975). However, higher fungi are very sensitive to mechanical movement and mixing of the substrate. Such conditions can destroy hyphal structure and lead to limitation of fungal growth. For these reasons we have not used the rotating drum reactor technology described by Giovannozzi-Sermanni *et al.* (1989). A 10-l glass vessel reactor with a rotary motor, normally used for liquid fermentation, was used for the production of culture inoculum (spawn) on wheat grain (Fig. 1). Fungal growth under rotating conditions was very slow. Such a system was not suitable for practical requirements. In this publication we drescribe further experiments using wheat straw and poplar wood chips as substrate for fermentation by *Pleurotus* spp. (Schuchardt and Zadrazil, 1982, 1988) in a 10,000-l reactor.

MATERIALS AND METHODS

Fungus

Various *Pleurotus* spp., from the collection of Dr. F. Zadrazil, and commercial strains from Italspawn (Italy) and Somycel (France) were used for large-scale experiments.

Substrate

Wheat straw and poplar wood chips (see particle size Fig. 2) were used after thermal and microbial pretreatment (see pretreatment of substrate in the Results section).

Fig. 1. Rotary drum for (semi) solid-state fermentation of wheat strain with *Pleurotus* spp. as technology for spawn production (Research Institute von Sengbusch, Hamburg, 1973). The side and front views of the reactor are shown in the top and bottom photographs, respectively. 1, Drum-reactor; 1a, 10-1itre; 1b, 1-litre; 2, Unit for moving drum reactor; 3, Air pump; 4, Air sterilization filter; 5, Oil-seal ring; 6, Speed control of rotary drum reactor.

Reactor

The 10,000-1 reactor, as described by Zadrazil *et al.* (1990) elsewhere in this volume.

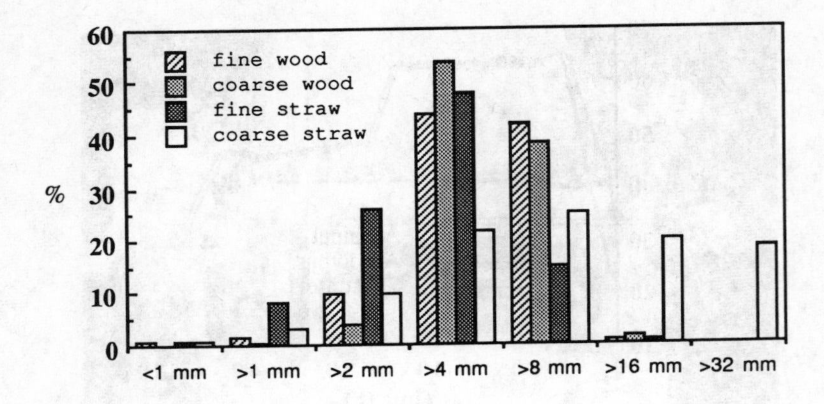

Fig. 2. Particle sizes of substrates (wheat straw and poplar wood chips) used for SSF.

Assay and measurement procedures

The particle sizes of substrate fractions were determined by sieving (Bosma and Dernedde, 1979). Each fraction is expressed as % of total substrate (Fig. 2). The water content of substrate was determined by drying samples at 105°C to constant weight. Lignin was determined as the residue after hydrolysis with HCl and H_2SO_4 (Halse, 1926). Ash content was determined by combustion at 550°C. *In vitro-* digestibility was determined using the method of Tilley and Terry (1963).

RESULTS AND DISCUSSION

Substrate pretreatment

Wheat straw and poplar wood were chopped (Fig. 2), watered and transferred to the first reactor for thermal and microbial pretreatment (Kalberer, 1974; Zadrazil and Kurtzman, 1982). The temperature of substrate, air and cooling liquid were measured (Fig. 3). Pretreatment time varied from 48 to 140 h. The O_2 content in the gas phase during pretreatment of the substrate has an influence on the development of thermophilic microorganisms thereon and also influences its properties and selectivity for growth of *Pleurotus* spp. and other cultivated fungi. An example of the air supply to the substrate during pretreatment is shown in Fig. 3. During this time the composition of the gas phase fluctuated. The CO_2 content increased for the first 12 h and then slowly decreased while the opposite pattern was true of the O_2 content (Fig. 3B, C).

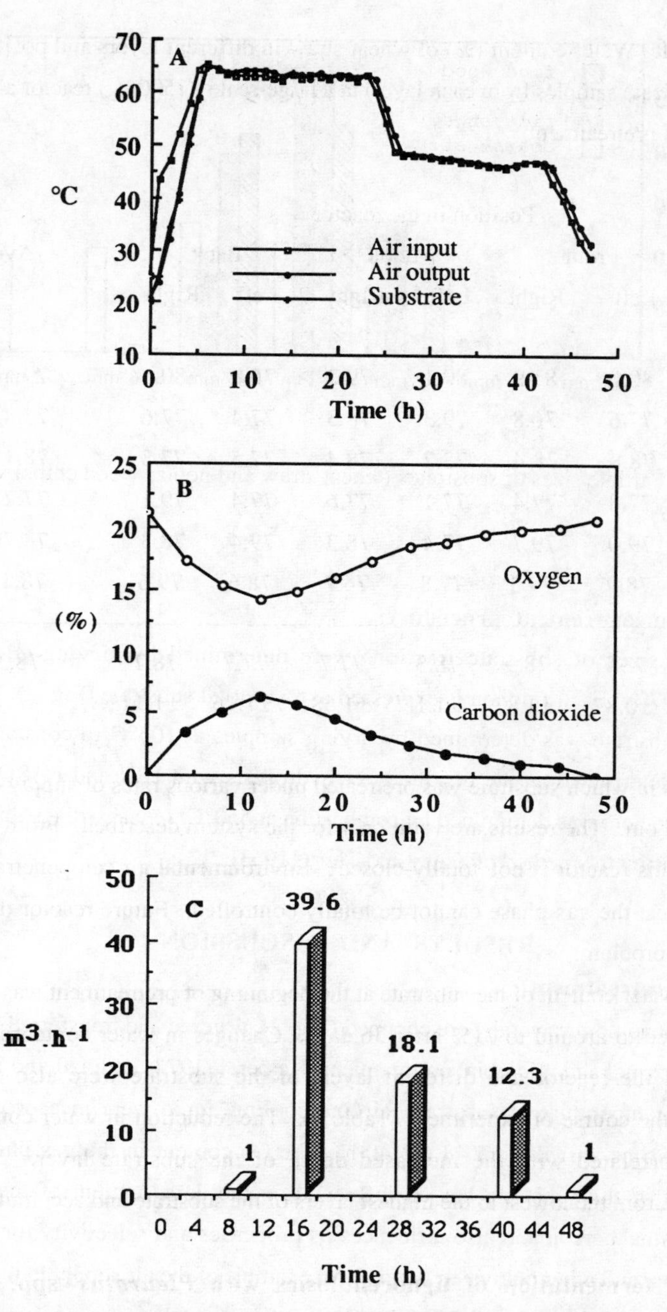

Fig. 3. Example of temperature control and air addition during pretreatment of wheat straw for SSF with white-rot fungi. A, Substrate temperature; B, Composition of gas phase; C, Air added during substrate pretreatment.

Table 1. Water content (%) of wheat straw in different layers and positions (2 replicate samples from each layer) in a large-scale (1500 kg) reactor after 6 days of pretreatment.

	Position in the reactor						
Layer	Front		Middle		Back		Average
(cm)	Left	Right	Left	Right	Left	Right	
0	80.0	81.9	80.1	79.2	76.0	80.1	79.4 ± 1.77
20	77.6	76.8	79.3	78.3	77.4	77.6	77.8 ± 0.85
40	78.8	78.4	78.2	78.4	77.5	77.3	78.1 ± 0.59
60	77.8	77.4	77.1	77.6	77.4	79.1	77.7 ± 0.71
80	79.0	79.7	77.4	78.3	79.4	78.3	78.7 ± 0.82
100	78.9	79.3	77.8	74.7	78.6	79.6	78.1 ± 1.81
Average	78.7 ±0.87	78.9 ±1.87	78.3 ±1.13	77.7 ±1.58	77.7 ±1.16	78.7 ±1.11	78.3 ± 0.55

Experiments in which substrate was pretreated under various rates of supply of fresh air were carried out. The results are valid only for the system described. From a technical viewpoint, this reactor is not 'totally-closed'. Environmental air can penetrate into the system. Thus, the gas phase cannot be totally controlled. Future reactor designs will address this problem.

The water content of the substrate at the beginning of pretreatment was about 80% and decreased to around to 71% after 36 days. Changes in water content in different locations of the reactor and different layers of the substrate were also determined throughout the course of experiment (Table 1). The reduction in water content of the substrate correlated with the increased depth of the substrate layer. Water was translocated from the lowest to the highest layers of the substrate and accumulated on the surface.

Solid-state fermentation of lignocellulosics with *Pleurotus* spp.

Inoculation of substrate

Pretreated substrate is moved with a special machine from one reactor to another (Fig. 4). During this step, the inoculum, either grain spawn or liquid culture, is added and mixed

Fig. 4. SSF reactors during transfer and inoculation of substrate. a, Reactor for substrate pretreatment; b, Reactor for substrate incubation; c, Machine for translocation and inoculation of substrate.

with the substrate. Inoculum was applied at different ratios of inoculum:substrate, e.g. 50 l of liquid inoculum (c. 100 g mycelium dry matter) or 30-50 kg grain spawn per 1000 kg cereal straw.

Incubation

Mycelial growth produces metabolic heat and gaseous metabolites. The most important problem associated with growth in the deeper layers of the substrate is in ensuring the uniform control of temperature, i.e. in the removal of heat from the system. The most promising way to exchange heat energy is to circulate conditioned gas through the substrate. The isothermal state of the substrate is controlled by the speed of gas movement, its temperature and humidity [Teifke and Bohnet (1990) in this volume].

Temperature

The temperature of the substrate after inoculation increases in proportion to the length of incubation (Figs. 5 and 6). The energy requirement for cooling of the substrate also increases with the incubation time. Due to limited gas circulation within the substrate, temperature peaks are observed (Fig. 5). Higher rates of gas circulation, or a reduction of the temperature of the gas phase, or both, are required for more uniform control of the temperature of the substrate.

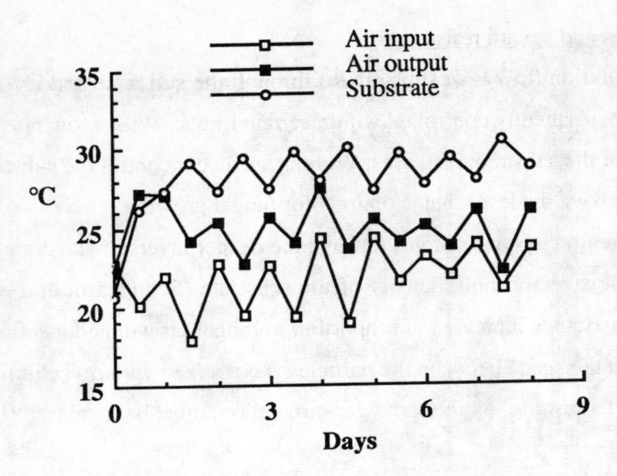

Fig. 5. Fluctuation of temperature in solid state reactor and substrate during solid-state fermentation.

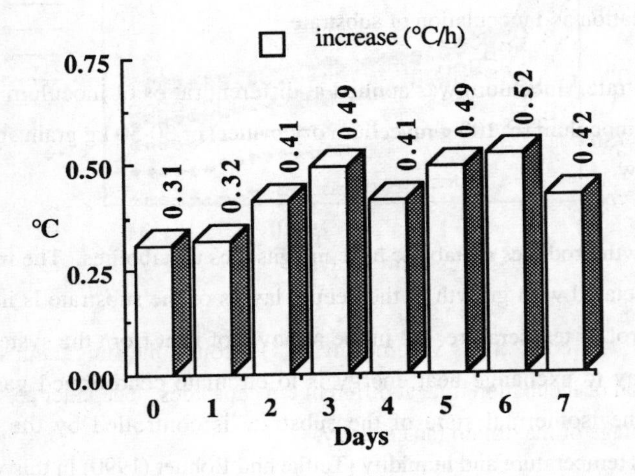

Fig. 6. Increase in temperature ($°C.h^{-1}$) of substrate during SSF relative to that at the beginning of the incubation.

The profile of increases in temperature in 1.5 tonnes of substrate are shown in Fig. 5. The differences between gas temperatures before entering and after leaving the substrate are easily controlled by varying the rate of circulation with gas. The substrate temperature during incubation is more difficult to control than that of the gaseous phase. At the onset of the incubation the temperature differences are low but increase with duration of incubation.

Air circulation and stream resistance

The gas circulation (m^3 gas/t substrate/h) through the substrate and its composition (0_2 and CO_2) is permanently controlled within certain limits. With a low rate of gas flow the temperature of the substrate increases. Without sufficient control the substrate temperature can increase above the level that is optimal for fungal growth.

The stream resistance of gas through the deeper layers of substrate depends on the shape, particle size and bulk density of the substrate (Schuchardt and Zadrazil, 1988). The stream resistance increases in proportion to fungal growth and gas flow rate. During fungal growth the space between the particles is colonized with mycelium and the energy for aeration of substrate, by the same gas-circulation, must be increased (Fig. 7).

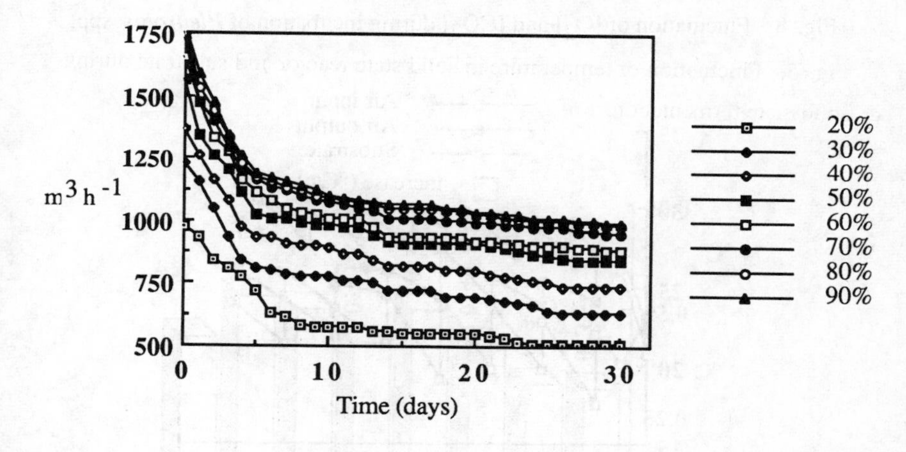

Fig. 7. Percolated gas volume ($m^3.h^{-1}$) through the fungal substrate as a function of incubation time at different fan capacities. The latter are given as a percentage of maximum fan capacity.

In one experiment, periodical control of fermentation conditions was tested (Fig. 5, 6, 8, 9) by allowing the gas to flow without control throughout the day. When the gas flow circulated in the system was not controlled, the temperature of substrate and the CO_2 concentration in the gas phase increased in proportion. When periodic environmental control of fermentation was used, the growth parameters fluctuated by unacceptably large amplitudes. Such a method for controlling fermentation conditions would not be ideal at the beginning of fungal growth but could, perhaps, be used at the later stages of incubation when the substrate is fully colonized.

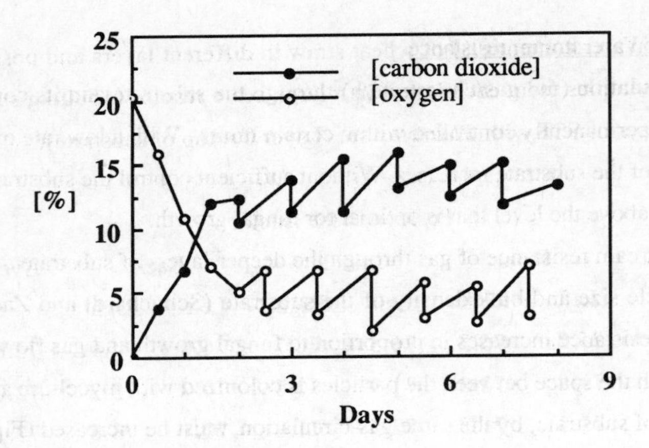

Fig. 8. Fluctuation of $[O_2]$ and $[CO_2]$ during incubation of *Pleurotus* spp.

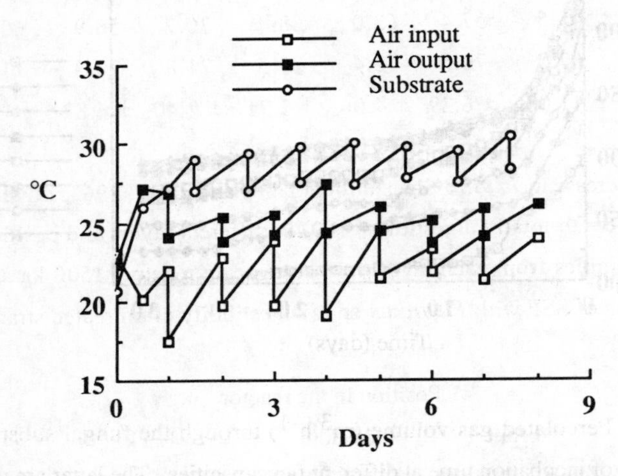

Fig. 9. Fluctuation of temperature in the gas phase and the substrate during SSF in which the system was controlled once per day.

Composition of the gas phase (O_2 and CO_2)

The present SSF-reactor cannot be assumed to be a hermetically-closed system as the concentrations of O_2 and CO_2 in the reactor are continually diluted with ambient air.

Water content in substrate and gas phase

Water tension in the gaseous phase is continually changed during gas circulation in the reactor. Presumably, the temperature differences in the whole system are very small as must, in consequence, be differences in the water tension values in gaseous phase (relative air humidity). The temperature of gas is reduced by the cooling system and the

Table 2. Water content (%) of wheat straw in different layers and positions (2 replicate samples from each layer) in a large-scale reactor (1500 kg substrate) after 38 days of SSF with *Pleurotus* sp. (Layer 0 = top of substrate).

Layer (cm)	Front Left	Front Right	Middle Left	Middle Right	Back Left	Back Right	Average
0	75.2	75.3	76.5	76.7	76.0	74.9	75.8 ± 0.73
20	73.9	72.3	71.9	70.5	70.2	73.2	72.0 ± 1.45
40	73.3	71.3	72.5	71.8	70.1	73.6	72.1 ± 1.29
60	73.7	72.7	73.3	73.2	70.7	73.4	72.8 ± 1.08
80	72.7	73.7	73.8	74.2	73.9	72.5	73.5 ± 0.71
100	45.5	57.4	63.0	68.4	70.2	56.9	60.2 ± 9.06
Av.	69.0	70.5	72.4	72.5	71.8	70.3	71.1 ± 3.81
	± 11.03	± 6.29	± 4.06	± 2.91	± 2.40	± 6.82	

Table 3. Increase in *in vitro* digestibility (Tilley and Terry, 1963) of wheat straw (indigestible = 0; totally digestible = 100) in different layers and positions (two replicate samples from each layer) in a large-scale reactor (1500 kg substrate) after 38 days of SSF with *Pleurotus* sp. (Digestibility of untreated straw = 40).

Layer (cm)	Front Left	Front Right	Middle Left	Middle Right	Back Left	Back Right	Average
0	19.4	11.8	16.0	20.6	21.6	22.8	18.7 ± 4.12
20	18.8	12.8	17.9	16.0	23.7	20.2	18.3 ± 3.71
40	11.9	12.1	16.9	15.7	23.3	14.3	15.7 ± 4.22
60	14.8	15.0	5.7	16.6	14.4	18.5	14.1 ± 4.41
80	8.3	9.5	10.9	11.9	7.8	10.2	9.8 ± 1.57
100	0.0	7.1	7.0	10.2	10.3	7.5	7.0 ± 3.73
Av.	12.2	11.4	10.7	15.2	16.8	15.6	13.8 ± 3.59
	± 7.30	± 2.73	± 5.29	± 3.70	± 6.98	± 5.95	13.7 ± 3.49

air humidity is increased to nearly 100%. After contact with warm substrate, the air temperature increased and water tension decreased. In the overlying substrate layers, the gas is able to take water from the substrate. On the surface of the substrate (i.e between the substrate and open space), the water again condenses and accumulates. *Pleurotus* mycelium on the surface produced yellow guttation droplets.

Evaporation of water from the substrate and condensation on the cold surfaces inside of reactor and gas cooling unit are disadvantages of the system. After termination of the experiment, the water balance and translocation were estimated.

During one experimental period, 106,151 m^3 of gas were moved through the substrate. In the same time, 291 litres of water were evaporated from the substrate and collected in the reservoir. Thus, one may calculate that each m^3 of gas, moved through the substrate, translocated 2.74 g of water. With parallel measurements of evaporation (Czeratzki, 1968), the same dependence was estimated.

The water content in the substrate decreased during incubation. After finishing the experiments, the water content at 36 different sites in the substrate were determined. The lowest content was found in the lowest layer of substrate and the highest on the surface. Correlation between content of water in different layers of substrate is shown in Table 2.

Digestibility and homogenity of product

The digestibility of substrate on fermentation increased on average by 13.8 digestibility units (Table 3). The highest increases (18.7 and 18.3 units) were found in the two layers near the substrate surface and the lowest (7.0 units) on the bottom layer. The observed increases in the digestibility of cereal straw after fungal treatment on a large-scale were comparable with results obtained by sodium hydroxide or ammonia treatment (Sundstøl and Owen, 1984).

After incubation, the substrate was also used for production of edible fungi. Colonized substrate was placed into a container for fructification. The yield of fruit bodies was satisfactory and was comparable to that obtained with other cultivation systems.

The final product (fungal substrate) differs in water content and digestibility. One may assume that differences between the water contents of different layers have an influence on the digestibility of substrate. We hope that this phenomenon can be eliminated by better control of the gas phase.

Infection by competitive microorganisms

The proposed system of SSF is based on the use of non-sterile culture conditions.

Selective propagation of thermophilic and mesophilic microorganisms during substrate pretreatment supports the saprophytic colonization by the cultivated fungus. Infection of substrate was not observed during colonization of substrate under these conditions. After 14 days of incubation, the substrate can be used for the production of edible fungi.

During the production of animal feed, colonies of *Trichoderma* sp. were observed on the surface of the substrate at the end of fermentation. This infection correlates with high substrate digestibility. Infection was frequently observed, when condition for the growth of *Pleurotus* sp. was suboptimal (e.g. temperature being too high).

CONCLUSION

Based on the studies reported here and elsewhere, one may conclude that there is a need for further research on several aspects of the SSF process. These include the following:

1. Development of new designs and constructions of SSF reactors.
    - 1.1 Requirements: Homogenous conditions during processing.
    - 1.2 Minimal differences between process parameters (growth conditions) in different layers of substrates.
2. Development of equipment and sensors for the control of the SSF process.
    - 2.1 Control of air speed in different parts of reactors.
    - 2.2 Control of air humidity (95-100% relative humidity).
    - 2.3 Control of water evaporation from substrate.
    - 2.4 Control of water translocation.
3. Development of strategy for process control.
4. Development of mathematical models of the SSF process.
5. Verification of results in laboratory and pilot-scale reactors.
6. Comparative economic studies with other lignocellulose-upgrading processes.

REFERENCES

Bosma, A.H. and Dernedde, W. (1979). Measuring the particle length of chopped forage. In *Proc. Conf. Forage Conservation in the 80's,* British Grassland Society, Brighton, pp. 331-334.

Czeratzki, W. (1968). Ein Verdungstungsmesser mit keramischer Scheibe. Landbauforschung Völkenrode, **18**, 93-98.

Eckstein, A. (1985). Herstellung von Cellulose aus Holz oder anderen lignocellulosehaltigen Pflanzen durch abbau der Lignocellulose. Europaische

Patenschrift No. 0 060 467.

Francescutti, B. (1972). Verfahren und Vorrichtung zum Anbau von Pilzen. Deutsches Patentamt, Offenlegungsschrift (1972).

Giovannozzi-Sermanni, G. and Perani, C. (1987). Solid-state fermentation in a pilot rotary bioreactor. Chimicaoggi, **3**, 55-57.

Giovannozzi-Sermanni, G., Bertoni, G. and Porri, A. (1989). Biotransformation of straw to commodity chemicals and animal feeds. In *Enzyme Systems for Lignocellulose Degradation*, ed. M.P. Coughlan, Elsevier Applied Science, London, pp. 371-382.

Hatakka, A.J. and Pirhonen, T.J. (1985). Cultivation of wood-rotting fungi on agricultural lignocellulosic material for the production of crude protein. Agricultural Wastes, **12**, 81-97.

Halse, O.M. (1926). Determination of cellulose and wood fibre in paper. Papier Journalen, **10**, 121-126.

Kalberer, P.P. (1974). The cultivation of Pleurotus ostreatus: Experiment to elucidate the influence of different culture conditions on the crop yield. Mushroom Sci. **9**, 653-661.

Kamra, D.N. and Zadrazil, F. (1985). Influence of oxygen and carbon dioxide on lignin degradation in solid-state fermentation of wheat straw with *Stropharia rugosoannulata*. Biotechnol. Lett. **7**, 335-340.

Kamra, D.N. and Zadrazil, F. (1986). Influence of gaseous phase, light and substrate pretreatment on fruit-body formation, lignin degradation and in vitro digestibility of wheat straw fermented with *Pleurotus* spp. Agricultural Wastes, **18**, 1-17.

Laukevics, J.J., Apsite, A.F., Viesturs, U.E. and Tengerdy, R.P. (1984). Solid-substrate fermentation of wheat straw to fungal protein. Biotechnol. Bioeng. **26**, 1465-1474.

Laukevics, J.J., Apsite, A.F., Viesturs, U.E. and Tengerdy, R.P. (1985). Steric hindrance of growth of filamentous fungi in solid-substrate fermentation of wheat straw. Biotech. Bioeng. **27**, 1187-1192.

Levonen-Munoz, E. and Bone, D.H. (1985). Effect of different gas environments on bench-scale solid state fermentation of oat straw by white-rot fungi. Biotechnol. Bioeng. **22**, 382-387.

Loske, D., Hüttermann, A., Majcherczyk, A., Zadrazil, F., Lorsen, H. and Waldinger, P. (1990). Use of white-rot fungi for the clean-up of contaminated sites. In

Advances in Biological Treatment of Lignocellulosic Materials, eds. M.P. Coughlan and M.T. Amaral Collaço, Elsevier Applied Science, London, pp. 311-322.

Majcherczyk, A., Braun-Lüllemann, A and Hüttermann, A. (1990). Biofiltration of polluted air by a complex filter based on white-rot fungi growing on lignocellulosic substrates. In *Advances in Biological Treatment of Lignocellulosic Materials,* eds. M.P. Coughlan and M.T. Amaral Collaço, Elsevier Applied Science, London, pp. 323-330.

Moo-Young, M., Moreira, A.R. and Tengerdy, R.P. (1983). Principles of solid substrate fermentation. In *Fungal Technology,* Vol. **4**, *Filamentous Fungi,* eds. J.E. Smith, D.R. Berry and B. Kristiansen, Edward Arnold, London, pp. 117-144.

Reid, J.D. and Seifert, K.A. (1982). Effect of an atmosphere of oxygen on growth, respiration and lignin degradation by white-rot fungi. Can. J. Bot. **60**, 252-260.

Schuchardt, F. and Zadrazil, F. (1982). Aufschluß von Lignocellulose durch hohere Pilze -Entwicklung eines Feststoff-Fermenters. In *Proc. 5th Symp Techn. Mikrobiologie "Energie durch Biotechnologie",* ed. H. Dellweg, Berlin, pp. 422-428.

Schuchardt, F. and Zadrazil, F. (1988). A 352-litre fermenter for solid-state fermentation of straw by white-rot fungi. In *Treatment of Lignocellulosics with White-rot Fungi,* eds. F. Zadrazil and P. Reiniger, Elsevier Applied Science, London, pp. 77-89.

Sundstøl, F. and Owen, E. (1984). *Straw and Other Fibrous By-products as Feed.* Elsevier, Amsterdam.

Teifke, J. and Bohnet, M. (1990). Modelling of physical process parameters of technical lignin degradation by *Pleurotus* spp. In *Advances in Biological Treatment of Lignocellulosic Materials,* eds. M.P. Coughlan and M.T. Amaral Collaço, Elsevier Applied Science, London, pp. 71-84.

Tengerdy, R.P. (1985). Solid-substrate fermentation. Trends Biotechnol. **3**, 96-99.

Tengerdy, R.P. (1987). Bioconversion of lignocellulose by solid substrate fermentation. In *Proc. Second Latin American Symposium on Biotechnology,* ed. R. Garcia, Guatemala, pp. 130-143.

Tengerdy, R.P., Wissler, M.D. and Murphy, V.G. (1983). Solid-state fermentation of wheat straw. In *Biochemical Engineering III,* eds. K. Venkatsubramanian, A.

Constantinides and W.R. Vieth, Ann. N.Y. Acad. Sci., **413**, 469-472.

Tilley, J.M.A. and Terry, R.A. (1963). A two-stage technique for in vitro digestion of forage crops. J. Br. Grassl. Soc. **18**, 104-111.

Ulmer, D.C., Tengerdy, R.P. and Murphy, V.G. (1981). Solid-state fermentation of steam-treated feedlot waste fibers with *Chaetomium cellulolyticum*. Biotechnol. Bioeng. Symp. **11**, 449-461.

Viesturs, U.E., Strikauska, S.V., Leite, M.P., Berzincs, A.J. and Tengerdy, R.P. (1987). Combined submerged and solid-substrate fermentation for the bioconversion of lignocellulose. Biotechnol. Bioeng. **30**, 282-288.

Yang, H.H., Effland, J.J. and Kirk, T.K. (1980). Factors influencing fungal degradation of lignin in a representative lignocellulosic, thermomechanical pulp. Biotechnol. Bioeng. **22**, 65-77.

Zadrazil, F. (1975). Influence of CO_2 concentration on the mycelium growth of three *Pleurotus* species. Eur. J. Appl. Microbiol. **1**, 327-335.

Zadrazil, F. (1979). Grundlagen für das Wachstum von höheren Pilzen in Schüttsubstraten. Mushroom Sci. **10** (1), 529-538.

Zadrazil, F. (1980). Conversion of different plant waste into feed by Basidiomycetes. Eur. J. Appl. Microbiol. Biotechnol. **9**, 243-248.

Zadrazil, F. (1985). Screening of fungi for lignin decomposition and conversion of straw into feed. Angewandte Botanik, **59**(5-6), 433-452.

Zadrazil, F. and Brunnert, H. (1980). The influence of ammonium nitrate supplementation on degradation and *in vitro* digestibility of straw colonized by higher fungi. Eur. J. Appl. Microbiol. Biotechnol. **9**, 37-44.

Zadrazil, F. and Brunnert, H. (1981) Investigation of physical parameters on solid-state fermentation by (saprophytic) white-rot fungi. Eur. J. Appl. Microbiol. Biotechnol. **11**, 183-188.

Zadrazil, F. and Kurtzman, R.H. Jr. (1982). The biology of *Pleurotus* cultivation in the tropics. In *Tropical Mushrooms,* eds. S.T. Chang and T.H. Quimio, The Chinese University press, pp. 277-298.

Zadrazil, F., Janssen, H., Diedrichs, M. and Schuchardt, F. (1990). Pilot-scale reactor for solid-state fermentation of lignocellulosics with higher fungi: Production of feed, chemical feedstocks and substrates suitable for biofilters. In *Advances in Biological Treatment of Lignocellulosic Materials,* eds. M.P. Coughlan and M.T. Amaral Collaço, Elsevier Applied Science, London, pp. 31-42.

The Role of the Soluble Lignocellulose Produced During Solid-state Conversion of Plant Material by White-rot Fungi

G. Giovannozzi-Sermanni, A. Porri, C. Perani, L. Badalucco and A.M. Garzillo

Dept. Agrobiologia e Agrochimica, Universitá Degli Studi Della Tuscia
01100 Viterbo, Via S. Camillo de Lellis, Blocco B, Italy

In this paper we present the results of studies on the utilization of a water-soluble lignocellulose copolymer (SLC) produced during solid-state and submerged fermentation by ligninolytic mycelia, and on the influence of the SLC on the physiology and biochemistry of the fungi. The results indicate that SLC is readily utilized as a carbon source by the mycelia and that the metabolic activities of *Lentinus edodes* and *Pleurotus ostreatus* change when grown on media containing SLC in addition to other organic compounds. The main changes concern cellular activity and the production of cell wall-degrading enzymes. This suggests that, during the degradation of cell walls, the presence of SLC or straw influences different aspects of fungal metabolism.

A water-soluble lignin-hemicellulose-cellulose complex may be of considerable help in studying growth and the mechanisms involved in degradation and utilization of cell wall material. The cellular content of ATP appears to be a particularly useful parameter for comparing mycelial masses of the same age but grown under different conditions of solid-state cultivation.

INTRODUCTION

Two problems must be faced when carrying out basic research on the physiological behaviour of ligninolytic fungi. These are the insolubility of lignocellulosic material, and our incomplete understanding of the physiological and biochemical mechanisms of lignin and cellulose degradation. These areas of research are currently of considerable interest and many workers are engaged in developing biotechnological processes for the utilization of plant materials. The data presented in this paper will be useful in process optimization of the 3500-l solid substrate bioreactor assembled in our LINA pilot plant service to produce feed, enzymes and soluble lignocellulose (Fig. 1).

The utilization of lignocellulose may occur via the formation of soluble polymeric lignocellulose (SLC), which in turn is depolymerized to small molecules that are taken up

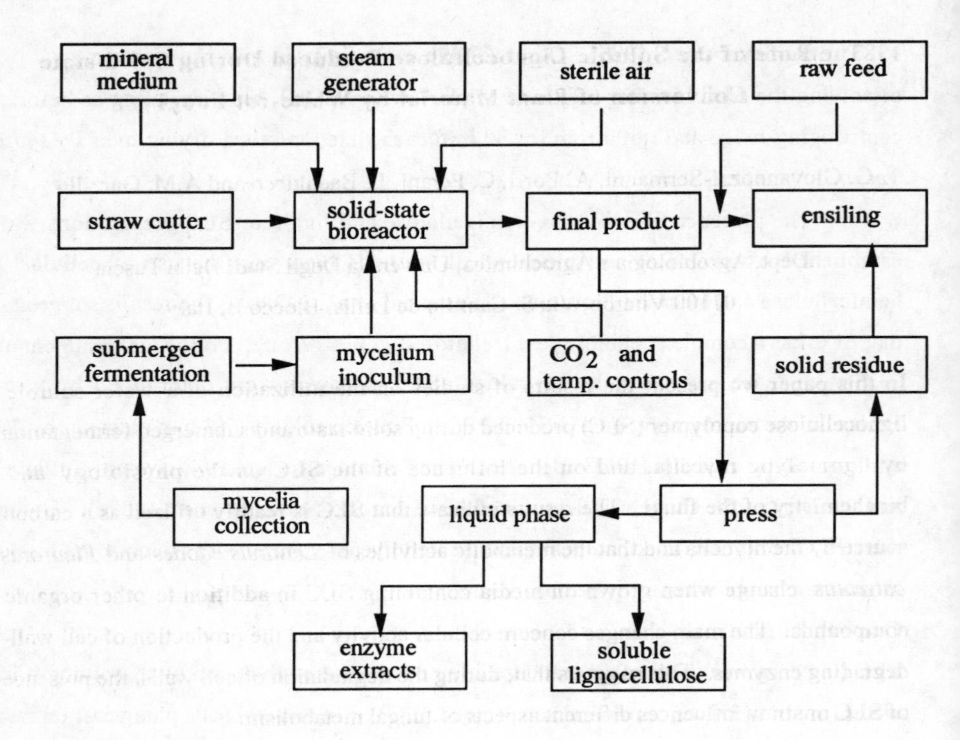

Fig. 1. Pilot-plant scheme.

by the mycelia (Giovannozzi-Sermanni *et al.*, 1989a, b). In order to ascertain whether this proposed chain of events is correct, we used SLC as a substrate for mycelial growth and for the production of cell wall-degrading enzymes during which time we monitored the alterations to the polymeric characteristics of the substrate. We have also examined the changes in the fungal growth pattern that occur when SLC is used in addition to other soluble or insoluble organic nutrients.

MATERIALS AND METHODS

SLC preparation

SLC was obtained after 10 days of solid-substrate fermentation of straw by white-rot fungi such as *Pleurotus* sp. or *L. edodes* (Giovannozzi-Sermanni *et al.*, 1989a). The lignocellulosic mush (containing 66% water by weight) was pressed and the dark-brown liquid was filtered to remove insoluble material. By adjusting the pH of the filtrate to 1 using 12 N HCl, brown material was precipitated from the water phase (Crawford *et al.*,

1983; Bergmeyer *et al.*, 1985). Other soluble coprecipitated material was removed by dissolving the SLC in water (adjusted to pH 7 with NaOH), reprecipitating as before, centrifuging twice at 11,000 rpm for 30 min each time, and then drying over P_2O_5 *in vacuo.*

The presence of cellulose, hemicellulose and lignin in SLC preparations was established by chemical and biochemical methods including the use of cellulase, hemicellulase and laccase (Giovannozzi-Sermanni *et al.*, 1990b). The results suggested that SLC has a complex cellulose-hemicellulose-lignin structure. Following treatment of SLC with these enzymes, the M_r values of the products were determined by size exclusion chromatography on a G-3000 PWXL Supelco column (7.8 mm diameter by 30 cm long) using a Varian HPLC apparatus. The column was eluted with water (1ml/min) and the A_{278} of the eluate was determined (Giovannozzi-Sermanii *et al.*, 1989a). Under the HPLC conditions used, no peaks ascribable to yeast extract, with which some liquid media were supplemented, were observed.

Growth conditions

SLC (0.1%, w/v) was used a a standard component of growth media by dissolving it in water (medium D), or in mineral salts (medium C) or in mineral salts plus yeast extract (medium B) as described by Giovannozzi-Sermanii *et al.* (1989a). Difco potato broth (medium A) was used as a reference medium. All media were sterilized at 120°C for 20 min and inoculated with *L. edodes* or with *P. ostreatus.* Eleven flasks, each containing 250 ml of liquid medium, were incubated at 28°C for 10 days with orbital shaking (70 rev/min). The broths obtained following centrifugation at 11,000 rpm for 30 min were used for size exclusion chromatography studies. Medium C was used as the basic growth medium for submerged cultures, but the carbon source (SLC, straw or glucose) was varied as described below.

Enzyme assays

Phenol oxidase activity in culture broth, clarified by centrifugation at 10,000 rpm for 15 min, was assayed using 2,2'-azino bis(3-ethyl-thiazoline-6-sulphonic acid) (ABTS) or 2,6-dimethoxyphenol (DMP) as substrate. With ABTS as substrate, activity was measured by monitoring the increase in A_{420} in a reaction mixture (1 ml) containing an aliquot of enzyme and 2 mM ABTS in 100 mM glycine-HCl buffer, pH 3.0. Activity with DMP as substrate, was measured by monitoring the change in A_{468} following addition of an aliquot of enzyme to a reaction mixture (final volume 3 ml) containing 1 mM DMP in 100 mM citrate-phosphate buffer, pH 6.0.

Endocellulase activity was measured using carboxymethylcellulose (CMC) as substrate. The reaction mixtures (final volume 0.5 ml) consisted of 0.25 ml of culture broth and 0.25 ml of 1% (w/v) CMC in 50 mM citrate buffer, pH 3.7. After 1 h incubation at 50°C, the amount of reducing sugars liberated was determined by the method of Robyt and Whelan (1968) using glucose as standard.

Exocellulase activity was tested by incubation of an aliquot of enzyme with 1% (w/v) Avicel in 1 ml of 100 mM acetate buffer, pH 5.0, for 2 h at 50°C. After clarification by centrifugation, the reducing sugar content of the supernatant was determined as above.

ATP determination

ATP was extracted from mycelial samples by using a modification of the procedure of Jenkinson *et al.* (1979) for extracting ATP from soil. To 25-250 mg of freshly harvested mycelia, or 250-500 mg of lignocellulosic mush (inoculated with mycelia), were added 25 ml of extracting solution (0.5 M trichloroacetate, 0.25 M $Na_2HPO_4.12H_2O$ and 2 mM Na_2EDTA). The extraction was carried out for 3 min at 0°C using a homogenizer operating at high speed. The extract was filtered through Whatman No. 42 paper and stored at -18°C. The extract was diluted 10-fold with 0.1 M Tris-acetate buffer, pH 7.75, containing 2 mM Na_2EDTA. An aliquot (0.2 ml) of this solution was assayed using the 1243-107 ATP bioluminescence assay kit (luciferin-luciferase system) and the 1250 Luminometer, both supplied by LKB-Wallac.

ATP was not detectable in extracts of wet mycelium or inoculated straw that had been stored at -18°C for a few days prior to extraction. Native straw (used as a control) did not contain measureable amounts of ATP.

RESULTS AND DISCUSSION

Characteristics of SLC

The SLC fraction, obtained by acidic precipitation from the liquid phase of lignocellulosic mush after colonization by *L. edodes,* is a lignin-hemicellulose-cellulose complex comprised of macromolecules with M_r values ranging from 2,000 to approximately 100,000 (Giovannozzi *et al.*, 1990b). It contains 3% nitrogen by weight as determined by conventional micro-Kjeldahl procedures. As yet, we are uncertain as to whether the nitrogen compounds are adsorbed to the copolymer or whether they are linked to it by chemical bonds, as is the case with humic matter.

SLC as a substrate for mycelial growth

We have used solid-state and submerged fermentation procedures to study the bioconversion of lignocellulose. The form of wheat straw used as growth substrate depended on the cultivation conditions: (i) particle size - normally short sticks for solid-state fermentation (SSF) and crushed material for submerged cultivation; (ii) the lignocellulose:water ratio - for SSF this was usually 300:1000 and corresponded to an SLC:water ratio of 1.5:1000 (0.45% SLC in durum wheat straw). In submerged cultivation the SLC:water ratio was 1:1000; (iii) the specific weight of the substrate - under SSF conditions 1000 cm^3 weighed 220 g (i.e. 73 g dry weight straw) while in submerged cultivation 1000 cm^3 weighed about 1000 g. Therefore, the total lignocellulose concentrations and, hence, the exchange of O_2 and CO_2 between liquid and gas phases in the two cultivation procedures differed significantly.

Consequently, when studying the bioconversion efficiency, one must take account of the mycelial biomass produced under both conditions of cultivation. In the literature, various parameters (e.g. acetylglucosamine, nucleic acid, ergosterol or protein contents or enzyme activities) have been recommended for use in the measurement of mycelial biomass. These recommended procedures attempt to overcome the problems associated with the great variety in amounts of soluble and insoluble organic compounds in solid-state cultures. However, each analysis is limited by the lack of constancy in the ratio of the quantity of biomass and the measured parameter during the growth cycle. Mycelial ATP content could be an additional parameter for monitoring of the metabolic conditions within the cells and, hence, of correlating mycelial growth with culture conditions.

In our hands, the maximum values obtained in submerged cultures was 0.51 ± 0.034 µmol $ATP.g^{-1}$ dry weight of mycelium, while that for solid-state cultures was 0.04 ± 0.002 µmol $ATP.g^{-1}$ dry weight of straw. These values are lower than those generally reported for other organisms. Determinations on a wide range of growing cells show that the ATP content of prokaryotes is 5.0 ± 2.4 $µmol.g^{-1}$ dry biomass, while that of eukaryotes is 5.9 ± 1.8 $µmol.g^{-1}$ dry biomass (Knowles et al., 1977). Other authors have reported an average value of 3.63 µmol.g-1 dry biomass (Lee *et al.*, 1971).

In our experiments, the cellular ATP content changed during mycelial growth, with the maximum values being obtained after 4-5 days of submerged cultivation. At this time other metabolic parameters also reach maximum values - a fact, that has also been reported for other types of fungal mycelia (see below). In order to establish a correlation

Table 1. ATP content and biomass production during submerged cultivation of *L. edodes* for 10 days.

Medium	Growth rate (g dry wt.l^{-1})	[ATP] μmol.l^{-1}
A. Potato broth	1.939	2.8900
B. 0.1% SLC + mineral salts + yeast extract	1.098	1.5300
C. 0.1% SLC + mineral salts	0.135	0.0634
D. 0.1% SLC	0.184	0.0699

between ATP content and biomass production after the same growth period, *L. edodes* was grown for 10 days in submerged cultivation as described in Table 1. The mycelium grew at different rates as sphere-like agglomerates with invariable water content (94.54%) but with differences in size, morphology, colour and ATP content (Fig. 2).

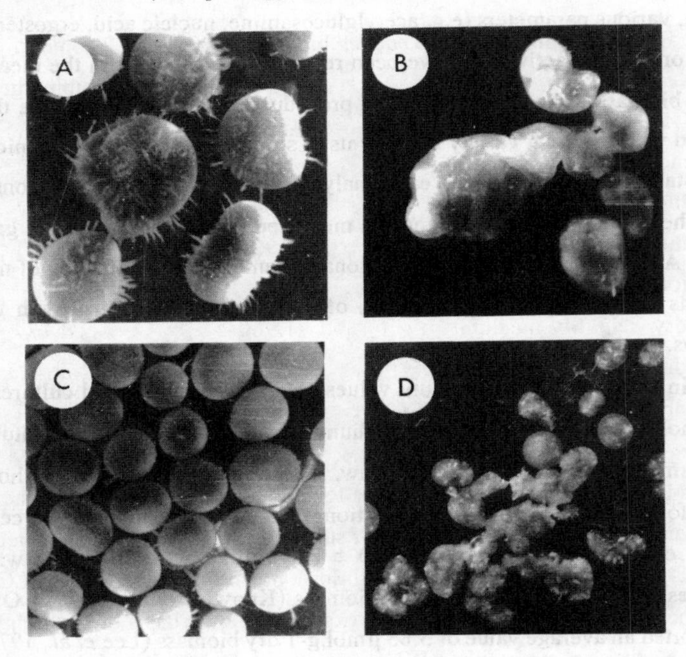

Fig. 2. Mycelial agglomerates produced on submerged cultivation. A, potato starch; B, SLC + mineral salts; C, SLC + mineral salts + yeast extract; D, SLC alone.

From the above data, the following linear regression between biomass and ATP content was obtained:

Biomass (g. dry wt.l^{-1}) = 0.63145 (ATP µmol.l^{-1}) + 0.1202 (r^2 = 0.999)

Thus, at a given culture age, ATP content appears to be a reasonable index of biomass production under different conditions of growth. Since ATP content is dependent on the availability of nutrients, its determination could give information on mycelial growth rate. Moreover, since mycelial growth rate is strongly dependent on ATP, if all other parameters are kept constant, the determination of ATP content could be utilized to compare the the effects of various treatments particularly in the unfavourable analytical situations presented by solid-state cultures.

By using the above results, the ATP contents in 10-day samples of solid-state and submerged cultures of *L. edodes* were determined in order to estimate mycelial biomass as a function of the volume of the medium, the amount of straw used or the amount of SLC used.

Table 2. Estimation of *L. edodes* biomass from the ATP content.

	Solid-state	Submerged
ATP concentration		
µmol.l^{-1} medium	2.920	0.021
µmol.g^{-1} straw	0.040	-
µmol.g^{-1} SLC	2.920	0.021
Biomass production		
g dry weight.l^{-1} medium	1.960	0.134
g dry weight.g^{-1} straw	0.145	-
g dry weight.g^{-1} SLC	1.960	0.134

The results in Table 2 show that the values for biomass production under the solid-state conditions used, given the concentration of substrate, were the same whether expressed on the basis of the volume of the lignocellulosic mush or the weight of SLC used. By contrast, the value was much lower when expressed on the basis of the weight of straw used. If one considers straw, like SLC, to be a readily available carbon source, one may note that the biomass value obtained following solid-state fermentation on straw was the same as that obtained following submerged cultivation on SLC alone. Indeed, if one considers the SLC content of straw, the biomass content of submerged cultures should be

10-fold higher than that obtained by solid-state cultivation. Thus, it appears that solid-state conditions allow of a more efficient conversion of the lignocellulosic matter.

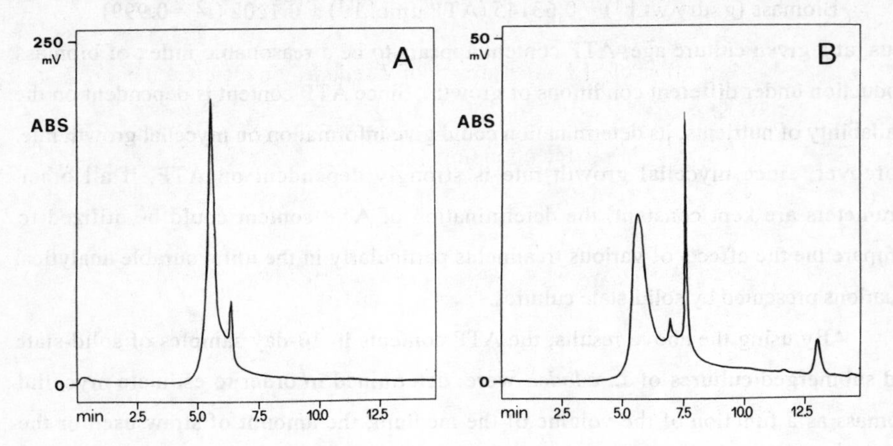

Fig. 3. Size exclusion chromatographic pattern of SLC before (A) and after (B) 10-days fermentation by *L. edodes* supplemented with yeast extract.

The pattern of M_r values of soluble lignocellulose, as determined by size exclusion chromatography, is characterized by 2 distinct peaks (Fig. 3). After 10-days growth, the total amounts of SLC declined quickly in all media even though fractions with the same retention times were still detectable (Table 3). In the presence of yeast extract, the mode of degradation of SLC was clearly modified since 2 new peaks were observed in the chromatographic patterns. The peak with a retention time of 13.3 min was also observed after treatment of SLC with cellulase (Giovannozzi *et al.*, 1990b).

The above results suggest that this polymeric material can be considered to be a water-soluble plant cell wall, that it is useful in preparing nutritive media with a defined lignocellulose content, and that the path of degradation of SLC may differ depending on the presence or absence of other organic compounds in the medium.The composition of the medium used also affects the production pattern of phenol oxidase activity. When glucose or SLC was used as carbon source, maximal expression of laccase-like activity was observed on the 4th day of growth. However, when glucose was replaced by powdered straw, mycelial growth and enzyme production were greatly depressed at 4-5 days. However, production of laccase-like activity reached an even greater maximum (than with glucose as growth substrate) at 8-10 days with absolute values comparable to those obtained with SLC as substrate (Fig. 4). Thus, it appears that SLC is a readily available carbon source for mycelial growth and that it induces greater synthesis of laccase-like activity than do other individual organic compounds.

Table 3. M_r profiles of SLC after growth of *L. edodes* on different media.

Medium	Peak retention time (min)				
	5.64	6.42	7.76	13.3	
	5.79	7.01	7.63		
	Peak area as % of total peak area				
Control*	82	18	-	-	100.0
B	61	12	21	5	23.4
C	73	26	-	-	10.5
D	84	15	-	-	28.2

*Uninoculated medium

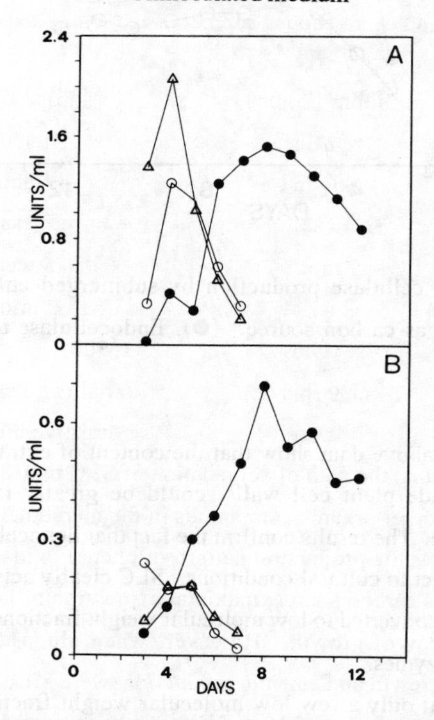

Fig. 4. Extracellular phenol oxidase activities of submerged cultures of *P. ostreatus* grown on glucose (O), SLC (Δ) or straw (●). A, ABTS as substrate; B, DMP as substrate.

On the other hand, induction of cellulase synthesis occurred only when the native structure of lignocellulose (straw powder) was present and the other ready-to-use carbon sources were absent. No cellulase activity was detectable when glucose or SLC was used as carbon source, whereas a large amount of such activity was excreted to the culture fluid when powdered straw was the only substrate present (Fig. 5).

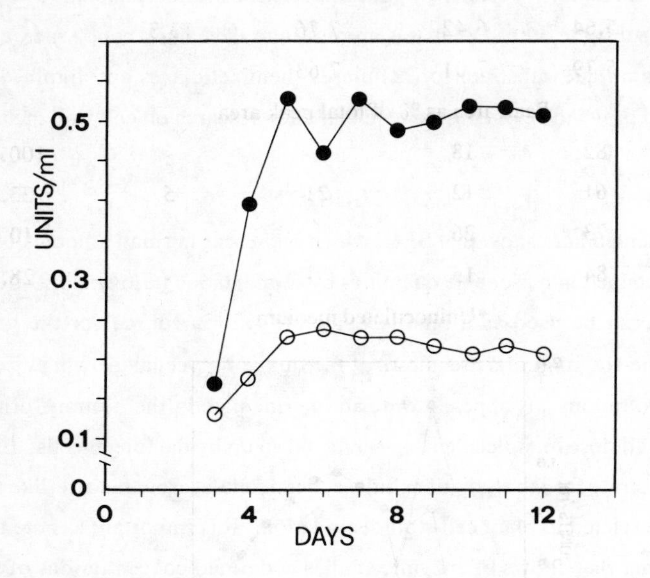

Fig. 5. Extracellular cellulase production by submerged cultures of *P. ostreatus* with straw as carbon source. (●), Endocellulase activity (O), exocellulase activity.

One may consider that the above data show that the content of extracellular enzymes, probably those that degrade plant cell walls, could be greatly influenced by the accessibility of the substrate. The results confirm the fact that lignocellulose degradation is a complex problem subject to cultural conditions. SLC clearly acts as a substrate for mycelial growth, since it is converted to low molecular weight fractions in the presence of plant cell wall-degrading enzymes.

It is noteworthy that only a few low molecular weight fractions appear. This could be due to the fact that some regions of SLC are more subject to enzymic attack than others. Indeed, the proposed lignin-hemicellulose-cellulose structure of SLC and the fact that some regions may be more resistant than others to enzymic attack may suggest the presence of repeat units in the complex.

One should keep in mind that the substrates normally used to study cell wall-degrading enzymes are chemically-modified celluloses and water-soluble polyphenols (as lignin models). Moreover, the sizes and structures of these artificial substrates differ substantially from naturally-occurring lignocellulosic substrates. Thus, the availability of a water-soluble substrate, i.e. SLC, that is more akin to the natural copolymer, and which can be used as a unique substrate for cellulases, hemicellulases and lignin-degrading enzymes, should prove to be of considerable benefit in research on cell wall metabolism.

CONCLUSION

The results presented here show that SLC, which is present in small amounts in wood or straw, can be obtained in reasonable quantities by fermentation (Giovannozzi-Sermanii *et al.*, 1989a). It can be used as a readily available carbon source for the growth of ligninolytic white-rot fungi. Its use allows of the study of mycelial growth rate at defined substrate concentrations. It appears to be an intermediate in the biotransformation of insoluble lignocellulose to molecules that can be taken up by the fungal cells. It may also be used as a substrate for the study of cellulase, hemicellulase and laccase-like activities, all of which are related to plant cell wall degradation. It is important to note that it is a natural copolymer that allows of enzymic studies at defined concentrations of substrate. Indeed, by using this substrate, one should be able to gain a better understanding of the biochemistry of cell wall degradation, of which ligninolysis and cellulolysis are the main processes.

The determination of ATP concentration appears to be a suitable procedure for the determination of biomass content in solid-state cultures and for the comparison of biomass production under different solid-state conditions.

Our increased understanding of the utilization of SLC and of the value of determining cellular ATP concentration will be of considerable assistance in the development of applications in the paper industry, feed production and in the metabolism of aromatic compounds such as pesticides.

REFERENCES

Bergmeyer, J.R. and Crawford, D.L. (1985). Production and characterization of a polymeric lignin degradation intermediate from two different *Streptomyces* sp. Appl. Environ. Microbiol. **49**, 273-278.

Crawford, D.L., Pometto, H.L. and Crawford, R.L. (1983). Lignin degradation by *Streptomyces viridosporus:* Isolation and characterization of a new polymeric lignin degradtion intermediate. Appl. Environ. Microbiol. **45**, 898-904.

Giovannozzi-Sermanni, G. and Perani, C. (1987). Solid-state fermentation in a pilot rotary bioreactor. Chimicaoggi, **3**, 55-57.

Giovannozzi-Sermanni, G. and Porri, A. (1989a). The potential of solid-state biotransformation of lignocellulosic materials. Chimicacoggi, **7**, 15-19.

Giovannozzi-Sermanni, G., Bertoni, G. and Porri, A. (1989b). Biotransformation of straw to commodity chemicals and animal feeds. In *Enzyme Systems for Lignocellulose Degradation,* M.P. Coughlan (ed.), Elsevier Applied Science, London, pp. 371-382.

Giovannozzi-Sermanni, G., Perani, C. and Porri, A. (1990a). Biodelignification and metabolite production under different solid-state fermentation conditions. In *Proc. Fourth Int. Symp. Biotechnology in the Pulp and Paper Industry,* H.-M. Chang and T.K. Kirk (eds.), in press.

Giovannozzi-Sermanni, G., Perani, C., Porri, A., De Angelis, F., Barbarulo, M.V., Mendola, D. and Nicoletti, R. (1990b). Solid-state bioconversion of plant cell wall macromolecules by means of *Lentinus edodes.* Biomass, submitted.

Jenkinson, D.S. and Oades, J.M. (1979). A method for measuring adenosine triphosphate in soil. Soil. Biol. Biochem. **11**, 193-199.

Knowles, C.J. (1977). Microbial metabolic regulation by adenosine nucleotide pools. Symp. Soc. Gen. Microbiol. **27**, 241-283.

Lee, C.C., Harris, R.F., Williams, J.D.H., Syers, J.K. and Armstrong, J.K. (1971). Adenosine triphosphate in lake sediments. I. Soil Soc. Amer. Proc. **35**, 82-86.

Robyt, J.F. and Whelan, W.J. (1968). The α-amylases. In *Starch and its Derivatives,* S.A. Radley (ed.), Chapman and Hall, London, pp. 431-432.

MODELLING OF THE PHYSICAL PROCESS PARAMETERS OF TECHNICAL LIGNIN DEGRADATION BY *PLEUROTUS* SPP.

J. Teifke and M. Bohnet

Institute of Process Technology
Technical University of Braunschweig
Langer Kamp 7, D-3300 Braunschweig, FRG

The phenomenon of lignin-decomposition by fungi is regarded as a heat and mass transfer process. Physical, chemical and biological fundamentals have to be taken into account by formulating the parameters that describe the process. In the first step of the description, far-reaching and simplified assumptions are necessary. The aim of modelling in the first instance is to derive a set of equations that describe the process well and which can be developed by further consideration of physical, chemical and biological details. In the Federal Research Centre for Agriculture at Braunschweig, a solid-state fermentation technique, based on the use of white-rot fungi and wheat straw, was developed. In this context, modelling is restricted to describing the steady-state and transient flow behaviour. The latter is an important factor governing heat and mass transfer.

INTRODUCTION

The economic importance of technical processes using microorganisms that convert lignocelluloses into products useful to man has been emphasized in the literature (e.g. Smith *et al.*, 1988). White-rot fungi are most effective microorganisms for the biological degradation of lignocellulosic materials. Schuchardt and Zadrazil (1988) regard the aims of such treatment to be the production of material of higher digestibility for ruminants and to produce edible mushrooms (e.g. *Pleurotus* sp.). Development of suitable fermentation facilities is assisted by the availability of a theoretical model that describes the process parameters. However, the formulation of such a model has to be based on experimental data. But when the accuracy of the model is satisfactory, information about the influence of a number of parameters may be obtained much more easily. In this way, the number of costly experiments required can be reduced considerably.

MATERIALS AND METHODS

Experimental fermenter equipment

Two experimental fermenter plants have been constructed at the Federal Research Centre for Agriculture, Braunschweig. One, described by Schuchardt and Zadrazil (1988), has a net volume of 352 litre and the other a net volume of 10.5 m^3. Both are based on the solid-state fermentation (SSF) of wheat straw by *Pleurotus sajor caju.*

The cylindrical 352-l fermenter consists of PVC and has a filling height of 2 m and a diameter of 0.5 m. The fermenter is hermetically-sealed by 2 lids with openings for gas circulation pipes. Gas is circulated by way of a pump with a capacity of up to 40 $m^3.h^{-1}$. A humidifier (injection of heated water by nozzles), a cooler, a heater, and facilities for the addition of nitrogen (or air), CO_2 and O_2 were also installed in the gas circulation loop. An electronic control system maintained constant temperature, humidity, CO_2 and O_2 concentration in the gas phase entering the fermenter. Exhaust gas left the fermenter through a water column or a hand valve. The flow rates of the circulating gas, the added gas and the exhausted gas were measured by gas meters. The temperatures of the inlet and outlet gases were measured by a PT 100 resistance thermometer, as was the temperature at 9 different levels in the fermenter. The CO_2 concentration was measured using an infrared method and that of O_2 with a paramagnetic measurement instrument. The fermenter was enclosed by a temperature-controlled chamber in order to maintain adiabatic conditions. The pressure loss of the fermenter was measured by a U-tube manometer.

The 10.5-m^3 fermenter has a rectangular basal surface (width 2.3 m, depth 2.1 m, height 2.16 m) and a filling height of 1.8 m. Gas passes through a slatted floor before flowing through the substrate. An air-cooler and a centrifugal fan have been installed in the circulation loop. Below the slatted floor there is a basin to collect the water that leaves the substrate during fungal growth. CO_2, O_2 or air can be added to the circulated gas, the flow rate being measured by means of the dynamic pressure and a calibration curve. Substrate temperature can be measured at two levels located 0.2 m and 0.6 m below the surface.

PRESSURE LOSS THROUGH THE PACKED SUBSTRATE

Stationary conditions

To simplify the modelling the packed substrate, i.e. chopped-milled wheat straw, can be regarded as a bed of granular solids. Ergun (1952) derived an equation to predict the pressure loss during the flow of fluids through such beds:

$$\Delta p = 150 \frac{[1-\varepsilon]^2}{\varepsilon^3} \frac{H}{d_p^{\,2}} \eta w + 1.75 \frac{[1-\varepsilon]}{\varepsilon^3} \frac{H}{d_p} \rho w^2 \qquad \text{(i)}$$

where w is the superficial fluid velocity; ρ, is the density of fluid; η, is the dynamic viscosity ; H, is the filling height; ε, is the fractional void volume in bed; d_p, is the effective diameter of particles

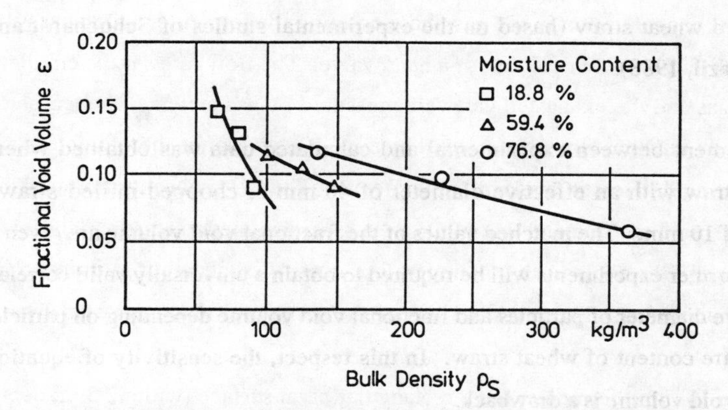

Fig. 1. Calculated fractional void volume versus bulk density for chopped wheat straw (based on the experimental studies of Schuchardt and Zadrazil, 1988).

As an analogous equation for beds of chopped-milled wheat straw does not exist, equation (1) was used. To get a complete equation the unknown parameters, viz. fractional void volume and effective particle diameter, were determined by matching the equation with some experimental results. In these studies (using the 352-l fermenter) of flow resistance behaviour, the superficial gas velocity, the bulk density of the chopped-milled wheat straw, the moisture content, and the particle length were varied. Some of the results were reported by Schuchardt and Zadrazil (1988).

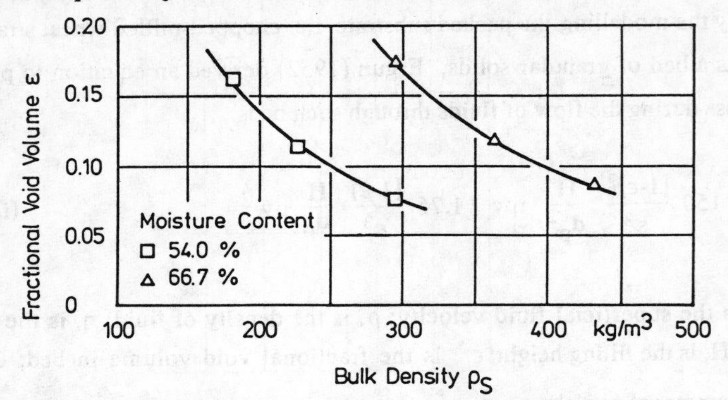

Fig. 2. Calculated fractional void volume versus bulk density for chopped-milled wheat straw (based on the experimental studies of Schuchardt and Zadrazil, 1988).

Best agreement between experimental and calculated data was obtained when using chopped straw with an effective diameter of 40 mm or chopped-milled straw with a diameter of 10 mm. The matched values of the fractional void volume are given in Figs. 1 and 2. Further experiments will be required to obtain a universally valid correlation for the effective diameter of particles and fractional void volume depending on particle length and moisture content of wheat straw. In this respect, the sensitivity of equation (i) to fractional void volume is a drawback.

Growth period

The pressure loss through the substrate increases during growth because the growing mycelial fill more and more of the pore spaces. While the fractional void volume decreases the flow resistance increases. As the rate of change is low, the pressure loss can be calculated each time to be quasi-steady state.

The pressure loss during the growth period of a 24-day fermentation experiment is shown in Fig. 3. The substrate temperature was 24 to 30°C, i.e. within the range for growth of *Pleurotus* spp. The O_2 concentration was controlled at 5 vol-% by the addition of air.. Accordingly, the CO_2 concentration remained constant at a value of about 15 vol-%. Before growth started the CO_2 concentration was about 80 vol-%. Hence, the reason that growth of *P. sajor caju* was prevented.

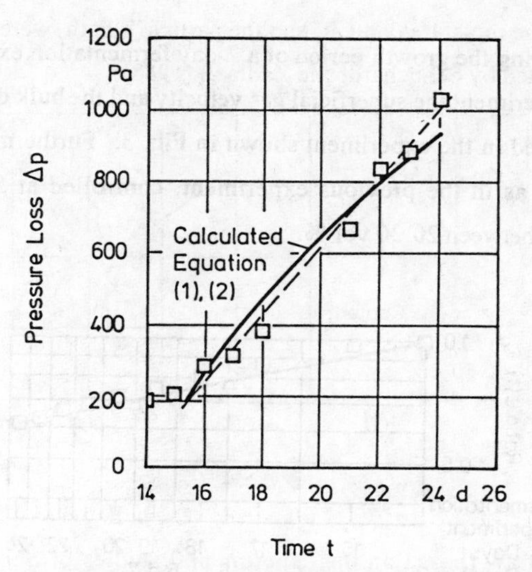

Fig. 3. Pressure loss during the growth period of a 24-day cultivation of *P. sajor caju* on chopped wheat straw. Filling height, 2 m; moisture content 68%; bulk density, 284 kg.m^{-3}; superficial velocity, 0.0163 m.s^{-1}.

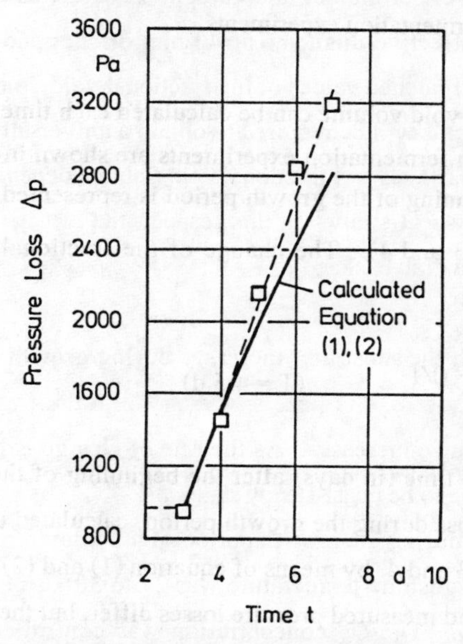

Fig 4. Pressure loss during the growth period of a 7-day cultivation of *P. sajor caju* on chopped-milled wheat straw. Filling height, 2 m; moisture content, 68%; bulk density, 341 kg.m^{-3}; superficial velocity, 0.0325 m.s^{-1}.

The pressure loss during the growth period of a 7-day fermentation experiment is shown in Fig. 4. In this experiment the superficial gas velocity and the bulk density values were higher than those used in the experiment shown in Fig. 3. Furthermore, while the O_2 concentration was, as in the previous experiment, controlled at 5 vol-%, the CO_2 concentration varied between 20-30 vol-%.

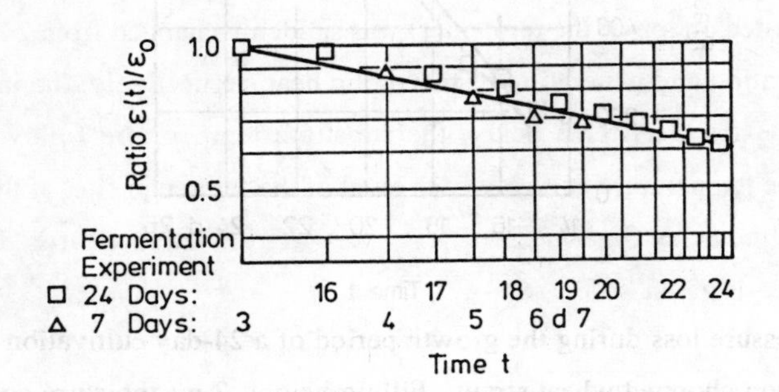

Fig. 5. Changes of the fractional void volume during the growth period of the two different fermentation experiments.

The unknown fractional void volume can be calculated each time by means of equation (1). The results for both fermentation experiments are shown in Fig. 5. The fractional void volume at the beginning of the growth period is represented by ε_o (see location of the beginning in Figs. 3 and 4). The change of the fractional void volume can be approximated as follows:

$$\varepsilon(t) = \varepsilon_o . \varepsilon^{-\ln(1+t)/T} \qquad (T = 4.5\ d) \qquad \text{(ii)}$$

In this equation, t is the time (in days) after the beginning of the growth period. The curves of the pressure loss during the growth period, calculated using equations (i) and (ii), are shown in Figs. 3 and 4. by means of equation (1) and (2) are shown in Figure 3 and 4. The calculated and measured pressure losses differ, but the error is less than 15%, and it is not expected that these different fermentation experiments can be described by one equation for the fractional void volume.

TRANSIENT BEHAVIOUR OF THE PACKED SUBSTRATE

Temperature distribution in the substrate in response to discontinuity in heat supply

A sudden change in gas temperature was introduced in order to investigate the transient behaviour of the packed substrate. The conditions are given in Fig. 6. Until time t = 0, the whole of the packed substrate had a uniform temperature of 24°C. The temperature of the inlet gas (located on top of the fermenter) was suddenly changed from 24°C to 38°C. To avoid evaporation and generation of metabolic heat, respectively, the inlet gas was saturated and O_2-free. After a delay, the substrate temperature followed the gas temperature. The temperature curves at each level of the fermenter (i.e. of the substrate) were similar but the delay times differed. The temperature curve measured in position 209, for example, is given in Fig. 6.

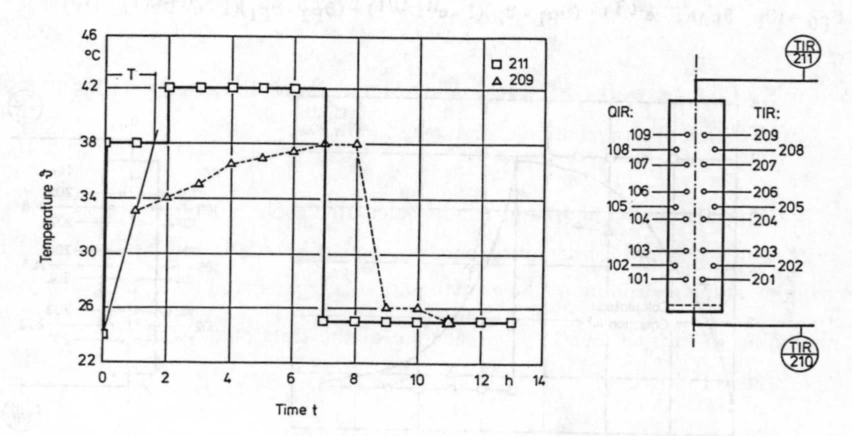

Fig. 6. Temperature curve of of substrate (at position 209) when the gas temperature was suddenly changed. The gas inlet was located on the top of the fermenter. The substrate was chopped wheat straw. The moisture content was 71%; bulk density, 341 kg.m^{-3}; volume of flow rate 40 m^3.h^{-1}. The gas composition was: O_2, 0%; CO_2, 16-17%; nitrogen, 83-84%.

After a further 2 h, the gas temperature was again changed suddenly, this time up to 42°C. Five h later the temperature was reduced suddenly to 25°C. As expected, changes in substrate temperature occurred after a delay time. Generally, the transient behaviour of the substrate temperature can be regarded as proportional response with first-order delay.

From this point of view, all levels in the fermenter have different delay times T_t. The time constant T for the regarded packed substrate investigated packing is assumed to be 1.55 h. Proportional response with first-order delay can be described by the equation:

$$\vartheta_S - \vartheta_{Fo} = (\vartheta_F - \vartheta_{Fo})(1 - e^{-t/T}) \qquad \text{(iii)}$$

where ϑ_S is the substrate temperature; ϑ_{Fo}, the gas temperature before the sudden change; ϑ_F, the gas temperature after the sudden change; t, the time; T, the time constant. When there was a further sudden change to the temperature ϑ_{F1} at the time $t = t_1$, and another to the temperature ϑ_{F2} at the time $t = t_2$, the following complete equation described the temperature curve:

$$\vartheta_S - \vartheta_{F0} = (\vartheta_F - \vartheta_{F0})(1 - e^{-t/T}) + (\vartheta_{F1} - \vartheta_F)(1 - e^{(t_1-t)/T}) + (\vartheta_{F2} - \vartheta_{F1})(1 - e^{(t_2-t)/T}) \qquad \text{(iv)}$$

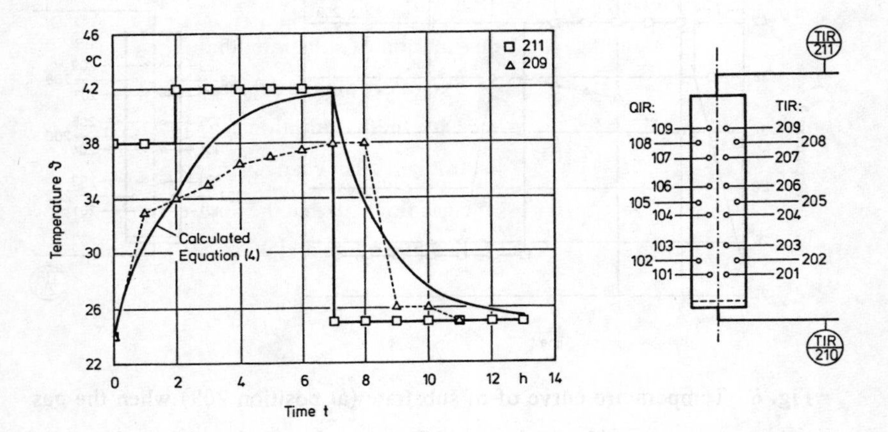

Fig. 7. Comparison between measured and calculated substrate temperatures (according to the conditions of Fig. 6).

Comparison between theory and experimental data

The observed and calculated (by eqn. iv) substrate temperature curves at position 209 are compared in Fig. 7. The value of the time constant, T = 1.55 h, was taken from the data in Fig. 6. The theoretical equation for the time constant T is:

$$T = m_S \, c_{pS} / (\alpha F) \qquad (v)$$

where m_S is the mass of substrate; c_{pS}, the specific heat capacity of the substrate; α, the heat transfer coefficient; F, the surface of the substrate.

As the exact specific heat capacity of the substrate is not known, equation (v) cannot be used to calculate the time constant. The calculated curve agrees with the experimental data for the first 2 h, but, thereafter, increases more than the measured temperatures. This may indicate that delay time effects occur after 2 h. By taking a delay time into account, the agreement between calculated and measured substrate temperatures can be improved. However, to do this work, and to find generally valid expressions for the time constant and the delay time, further experiments have to be carried out. The existing results show that the method described is useful.

HEAT AND MASS TRANSFER IN THE GROWTH PERIOD

Theoretical approach

The first step in predicting the local distribution of substrate temperatures is to derive fundamental equations for the heat and mass transfer in the substrate (Fig. 8). Therefore, the real conditions have to be idealized by the assumptions: a, that conditions are stationary; b, that substrate packing is isotropic; c, that changes in temperature within the substrate occur only in the direction of the gas flow; d, that the physical characteristics of the substrate do not change with changes in temperature and pressure.

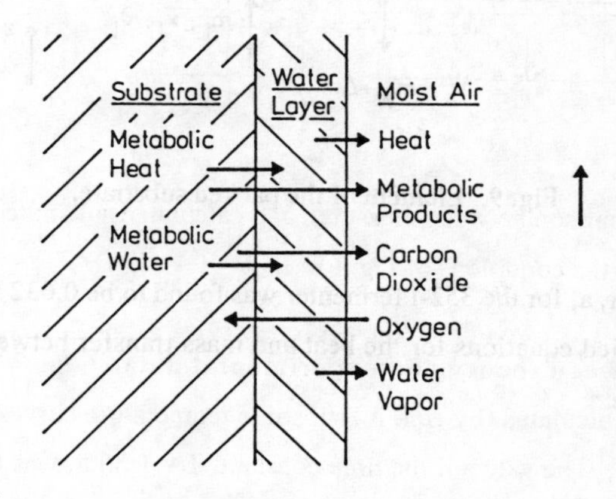

Fig. 8. Idealized surface conditions in a section of the packed substrate.

As the gas composition was controlled, no further account was taken of the mass-flow rates of O_2, CO_2 and the other metabolic products through the phase surfaces. The equations may be derived by dividing the whole of the packed substrate into elements of height Δz, according to Fig. 9.

To describe the effect of the metabolic heat a source of heat is assumed for each element as follows:

$$\Delta\dot{\Phi} = a/n\,[\vartheta_{S,\,max} - (\vartheta_S + 0.5\,\Delta\vartheta_S)] \qquad \text{(vi)}$$

where n represents the number of elements; a, an empirical factor valid for the whole substrate packing; $\vartheta_{S,\,max}$, the final temperature of the substrate without cooling by gas (for *Pleurotus* spp. this was about 40°C).

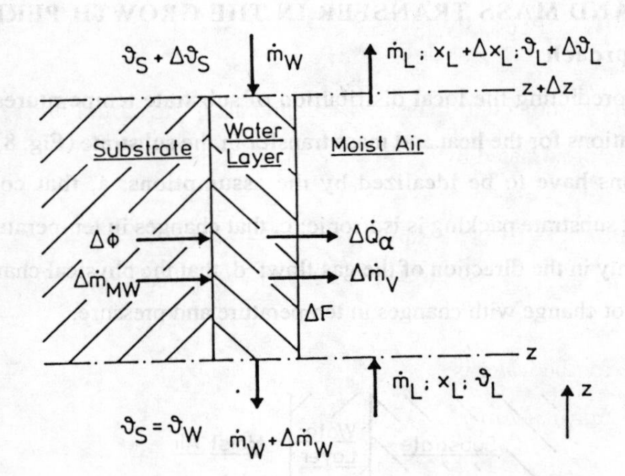

Fig. 9. Element of the packed substrate.

The empirical factor, a, for the 352-l fermenter was found to be 0.032 kW.K^{-1}.

The simplified equations for the heat and mass transfer between the water layer and the moist air are:

$$\Delta\dot{Q}_\alpha = \alpha\,\Delta F\,(\vartheta_S + 0.5\,\Delta\vartheta_S - \vartheta_L - 0.5\,\Delta\vartheta_L) \qquad \text{(vii)}$$

$$\Delta\dot{m}_V = \sigma\Delta F\,(x_S - x_L - 0.5\,\Delta x_L) \qquad \text{(viii)}$$

$$\Delta\dot{m}_V = \dot{m}_L\,\Delta x_L \qquad \text{(ix)}$$

where $\Delta\dot{Q}_\alpha$ is the heat flux within the element; $\Delta\dot{m}_V$, the mass-flow of water due to evaporation; α, the heat transfer coefficient; σ, the mass transfer coefficient; x_L, the ratio mass of water vapor to mass of dry air; x_S, the ratio mass of water vapor to mass of dry air at saturation; $\dot{m}_L$, the mass-flow rate of moist air.

According to the proposal of Löffler (1989), x_S can be written as:

$$x_S = \frac{M_W}{M_L} \cdot \frac{p_{W,S}(\vartheta_S + 0.5\Delta\vartheta_S)}{p_{ges} - p_{W,S}(\vartheta_S + 0.5\Delta\vartheta_S)} \qquad \text{(x)}$$

with M_W and M_L as molecular mass of water and air. The partial pressure of water at saturation can be approximated by:

$$p_{W,S}\,(\vartheta_S + 0.5\Delta\vartheta_S) = 0.018 \text{ bar} \exp\left(18.33 - \frac{5300}{273.15 + (\vartheta_S + 0.5\Delta\vartheta_S)/°C}\right) \qquad \text{(xi)}$$

The mass balance for the water is:

$$\Delta\dot{m}_{MW} = \Delta\dot{m}_W + \Delta\dot{m}_V \qquad \text{(xii)}$$

If one considers only the space for the moist air, the energy balance yields:

$$\Delta\vartheta_L\,\{\dot{m}_L\,(c_p\overset{\circ}{L} + x_L\,c_p\overset{\circ}{W}) + (\dot{m}_L\Delta x_L\,c_p\overset{\circ}{W} + 0.5\,\alpha\Delta F\} - \Delta\vartheta_S\,\{\dot{m}_L\Delta x_L\,0.5\,c_p\overset{\circ}{W} + 0.5\,\alpha\Delta F\}$$
$$= \alpha\Delta F\,(\vartheta_S - \vartheta_L) + \dot{m}_L\Delta X_L\,c_p\overset{\circ}{W}\,(\vartheta_S - \vartheta_L) \qquad \text{(xiii)}$$

and considering the space for moist air and substrate:

$$\Delta\vartheta_L\,0.5\,\alpha\Delta F - \Delta\vartheta_S\,\{0.5\,\alpha\Delta F + \alpha/(2n) + 0.5\,\dot{m}_L\Delta x_L\,c_p\overset{\circ}{W} - 0.5\,\Delta\dot{m}_{MW}\,c_{pW}^{f\,1} - \dot{m}_W\,c_{pW}^{f\,1}\}$$

$$= \alpha\Delta F\,(\vartheta_S - \vartheta_L) - a/n\,(\vartheta_{S,max} - \vartheta_S) + \dot{m}_L\Delta x_L(h_V + c_p\overset{\circ}{W}\,\vartheta_S - c_{pW}^{f\,1}\,\vartheta_S) \qquad \text{(xiv)}$$

where $\overset{\circ}{c}_{pL}$ is the specific heat capacity of dry air; $\overset{\circ}{c}_{pW}$, the specific heat capacity of water vapor; c^{fl}_{pW}, the specific heat capacity of (liquid) water; h_V, the evaporation enthalpy at 0°C.

By using equations (vi) - (xiv), the change in substrate temperature within one element can be calculated by iteration. When the inlet properties of the lowest element are known, the local distribution of the substrate temperature can be calculated by considering all elements. Before the calculation can be started, 4 further parameters have to be fixed by experiments or approximations for the special conditions. For a 24-day fermentation experiment (conditions specified in Fig. 3), the parameters for the growth period are found to be:

$$n \cdot \dot{m}_{MW} = 1.8.10^5 \text{kg/s (estimate value)}$$

$$(\dot{m}_W)_{z=0} = 4.4 \cdot 10^6 \text{kg/s (estimate value)}$$

$$\alpha F = 0.0548 \text{ kW/K (calculated with data from another experiment)}$$

$$\sigma F - 0.0527 \text{ kg/s (calculated with data from another experiment)}$$

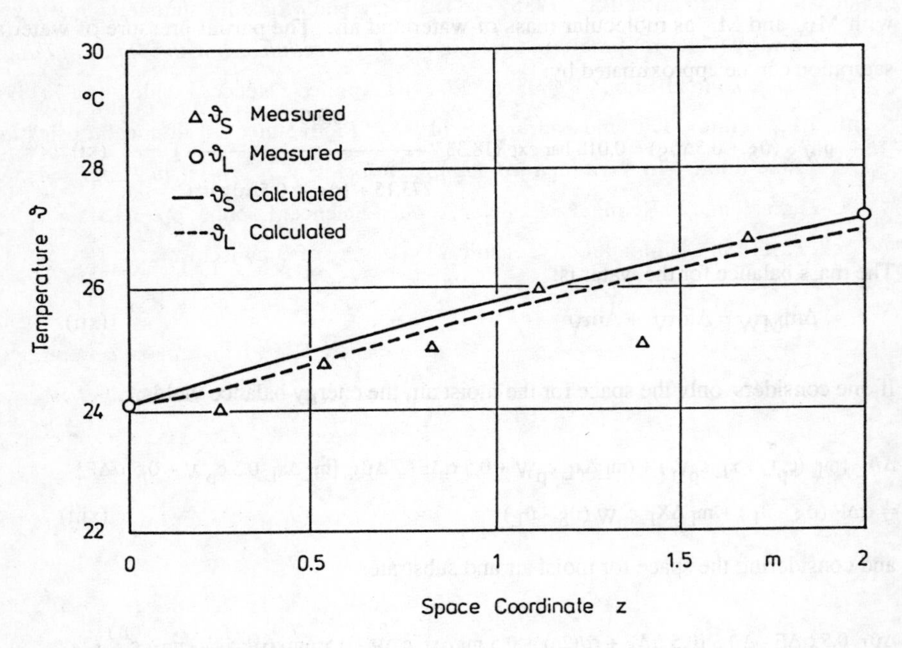

Fig. 10. Comparison between calculated and measured temperatures.

Comparison between theory and experimental data

A comparison between calculated and measured temperatures of substrate and moist air for the 24-day fermentation experiment specified in Fig. 3 is presented in Fig. 10. The agreement between calculated and measured temperatures shows that the method chosen to describe the heat and mass transfer is useful. To develop the method to a generally valid instrument further experiments and modifications of the theory have to be carried out.

Further information on the theoretical approaches and the calculations presented are given in Usbeck (1988).

REFERENCES

Ergun, S. (1952). Fluid flow through packed columns. Chem. Eng. Prog. **48**(2), 89-94.

Löffler, H.-J. (1989). Lecture Course in Thermodynamics (II), Technical University of Braunschweig.

Schuchardt, F. and Zadrazil, F. (1988). A 352-l fermenter for solid-state fermentation of straw by white-rot fungi. In *Treatment of Lignocellulosics with White-rot Fungi,* eds. F. Zadrazil and P. Reiniger, Elsevier Applied Science, London, pp. 77-89.

Smith, J.F., Fermor, T.R. and Zadrazil, F. (1988). Pretreatment of lignocellulosics for edible fungi. In *Treatment of Lignocellulosics with White-rot Fungi,* eds. F. Zadrazil and P. Reiniger, Elsevier Applied Science, London, pp. 3-13.

Usbeck, E. (1988). Diplomarbeit, Technical University of Braunschweig.

CHAIRMAN'S REPORT ON SESSION I

Solid-state fermentation of plant residues with white-rot fungi

As implied by the title of this session, solid-state fermentation of lignocellulosics, specifically with white-rot fungi, is considered to be among the most practical and economically feasible means of exploiting the potential of these materials. Gerrits and van Griensven spoke about the new developments in composting of agricultural wastes for use as a substrate for the cultivation of edible mushrooms. In particular, the authors discussed the several advantages the indoor 'tunnel' process has over the traditional outdoor processes. One such advantage, likely to be acclaimed by environmentalists, is that the tunnel process obviates or minimizes the offensive odour that is characteristic of outdoor composting processes, while at the same time providing a more suitable material for mushroom growth. Giovannozzi-Sermanni and colleagues presented results on the 'soluble' lignocellulosic (SLC) polymer that is produced during growth of ligninolytic fungi on lignocellulosic materials. They have partially characterized SLC and have examined the influence this material has on the physiological and biochemical aspects of the relevant fungi. It would seem that SLC may be of considerable importance as a model 'native' substrate for the study of growth of ligninolytic species and for investigation of the mechanisms involved in the degradation and utilization of cell wall materials. One should note that currently used substrates do not adequately represent native materials - a problem the use of SLC may overcome.

Much of the work carried out at the Institut für Bodenbiologie in Braunschweig has to do with the solid-state fermentation of agricultural wastes, residues and surpluses with *Pleurotus* species. The aims of these studies include the production of animal feeds, chemical feedstocks, and substrates suitable for the cultivation of edible mushrooms or for use as biofilters. One of the papers presented by this group dealt with the theoretical and practical aspects of pilot-scale reactors for solid-state fermentation and the lessons to be learned from work at the pilot-scale for the setting up of industrial-size processes. The latter was the subject matter of the second paper from the Braunschweig group. Much of the discussion centered on characterization of the large-scale process currently operating and the problems attendant on its operation. While the process has shown considerable promise, the authors emphasized the fact that a number of questions, which they outlined, must still be resolved before the operation can be considered to be ready for industrial application. In the final paper, Teifke and Bohnet presented a model of the process of lignin degradation by *Pleurotus* species. In their presentation, they paid particular attention to the various physical parameters that influence the overall process.

M.P. Coughlan
for F. Zadrazil

SESSION II

Changes in lignocellulosic materials during biological treatment

NEAR AND MID-INFRARED SPECTROSCOPY AND WET CHEMISTRY AS TOOLS FOR THE STUDY OF TREATED FEEDSTUFFS

James B. Reeves III,[1] Guido C. Galletti, [2] J. George Buta[3] and Frantisek Zadrazil[4]

[1]Ruminant Nutrition Lab., USDA, Beltsville, MD 20705, USA
[2]CNR, Via Filippo Re 8, 40126-Bologna, Italy
[3]Plant Hormone Lab., USDA, Beltsville, MD 20705, USA
[4]Inst. für Bodenbiologie, Bundesforschungsanstalt für Landwirtschaft
Bundesallee 50, D-3300 Braunschweig, FRG

The possibility of using near-infrared and mid-infrared spectroscopic techniques to study treated feedstuffs was examined. The central objective was to determine the usefulness of each spectral region in determining both qualitative and quantitative changes occurring in feeds during treatment. The role of wet chemistry in determining the usefulness and interpretation of spectral results was also examined. The fibre and lignin composition of sodium chlorite-treated feeds, intact lignin in wheat straw degraded by the white-rot fungus, *Stropharia rugosoannulata,* were examined as were different methods of determining and extracting lignin. Results indicated that both near-infrared and mid-infrared spectroscopy have the potential for the rapid determination of the quantitative changes in fibre digestibility and, perhaps, lignin composition in treated feeds. However, due to better spectral definition, mid-infrared spectroscopy appears to be more useful than near-infrared in terms of qualitative examinations. The combination of studies in the two spectral regions in conjunction with chemical methods is an area of great potential.

INTRODUCTION

The use of treatments (biological, chemical and/or physical) to improve the digestibility or quality of animal feedstuffs has been investigated by many researchers over the years. However, lack of understanding as to how these treatments operate has always been a problem. Consequently, the results of one study cannot easily be extended to other feeds and/or treatments. The multitude of papers dealing with one or two feeds treated in some manner is evidence of this. Even when dealing with only a few feeds and chemicals, it is possible to generate samples that can take years to analyze by standard chemical procedures and a lifetime if using animal-feeding trials. Nevertheless, because of the myriad combinations of reagents, treatment conditions and feeds, large numbers of samples must be analyzed in order to determine the most effective treatment. A less time-consuming non-chemical method, such as near-infrared reflectance spectroscopy (NIRS), to analyze the treated samples might be an ideal tool for selecting promising treatments.

In this paper, we show that the combined use of spectroscopic methods near-infrared (both Fourier, FTNIR and grating type instrumentation, NIRS) and Fourier transform mid-infrared (FTIR) with chemical and *in vitro* procedures can be used to: (i) gain insights into the effects of treatments on feeds; (ii) reduce the time needed to determine the results of treatments; (iii) reveal the weaknesses of chemical and/or spectroscopic procedures.

NIRS is an instrumental technique used to measure the composition and quality of a wide variety of materials, including feedstuffs, food, and industrial products. It is a diffuse reflectance technique that, through spectral differences in the stretching and bending of hydrogen bonds (O-H, C-H, N-H, etc.), is capable of measuring the composition of lignocellulosic materials present in the plant constituents (Marten *et al.*, 1985). It has been successfully used to measure the composition and quality of single species (Norris and Williams, 1979; Burdick *et al.*, 1981; Bengtsson and Larsson, 1984) and mixed species forages (Coleman *et al.*, 1985), as well as to determine animal response (Eckman *et al.*, 1983). Various reports show that, with proper calibration, NIRS is capable of measuring the quality and composition of forage materials to an accuracy comparable to the more time-consuming and more costly chemical analyses (Moe and Carr, 1984; Marten *et al.*, 1985).

NIRS spectra consist of overtones and combinations of fundamental infrared absorption frequencies (Osborne, 1981; Marten *et al.*, 1985). Because of this and because of the small spectral differences involved, extensive mathematical treatments are used to extract useful information from the NIRS spectra (Marten *et al.*, 1985). In addition to the proper mathematical treatment of such spectra, selection of the calibration set is extremely important. It is necessary to select samples that represent the range of constituent variation found in the sample set to be analyzed (Marten *et al.*, 1985).

Because of misunderstandings of the exact nature of the technique, considerable misinformation about its use exists. Some naive potential users believe that all one need do to determine the chemistry of feed samples is to buy the instrument and proper equations, grind and scan the sample and the answer is forthcoming. For some materials of consistent composition this may work. However, the variability in the composition of animal feedstuffs can present problems. In the case of wet chemistry, and to some extent in mid-infrared spectroscopy, some component of the feed material, such as nitrogen or a specific wavelength, is directly measured by a test specific for that entity. By contrast, NIRS is an indirect method in that spectral properties of the material are matched against results obtained from wet chemistry. The analogy might be to determine peoples' weights by analyzing their shadows. The spectrum of the sample, like the shadow, is based on the sample composition, but, at present, not enough is known to be sure that two samples with different composition will always give different spectra. Because it is based on a calibration set, NIRS often fails with samples of unfamiliar composition, viz. different varieties, growing conditions and so on (Blosser *et al.*, 1988). Statistical tests to pinpoint such samples are not always reliable.

While mid-infrared spectra can be used like their NIRS counterparts, they are based more on the fundamental vibrations and stretchings of both of those groups mentioned for the near infrared, and also on others, such as carbonyls and carbon-carbon (Socrates, 1980). The peaks tend to be narrower and the absorptions much stronger than for the near-infrared. The advantage of this spectral region is that much more is known about the basis of the spectra and, therefore, one can extract more information about the structures of the samples under study, i.e. the results are not based solely on correlations between spectra and calibration sets. The main disadvantages are that because of the strong nature of the absorptions, samples must be diluted and the presence of water and carbon dioxide interfere.

Finally, results obtained using wet chemistry are often the foundations upon which all our spectral interpretations (and the uses to which they are put) are based. Frequently, assumptions made about such results are forgotten or are plainly false. Work with the near-infrared has, perhaps more than anything else, pointed out the need for more reproducible and more specific wet chemical methods

MATERIALS AND METHODS

All determinations of fibre, permanganate- and sulphuric acid-lignins, and digestibility were performed according to the methods of Goering and van Soest (1970). *In vitro* digestibilities were determined, using ruminal fluid from a steer fed orchard grass, with 48-h incubations. Chlorite-lignins were determined by the methods of Collings *et al.* (1978) and acetyl bromide-lignins by the methods of Morrison (1972a, b). Protein (CP) concentration was calculated on the basis of Kjeldahl nitrogen (Williams, 1984). The lignin composition profiles were determined by nitrobenzene oxidation (Cymbaluk and Neudoerffer, 1970; Reeves, 1986).

Sodium chlorite treatment of feeds was carried out at 11 different ratios of reagent to feed, using 16 different feeds (alfalfa [AL], tall fescue [FS], orchard grass [OG], red clover [RC], timothy [TY] and grass-legume [MH] hays; barley [BS] and wheat [WS] straws; corn cobs [CC]; two corn [CSI, CSII] and two soybean [SSI, SSII] stovers; peanut [PH], rice [RH], and soybean [SH] hulls), to give final concentrations of 0-0.394 g sodium chlorite per g feed (Reeves, 1985). The non-treated feeds (alfalfa, tall fescue and orchardgrass, and the vegetative parts of corn and wheat plants) used for comparison were collected at various stages of growth at Beltsville during the 1982 growing season (Reeves, 1987). All computations of fibre content and digestibility of the treated feeds were based on organic matter basis so as to avoid errors and misconceptions due to increases in soluble inorganic content with increasing concentrations of sodium chlorite.

Near-infrared reflectance analysis was carried out using a Neotec 6250 scanning monochromator (grating-type instrument) and a Digilab FTS-65 Fourier transform spectrometer. The latter was also used for the mid-infrared work. Data from the FTNIR and FTIR work was converted to a format similar to the grating data and all sets were analyzed by multi-term regression methods to allow of comparisons. Proc stepwise with the MAXR technique (SAS Users' Guide, Statistical Analysis System Inc., Gary, SC, USA) was used with a maximum of 9 wavelengths. Samples for the mid-infrared work were diluted to 5% (w/v) with KBr, while near-infrared samples were analyzed without dilution. The sample area scanned was also about 4-fold larger on the grating instrument due to sample presentation methods. All spectra were taken using diffuse reflectance.

Mid-infrared spectra of extracted lignin, 72% sulphuric acid-lignin, and wheat straw degraded by *Stropharia rugosoannulata* were taken on a Nicolet Model 60-SX FT-IR using 3 mm pellets at a resolution of 4 cm^{-1}.

RESULTS AND DISCUSSION

NIRS and chlorite-treated feeds

The fibre composition, protein content and *in vitro* digestibility of untreated and treated (at the highest ratio of chlorite to feed) samples of some of the 16 feedstuffs are given in Table 1. The data are presented to show the range of values to which NIRS was applied. It is interesting that for many of the low-protein materials, crude protein content (CP) increased with treatment (BS, CC, CSI, RH and WS),while for the higher quality materials (AL, MH, OG, RC and SH), it decreased. These changes were significant at the P<.05 level for CC, RH, WS, AL, OG and RC for feeds treated with the lowest and highest concentrations of chlorite. Apparently, in the low-protein materials, the CP was protected to some degree so that, as the lignin was destroyed by the sodium chlorite and as the total fibre present decreased, the CP represented a larger proportion of the remaining material. With the higher quality materials, it appears that CP, in addition to lignin, was destroyed to some extent.

Table 1. Fibre composition, *in vitro* digestibility, and protein content of sodium chlorite-treated feeds.

Feed[1]	$[NaClO_2]$[2]	NDF	ADF	LIG[3]	DNDF[4]	DDM[5]	CP
AL	.000	50.3	39.1	14.9	37.7	68.7	16.2
AL	.394	33.4	26.6	5.6	93.2	97.7	14.6
BS	.000	91.5	61.8	13.8	51.1	55.3	2.5
BS	.394	61.9	48.9	4.2	96.1	97.6	2.7
CSI	.000	84.5	60.9	12.3	51.4	58.9	2.9
CSI	.394	62.4	46.7	3.0	100.2	100.1	3.0
PH	.000	89.9	80.3	21.8	1.5	11.5	7.8
PH	.394	68.4	57.8	20.6	15.1	42.0	7.2
RC	.000	59.4	47.7	19.6	32.7	60.1	16.8
RC	.394	37.1	32.8	7.7	87.5	95.4	15.2
SSI	.000	81.9	69.9	21.0	37.9	49.1	6.2
SSI	.394	53.9	44.0	6.5	96.7	98.2	5.8
TY	.000	82.5	47.9	10.5	57.0	63.3	8.0
TY	.394	61.6	37.5	2.8	95.2	96.1	7.7
WS	.000	91.6	61.1	13.3	46.0	50.5	2.2
WS	.394	69.3	46.2	2.6	98.5	99.0	2.4

[1]Abbreviations have been described above: [2]$g.g^{-1}$ feed; [3]LIG, organic matter permanganate lignin; [4]DNDF, organic digestible NDF; [5]DDM, organic dry matter digestibility, based on NDF (neutral detergent fibre).

The results of the NIRS analyses are presented in Table 2. The same validation set was used for all determinations. The r^2s of determination and SEA remain fairly constant for the full, 3/4 and 1/2 calibration sets. In the case of DDM (digestible dry matter), a measure of prime concern when selecting a chemical treatment, the r^2s only varied by 0.02 between the three sets, although the SEA was more variable. For DNDF (digestible neutral detergent fibre) and DDM, r^2s values of 0.90 and 0.89 were achieved for the 1/4 set. Examination of the analyzed values (data not given) for the 1/4 calibration set for DDM showed that for AL, FS, MH, OG, RC and TY, the DDM was determined correctly to within five percentage units. The values for the other feeds were also within five percentage units except for a specific treatment. This was true for BS, CSII, RH and WS. Values for PH and SS (I and II) were particularly poor, with residuals of as much as 22 (PH). Examination of the results for DNDF also showed PH and SS (I and II) with large residuals, although the results in general were poorer for all samples.

Table 2. Rsquares and SEA[1] for validation sets using equations derived from calibration sets of various sizes.[2]

	Calibration set size.							
	Full		3/4		1/2		1/4	
	r^2	SEA	r^2	SEA	r^2	SEA	r^2	SEA
NDF	0.94	3.7	0.94	3.6	0.94	3.6	0.91	4.5
ADF	0.97	2.2	0.96	2.3	0.92	3.2	0.90	3.5
Lignin	0.94	1.4	0.92	1.6	0.88	1.9	0.89	1.9
DNDF	0.95	6.5	0.96	6.1	0.91	8.9	0.90	9.0
DDM	0.96	4.7	0.97	4.7	0.95	5.8	0.89	7.9

[1]SEA, standard error of analysis for the validation set. [2]The full set consisted of 131 samples (3/4 of all treated feeds); the 3/4 set consisted of 98 samples (3/4 of full set); the 1/2 set consisted of 65 samples (1/2 of full set) and so on. The remaining 1/4 of all treated feeds used as the validation set (43 samples) for all the various calibration sets shown.

For the 1/8 calibration set (data not shown), a 10% unit limit was necessary to avoid excluding most of the samples. Even then, BS, PH, RH, SS (I and II) had determined values for DDM far outside the imposed limit. For SSI and SSII, the determined values were off by as much as twenty nine percentage units.

However, examination of NIRS determinations (data not shown) based on the full calibration set did not show any feeds that were exceptions (as judged by the residuals of predicted versus actual DDM) as far as difficulty in analyzing DDM, although some were better than others. Thus, it appears that sample selection was of more importance in developing good regression equations than any inherent problems with specific feeds. The results from the treated feeds indicate that NIRS is basically quite capable of determining the composition and digestibility of sodium chlorite-treated feeds. The SEA values reported are similar to those found by us for other samples and to those reported by others (Goering and van Soest, 1970). The SEA values generally reflect the accuracy of the chemical determinations, as accurate determinations of NDF or ADF are more easily obtained than those for digestibility. The results shown above support this. Also, as the values decrease (LIG < ADF), the relative errors increase. Overall, these results indicate that the basis for NIRS analysis of sodium chlorite-treated feeds is not fundamentally different than that for non-treated feeds of naturally-occurring variable composition. Finally, in results not shown, the values for the chemically-treated feeds were adjusted to include the ash content (that present naturally plus that generated as a result of treatment). The results obtained using 1/3 as a validation set were virtually the same as those for the ash-free values shown.

During the course of the NIRS investigations, it was noticed that the chemistry being examined could have a great influence on the ability of NIRS to operate accurately. Protein and water contents are generally determinable with the greatest accuracy. It can readily be seen that these two components represent true chemical entities to a greater degree than do NDF, ADF or DDM. The exact natures of the last named three features are not known and may vary with the feed in question. For this reason, lignin oxidation products were examined, in the hope that, as they represent more definable entities, they would have more in common across a variety of feeds than would NDF or ADF. The results in the first part of Table 3 indicate that there is a common base of information between the two sets of data (treated and 1982 samples). However, the use of equations derived from the 1982 feeds to predict or to determine the treated feeds was entirely unsatisfactory.

Table 3. NIRS analysis of lignin oxidation products in sodium chlorite-treated (N = 174) and non-treated feeds (N = 67).

	All feeds (2/3 in calibration set)					
	Calibration		Validation		Determined	Predicted
Component[1]	r^2	SEA	r^2	SEA	Mean	Mean
pHB	.88	.024	.90	.021	.087	.087
VAN	.87	.042	.89	.038	.363	.353
MIX	.95	.027	.96	.025	.091	.093
VNA	.74	.032	.80	.032	.097	.090
	Calibrated set = non-treated; predicted set = all treated feeds					
pHB	.89	.015	.18	.160	.07	.*
VAN	.87	.020	.49	.280	.36	.48
MIX	.94	.006	.21	.170	.12	.02
VNA	.94	.008	.01	.075	.11	.04
	Calibrated set = non-treated + 1/4 of treated; predicted set = rest					
pHB	.88	.021	.85	.027	.07	.07
VAN	.90	.031	.86	.049	.36	.36
MIX	.96	.018	.92	.046	.12	.11
VNA	.73	.027	.69	.042	.11	.11

[1]Component: pHB, p-hydroxybenzaldehyde; VAN, vanillin; MIX, mixture of acetovanillone and 4-allyl-2,6-dimethoxyphenol; VNA, vanillic acid; *, negative predicted mean.

The addition of 1/4 of the treated feeds to the 1982 set resulted in a great deal of improvement, again indicating that these two sets of data do share common chemistries. The reasons for the lack of cross predictability shown previously is not known at present.

NIRS and lignin assays

As a result of the differences found using various chemistries, research aimed at finding chemistries with which NIRS works well was initiated. The ranges and standard errors of the means (SE) for the assays performed are given in Table 4. The ranges of the lignin values vary, with some assays giving similar results for feeds either high or low in lignin content, but very different values for forages with lignin at the opposite extreme. The lignin assay with the most precisely determined values appears to be the acetyl bromide-lignin determined on ADF fibre (ABAF), and the poorest, chlorite-lignin performed on ammonium oxalate-fibre (OXCL). These differences in lignin assays have also been seen in other studies (Collings et al., 1978; Giger, 1985) even though the same entity, i.e. lignin, is supposedly being measured.

Table 4. Lignin content (as %DM) as determined by chemical means * or by absorbance (i.e. absorbance.g DM^{-1})**.

Lignin	High value	Low value	SE
ABAF[1]	7.09	0.79	0.06
ABOR[2]	9.64	2.98	0.15
AFCL[3]	6.02	0.73	0.16
OXCL[4]	8.15	3.98	0.28
H_2SO_4[5]	7.67	1.01	0.10
$KMnO_4$[6]	14.00	2.76	0.18

----

[1]ABAF, acetyl bromide-lignin on ADF fibre; [2]ABOR, acetyl bromide-lignin on hot water-organic solvent-fibre; [3]AFCL, chlorite-lignin on ADF fibre; [4]OXCL, chlorite-lignin on ammonium oxalate-fibre; [5]H_2SO_4, sulphuric acid-lignins; [6]KMnO4, permanganate ADF lignins. *, % of DM except for ABAF and ABOR; **, absorbance.g^{-1}DM for ABAF and ABOR.

The first part of Table 5 shows that various lignin assays have little in common, even when the basic differences among feeds are removed (feed as a covariate). Despite these differences, NIRS predicted most assays fairly well, with the acetyl bromide assay on ADF fibre being best.

Comparison of FTIR, FTNIR and NIRS

The results of comparative studies on the treated samples using both grating and Fourier-based near-infrared and Fourier-based mid-infrared spectroscopy are given in Table 6. The rsquares presented represent the best results from two runs, one using a first derivative and one a second derivative mathematical treatment with all samples as a

Table 5. Regression of lignin v lignin, with (values in parentheses) and without feed as a covariate, and of NIRS using every third sample as the validation set.

Lignin v lignin

	ABOR	AFCL	$KMnO_4$	OXCL	H_2SO_4
ABAF	0.35	0.84	0.63	0.03	0.52
	(0.80)	(0.85)	(0.88)	(0.55)*	(0.94)
ABOR		0.39	0.08	0.04	0.01
		(0.71)	(0.80)	(0.44)	(0.86)*
AFCL			0.63	0.13	0.49
			(0.84)	(0.59)	(0.67)
$KMnO_4$				0.06	0.87
				(0.59)	(0.93)
OXCL					0.03
					(0.62)*

Validation set results from NIRS analysis

	Mean	SD	PMN	PSD	Bias	r^2	SEA
ABAF	3.03	1.37	3.08	1.35	-0.06	0.95	0.31
ABOR	6.63	1.16	6.70	1.21	-0.08	0.86	0.46
AFCL	2.72	1.10	2.75	0.88	-0.03	0.90	0.39
OXCL	5.85	0.96	5.99	0.65	-0.15	0.77	0.49
H_2SO_4	3.80	1.66	3.87	1.53	-0.07	0.91	0.50
$KMnO_4$	7.55	2.53	7.24	2.55	0.31	0.90	0.82

***Regression without feed as covariate not significant at the $P<0.05$ level. ABAF, acetyl bromide on lignin; ABOR, acetyl bromide on hot water-organic solvent fibre; AFCL, ammonium oxalate on entire crucible and fibre; PMN, predicted average mean; PSD, predicted standard deviation.**

Table 6. Rsquares for near-infrared and mid-infrared predictions of fibre and lignin composition of chlorite-treated feeds.

Assay	Grating NIRS	FTNIRS*		FTIR	
		4	16	4	16
NDF	0.96	0.90	0.86	0.91	0.94
ADF	0.96	0.91	0.91	0.94	0.94
KMnO4-lignin	0.92	0.86	0.87	0.88	0.88
DNDF	0.95	0.86	0.86	0.90	0.88
DDM	0.95	0.90	0.87	0.92	0.90
CP	0.99	0.92	0.90	0.95	0.96
pHB	0.90	0.84	0.74	0.84	0.88
VAN	0.92	0.86	0.88	0.87	0.87
MIX	0.97	0.86	0.86	0.90	0.91
VNA	0.80	0.65	0.62	0.84	0.81
SYAL	0.83	0.78	0.77	0.92	0.93
UNK	0.80	0.73	0.64	0.84	0.88
SA	0.60	0.48	0.42	0.55	0.74

*4 and 16 refer to the resolution used in data collection. pHB, p-hydroxybenzaldehyde; VAN, vanillin; MIX, mixture of acetovanillone and 4-allyl-2,6-dimethoxyphenol; VNA, vanillic acid; SYAL, syringaldehyde; UNK, unknown; SA, syringic acid.

calibration set and a maximum of nine wavelengths chosen by stepwise regression. The Fourier work was performed at two resolutions, 4 and 16 cm^{-1}; the 16 corresponding roughly to the capabilities of our grating instrument. After collection, all data were averaged across several wavelengths to give 700 wavelengths from roughly 400 to 4000 cm^{-1}, and 4000 to 10000 cm^{-1}, for the mid-infrared and near-infrared regions.

Examination of the results shows that, for fibre measurements, the grating instrument performs best of all. This is probably due to the larger sample area scanned and perhaps also to the higher signal-to-noise ratio for the grating spectrometer. It is interesting that, for these samples, the mid-infrared performed better than the near-infrared (Fourier instrument) and not that much worse than the grating instrument. It has been reported by others (Olinger and Griffiths, 1989) that the mid-infrared region was not useful for such predictions (protein in wheat) and, therefore, was restricted to qualitative type information.

In the case of the lignin components, the mid-infrared spectra showed indications of being better than the near-infrared for at least some components (SYAL, UNK and perhaps SA). Overall, the resolution did not affect the results greatly. This might be due to smoothing of the data. Alternatively, since all samples were scanned 64 times, it may be that the higher resolution was offset by higher noise.

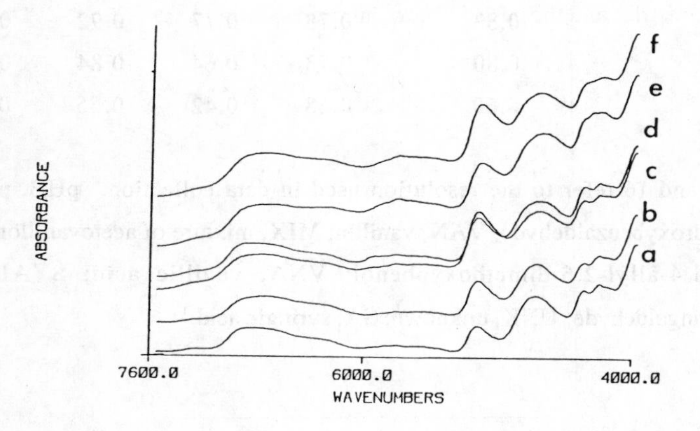

Fig. 1. FTNIR (near-infrared) absorbance spectra of untreated (a, c, e) and chlorite-treated (chlorite: feed ratio 0.394) feeds (b, d, f). Lettering denotes the following: (a, b) wheat straw; (c, d) peanut hulls; (e, f) alfalfa hay.

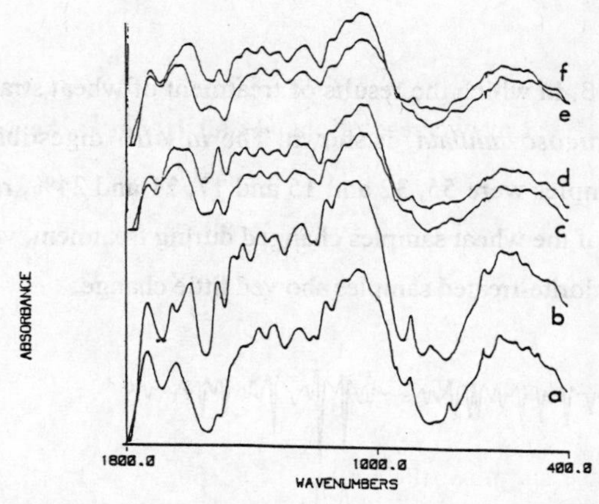

Fig. 2. FTIR (mid-infrared) absorbance spectra of untreated (a, c, e) and chlorite-treated (chlorite: feed ratio 0.394) feeds (b, d, f). Lettering denotes the following: (a, b) wheat straw; (c, d) peanut hulls; (e, f) alfalfa hay.

While work with the near-infrared region on extracted components of feeds has shown it to be difficult or almost impossible to reconstruct spectra or to do spectral subtraction (spectra of very different samples often look very similar, Fig. 1), these procedures are common in the mid-infrared (in which spectral differences are greater, Fig. 2). The mid-infrared appears to perform well with predictions on diffuse spectra and to be useful in spectral reconstruction. This gives a promising area for increasing our understanding of both feed chemistries and the basis for near-infrared spectra.

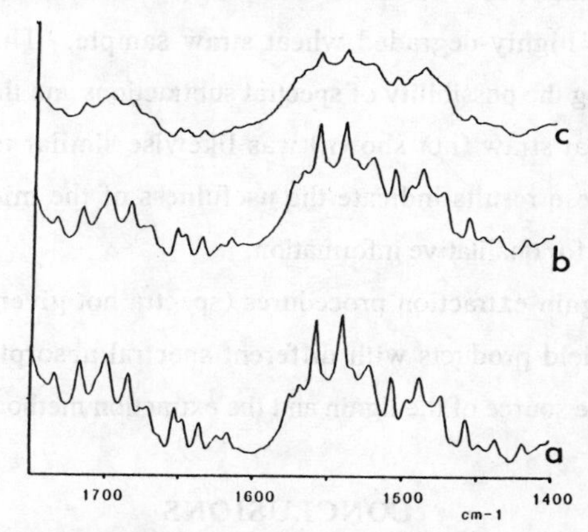

Fig. 3. Mid-infrared transmittance spectra of wheat straw subjected to treatment with white-rot fungus. The extents of degradation range from (a) highly-degraded to (c) virtually undegraded.

This can be seen in Fig. 3, in which the results of treatment of wheat straw by the white-rot fungus, *Stropharia rugosoannulata,* is shown. The *in vitro* digestibilities and lignin contents of the three samples were 55, 32 and 15 and 17, 21 and 24%, respectively. As can be seen, the spectra of the wheat samples changed during treatment, whereas the near-infrared spectra of the chlorite-treated samples showed little change.

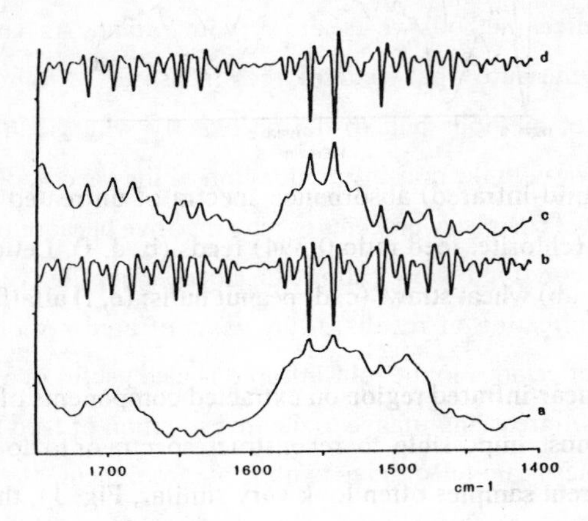

Fig. 4. Mid-infrared transmittance (a, c) and 2nd derivative (b, d) spectra of wheat straw (a, b) and 72% sulphuric acid-wheat straw lignin (c, d).

Fig. 4 shows the aromatic area of the mid-infrared spectra of sulphuric acid-extracted lignin and the most highly-degraded wheat straw sample. There are considerable similarities, indicating the possibility of spectral subtractions and the like. The spectra of the undegraded wheat straw (not shown) was likewise similar to the sulphuric acid-extracted lignin. These results indicate the usefulness of the mid-infrared regions in spectral comparisons for qualitative information.

Work on lignin extraction procedures (spectra not given) has shown that the different methods yield products with different spectral absorptions and definitions, depending on both the source of the lignin and the extraction method.

CONCLUSIONS

Our results indicate that NIRS can be used as a tool to scan large numbers of chemically-treated feeds so as to select those best suited for further study. That oxidized feeds can be evaluated by NIRS suggests that feeds altered by other treatments might also be evaluated by using this technique. The results indicate that chemical analyses required for adequate calibration can be reduced to less than 1/2 and perhaps 1/4 of the total number of samples, depending on how selective one might need to be and perhaps on the nature of the feeds treated. The number of calibration samples necessary for the large data sets that might be generated by a search for useful treatment schemes (sets of 1000 or more samples) has yet to be determined.

Since NIRS is at present based strictly on calibrations with samples of similar composition, the chemical methods we use are very important. As has been shown, these methods often do not measure what we think they measure. More research needs to be done to develop better methods and to define exactly what such methods actually measure. Finally, work with the mid-infrared region of the spectra offers great promise. On the one hand our wet chemistry procedures may improve because of a better definition of what they do. Secondly, by providing the needed qualitative information, our interpretation and application of results in the near infrared region may improve. It appears that diffuse reflectance in the mid-infrared is also useful in a quantitative sense. All of these areas are of great importance to the investigation of feed treatment, if we are to understand our studies more fully and to extrapolate from them.

REFERENCES

Bengtsson, S. and Larsson, K. (1984). Prediction of the nutritive value of forages by near-infrared reflectance photometry. J. Sci. Food. Agric. **35**, 951-958.

Blosser, T.H., Reeves, J.B. III. and Bond, J. (1988). Factors affecting analysis of the chemical composition of tall fescue with near-infrared reflectance spectroscopy. J. Dairy Sci. **71**, 398-408.

Burdick, D., Barton, F.E. II and Nelson, B.D. (1981). Prediction of bermuda grass composition and digestibility with a near-infrared multiple filter spectrophotometer. Agr. J. **73**, 399-403.

Coleman, S.W., Barton, F.E. II and Meyer, R.D. (1985). The use of near-infrared reflectance spectroscopy to predict species composition of forage mixtures. Crop Sci. **25**, 834-837.

Collings, G.F., Yokoyama, M.T. and Bergen, W.G. (1978). Lignin as determined by oxidation with sodium chlorite and a comparison with permanganate lignin. J. Dairy Sci. **61**, 1156-1160.

Cymbaluk, N.F., and Neudoerffer, T.S. (1970). A quantitative gas-liquid chromatographic determination of aromatic aldehydes and acids from the nitrobenzene oxidation of lignin. J. Chromatogr. **51**, 167-174.

Eckman, D.D., Shenk, J.S., Wangsness, P.J. and Westerhaus, M.O. (1983). Prediction of sheep responses by near-infrared reflectance spectroscopy. J. Dairy Sci. **66**, 1983-1987.

Giger, S. (1985). Review on lignin determination methods for forage analysis. Ann. Zoo Tech. **34**, 85-122.

Goering, H.K., and van Soest, P.J. (1970). *Forage Fibre Analyses. Agricultural Handbook* No. 379, USDA.

Marten, G.C., Shenk, J.S. and Barton , F.E. II (eds.) (1985). *Near-infrared Reflectance Spectroscopy (NIRS): Analysis of Forage Quality. Agricultural Handbook* No. 643, USDA.

Moe, A.J. and Carr, S.B. (1984). Laboratory analysis and near-infrared reflectance for predicting *in vitro* digestibility of rye silage. J. Dairy Sci. **67**, 1301-1305.

Morrison, I.M. (1972a). A semi-micro method for the determination of lignin and its use in predicting the digestibility of forage crops. J. Sci. Fd. Agric. **23**, 455-463.

Morrison, I.M. (1972b). Improvements in the acetyl bromide technique to determine lignin and digestibility and its application to legumes. J. Sci. Fd. Agric. **23**, 1463-1469.

Norris, K.H. and Williams, P.C. (1979). Comparing near-infrared reflectance with other methods of protein estimation in wheat. Cereal Foods World, **24**, 450.

Olinger, J.M. and Griffiths, P.R. (1989). Fourier transform infrared spectroscopic studies of wheat in the mid-infrared. In Proc. 7th Int. Conf. on Fourier Transform Spectroscopy Abs. No. P1.7.

Osborne, B.G. (1981). Principles and practice of near-infrared (NIR) reflectance analysis. J. Fd. Technol. **16**, 13-19.

Reeves, J.B.III. (1985). Lignin composition of chemically-treated feeds as determined by nitrobenzene oxidation and its relationship to digestibility. J. Dairy Sci. **68**, 1976-1983.

Reeves, J.B.III. (1986). Use of nitrobenzene oxidation for the study of lignin composition with an improved method for product extraction. J. Dairy Sci. **69**, 71-76.

Reeves, J.B.III. (1987). Lignin and fiber compositional changes in forages over a growing season and their effects on *in vitro* digestibility. J. Dairy Sci. **70**, 1583-1594.

Socrates, G. (1980). *Infrared Characteristic Group Frequencies.* Wiley, New York.

Williams, S. (ed.) (1984). *Official Methods of Analysis,* 14th edn., AOAC Inc. Arlington, VA.

RAPID METHODS FOR DETERMINATION OF SUBSTRATE QUALITY DURING SOLID-STATE FERMENTATION OF LIGNOCELLULOSICS

Guido C. Galletti and Roberta Piccaglia
Centro di Studio per la Conservazione dei Foraggi - C.N.R.,
Via Filippo Re, 8 - 40126 Bologna (Italy)
and
J. George Buta and James B. Reeves III
United States Department of Agriculture
Agricultural Research Service, Beltsville, Maryland 20705 (USA)

High performance liquid chromatography (HPLC), pyrolysis-gas chromatography-ion trap detector (PY-GC-ITD) and Fourier transform infrared spectroscopy (FT-IR) were tested as tools for the rapid monitoring of chemical changes in wheat straw subjected to solid-state fermentation (SSF) with *Stropharia rugosoannulata* under different degrees of oxygenation. Data on simple phenolic compounds found by HPLC in neutral and basic extracts, on pyrolysis fragments assignable to lignin or to carbohydrates, and on FT-IR absorption bands due to lignin will be discussed in relation to the different oxygen concentrations of the samples and, hence, to the different fungal activities. It will be shown that the results from the three techniques can be used both independently, as markers to check SSF efficiency, and in combination, to provide information on the molecular basis of the SSF process.

INTRODUCTION

Edible fungi may be grown on lignocellulosic wastes by solid-state fermentation (SSF). The procedure has the additional benefits of improving the digestibility of the residual lignocellulose as an animal feed and of facilitating its enzymic hydrolysis (Zadrazil and Reiniger, 1988).

The understanding of the mechanism of lignin degradation, the selection of optimal fungi, and ultimately the monitoring of SSF efficiency depend upon the use of suitable analytical methods.

The determination of biological, chemical and physical parameters can improve the control and reproducibility of the process, whereas the chemical composition of the lignocellulosic substrate is analyzed for the purpose of using the treated material for animal feed.

Analytical methods focus on the three main constituents of lignocellulose, namely cellulose, lignin and hemicellulose and on their derivatives.

Usually, the forage chemist studying the treated and untreated materials can choose from a number of analytical methods. These differ with respect to specificity, rapidity and simplicity and include measurements of total organic matter, *in vitro* digestibility, spectroscopic (IR, NMR) techniques and chromatographic analyses of simple and large molecules.

The present paper reports on the monitoring of some characteristic changes in lignocellulose during a laboratory-scale processing with the white-rot fungus, *Stropharia rugosoannulata.* Techniques used include high performance liquid chromatography (HPLC), pyrolysis-gas chromatography-ion trap detector (PY-GC-ITD) and Fourier transform infrared spectroscopy (FT-IR).

The analysis of phenolics in straw using HPLC is well established. Since the work of Hartley and Buchan (1984), a number of papers have been published about HPLC determination of phenolics in lignocellulose using UV and electrochemical detectors (Reeves, 1986; Chiavari *et al.*, 1988; Galletti *et al.*, 1988 and 1989).

Py-GC is a rapid and advantageous technique in the analysis of matrices in which the complexity of the natural macromolecules makes traditional analytical techniques difficult to use. Studies on lignin- and cellulose-related substances have been carried out recently by Faix *et al.* (1987) and Pouwels *et al.* (1989).

IR spectroscopy has been useful in studying lignin structure (Hergert, 1971). More recently, research has been carried out on lignin polymer models and on wheat lignin using FT-IR (Faix, 1986; Jung and Himmelsbach, 1989).

MATERIALS AND METHODS

Samples

Eight samples of fungal-treated straw (particle size < 1 mm) were taken from the SSF experiment performed by Zadrazil and Kamra (1989). These workers have published a detailed description of the experimental conditions. To summarize, 30 Erlenmeyer flasks containing wheat straw and *Stropharia rugosoannulata* were interconnected in a chain with polyethylene tubing and flushed with oxygen (2 l per day). Samples, representative of lignin degradation, were taken from flask numbers 1, 3, 5, 7, 9, 13, 21 and 30.

FT-IR

The ground samples were incorporated into KBr (2:50 w/w) and pressed into a 3 mm pellet using a Qwick Handi-Press. The infrared spectrum from 4000 to 700 cm^{-1} was obtained using a Nicolet Model 60-SX FT-IR spectrometer operating at a nominal resolution of 4 cm^{-1}. The spectrometer was equipped with a globar source and a liquid-nitrogen-cooled Hg-Cd-Te (MCT-A) detector. Thirty-two interferograms were collected, phase-corrected and apodised by the Happ-Genzel function prior to Fourier transformation. Duplicate experiments were performed. First and second derivative spectra were obtained using Nicolet DR1 and DR2 software, respectively.

PY-GC-ITD

Triplicate samples of straw, finely pulverized, were subjected to pyrolysis in a quartz holder. A CDS Pyroprobe 100 heated filament pyrolyser was used, fitted with a platinum coil probe. Pyrolysis was carried out at 600°C for 5 seconds. The probe was coupled directly to a Carlo Erba HRGC with a flame ionization detector. An SE 54 capillary column (25 m x 0.32 mm I.D., film thickness 0.25 μm) was operated at a temperature programmed from 50°C (10 min) to 250°C at 5°C/min. An ITD (Finnigan MAT) model 600 set at 70 eV and equipped with the software Release 3.0 was employed for mass spectral analyses.

HPLC

Samples were sequentially extracted in buffer (pH 7), 0.1 M NaOH and 2 M NaOH-nitrobenzene (100/2, v/v) to yield three phenolic fractions, namely free phenolics, alkali-labile lignin and alkali-resistant lignin. These were subsequently analyzed by HPLC. Details of the extraction are given elsewhere (Galletti *et al.*, 1988).

A reversed-phase column (120 x 4.6 mm) Viosfer C6, 5 μm (Violet, Rome, Italy) was used under isocratic conditions with methanol-0.1% perchloric acid in water (15:85, v/v) using an M 45 Waters pump, a Rheodyne 7010 injector and/or an ESA coulochem model 5100 A (analytical cell 5011 set at 0.80 V) or a 440 Waters UV detector.

RESULTS AND DISCUSSION

HPLC

Fig. 1 shows the trend of the total amount of the free phenolic fraction in the 8 samples. Higher quantities of free phenolics were contained in the first samples in which *S. rugosoannulata* was more active due to the higher oxygen concentration. Differences in the single compounds were marked and statistically significant ($P < 0.001$): o-hydroxyphenylacetic acid concentration ranged from 36.78 mg/100 g in sample 1 to zero in sample 30; p-hydroxybenzoic acid from 7.48 to 0; p-hydroxybenzaldehyde from

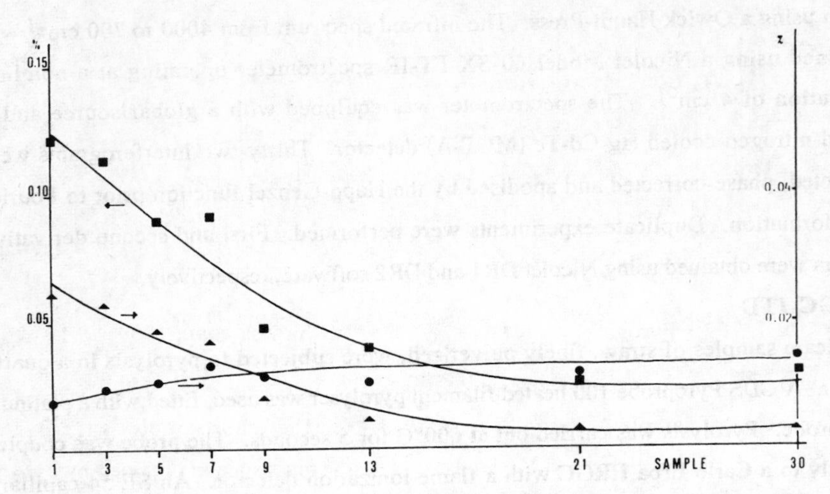

Fig. 1. HPLC trends of total free phenolics (■, left axis), syringic acid (▲, right axis) and p-coumaric acid (●, right axis) in the free phenolic fraction.

3.79 to 1.95; syringic acid from 24.39 to 2.05; while p-coumaric acid showed an opposite trend (7.80 mg/100 g in sample 1 and 12.11 in sample 30 (Fig. 1). A tentative explanation for this result is that the C (α) - C(ß) bond is cleaved by fungal activity. The relationships between the quantities of free phenolics and the sample position in the chain is well described by square functions with R^2>0.9.

The phenolic compounds from the alkali-labile lignin (data not reported) showed the same highly-significant, although less evident, trend shown by free phenolics. Also in this fraction, the phenolics having a propanoic chain on the ring (p-coumaric and ferulic acids) were relatively less abundant in the first flasks than in the last flasks, probably for the same reasons mentioned for the free phenolics. Differences between vanillin and syringaldehyde concentrations in the first and the last flasks were slightly significant and non-significant, respectively. This is consistent with the fact that these compounds develop with the lignification process and, therefore, can be more resistant to fungal attack.

With respect to the quantitative composition of phenolics in the last fraction (alkali-resistant lignin), lower values were found in the most degraded samples (first flasks) and higher values in the least degraded samples (last flasks). This result is consistent with and complementary to the other two fractions.

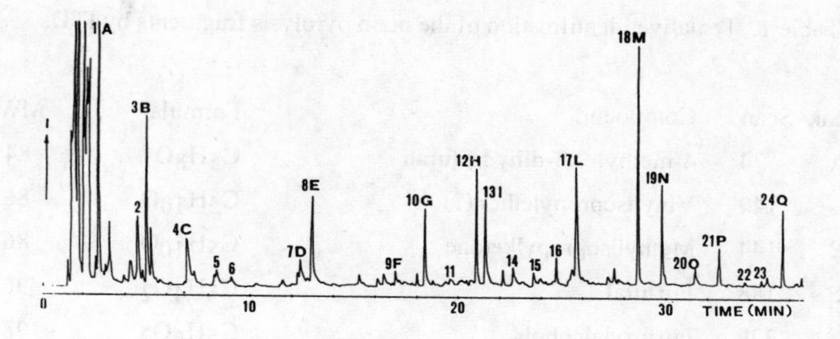

Fig. 2. Pyrogram of straw sample No. 1 (Chiavari *et al.*, 1989).

PY-GC-ITD

The pyrogram of sample No. 1 is shown in Fig. 2. The other samples gave rise to qualitatively similar pyrograms. The comparison of the mass spectra obtained by ion trap detection (ITD) with known spectra and the analogical interpretation of the Finnigan library software led to the tentative identification of most of the main fragments (Table 1).

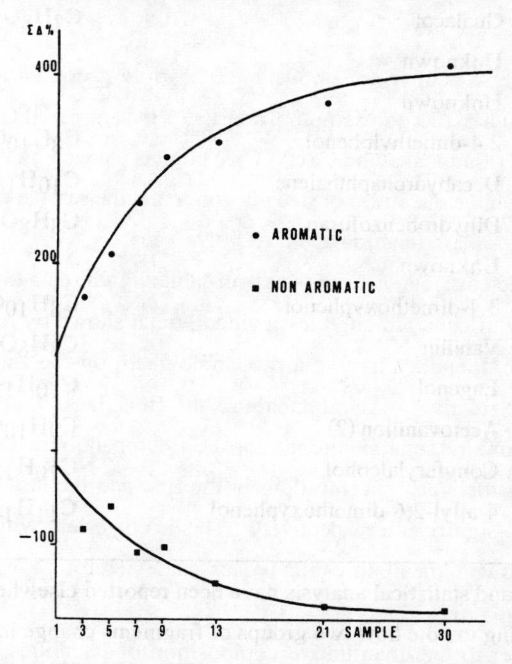

Fig. 3 Trends of the sum of the pyrolysis fragments of aromatic (●) and non-aromatic (■) nature expressed as % difference with respect to sample No. 1 (Chiavari *et al.*, 1989).

Table 1. Tentative identification of the main pyrolysis fragments by ITD.

Peak	Scan	Compound	Formula	MW
1A	74	4-methyl-2.3-dihydrofuran	C_5H_8O	84
2	140	Vinylisopropylether (?)	$C_5H_{10}O$	86
3B	144	Methylisopropylketone	$C_5H_{10}O$	86
4C	188	Furfural	$C_5H_4O_2$	96
5	229	Furfurylalcohol	$C_5H_6O_2$	98
6	248	Furfurylalcohol	$C_5H_6O_2$	98
7D	363	Dihydroxypyran	C_5H_8O	84
8E	380	Piperidone	C_5H_9ON	99
9F	515	Unknown	-	-
10G	562	Methylcyclohexanone	$C_7H_{12}O$	112
11	596	O-cresol	C_7H_8O	108
12H	636	Guaiacol	$C_7H_8O_2$	124
13I	679	Unknown	-	-
14	705	Unknown	-	-
15	711	2,4-dimethylphenol	$C_8H_{10}O$	122
16	757	Decahydronaphthalene	$C_{10}H_{18}$	138
17L	794	Dihydrobenzofuran	C_8H_8O	120
18M	883	Unknown	-	-
19N	920	3,4-dimethoxyphenol	$C_8H_{10}O_3$	154
20O	968	Vanillin	$C_8H_8O_3$	152
21P	1004	Eugenol	$C_{10}H_{12}O_2$	164
22	1043	Acetovanillon (?)	$C_9H_{10}O_3$	166
23	1102	Conyferylalcohol	$C_{10}H_{12}O_3$	180
24Q	1206	4-allyl-2,6-dimethoxyphenol	$C_{11}H_{14}O_3$	194

While quantitative and statistical analysis have been reported elsewhere (Chiavari *et al.*, 1989), it is interesting to note how two groups of fragments change in the chain of flasks as percentage difference relative to flask No. 1 (Fig. 3). The group formed by A-G fragments, possibly originating from carbohydrates, is slightly more abundant in the first flasks in which the fungal activity had been higher. The group formed by aromatic fragments H, L-Q, possibly assignable to lignin, shows an opposite trend, i.e. low in the first flasks, high in the last flasks.

This last observation is consistent with the decrease in aromatic absorption observed with FT-IR analyses in the following section and with the trends of phenolics determined by HPLC analysis discussed in the previous section.

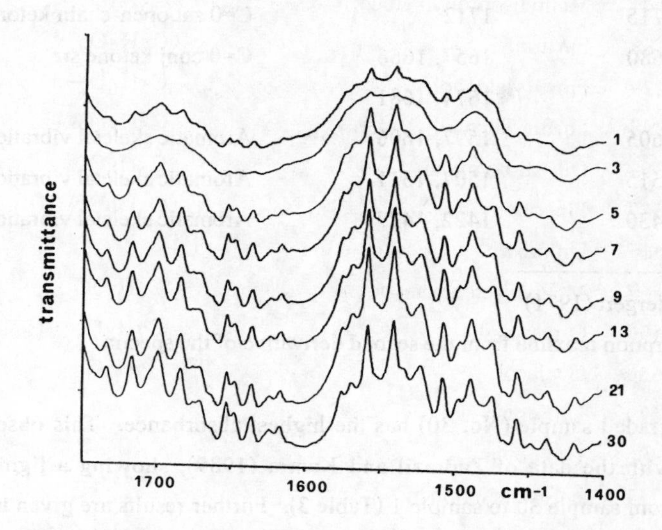

Fig. 4 FT-IR spectra of sample No. 1-30. Spectra are stacked to allow comparison of relative absorbances (Buta *et al.*, 1989).

FT-IR

The FT-IR spectra of the eight straw samples (Fig. 4) showed absorption maxima corresponding to those assignable to lignin (Hergert, 1971; Table 2). The determination of the interesting bands was performed using second-derivative spectra and a standard lignin prepared by extraction with 72% H_2SO_4.

The region from 1800 to 4000 cm^{-1} did not show any useful information other than broad hydroxyl and aliphatic C-H absorption. Similarly, spurious bands, possibly due to water vapour, in the 1800-1400 cm^{-1} region were not considered. The most significant features of the spectra are the relative changes of the aromatic absorptions due to lignin. These apparently dependent upon the degree of degradation caused by the different degree of oxygenation, the fungus being more active when more oxygen is present. Therefore, the most degraded sample (No. 1) has the smallest absorbance and

Table 2. Assignments of FT-IR absorption bands (cm^{-1})

lit [a]	sample 1 [b]	Assignment
1720	1729	C=0 aliph carboxyl str
1710-1715	1712	C=0 sat open-chain ketone str
1660-1680	1657, 1666	C=0 conj ketone str
	1674, 1681	
1595-1605	1597, 1606	Aromatic skeletal vibrations
1505-1515	1504, 1511	Aromatic skeletal vibrations
1425-1430	1422, 1427	Aromatic skeletal vibrations

[a] See Hergert (1971)

[b] Absorption maxima from the second derivative of the spectra.

the least degraded sample (No. 30) has the highest absorbance. This observation is consistent with the data of Zadrazil and Kamra (1989), showing a lignin content decreasing from sample 30 to sample 1 (Table 3). Further results are given in Buta and Galletti (1989).

Table 3. Lignin contents as reported by Zadrazil and Kamra (1989).

Sample	Lignin content *
1	16.97
3	19.11
5	20.64
7	21.38
9	24.08
13	24.96
21	25.07
30	23.61

* % of dry matter

CONCLUSION

To conclude, all the three methods investigated provided evidence for changes in some chemical parameters due to different degrees of oxygenation in laboratory-scale solid-state fermentor, while yielding results consistent with each other: FT-IR showed that a decrease in aromatic absorption bands was directly related to the oxygen concentration (and hence to fungal activity). PY-GC-ITD showed a relative increase in carbohydrate-related fragments and a simultaneous decrease of lignin-derived fragments with increasing fungal activity.

HPLC determination of phenolics showed higher quantities of free phenolics in the most degraded samples with respect to the least degraded ones. The same, although less marked trend, was shown by the phenolics from the alkali-labile lignin fraction, whereas the alkali-resistant lignin showed an opposite trend.

The results show what molecules are affected by the metabolism of *S. rugosaonnulata* in the solid-state fermentation of wheat straw. Meanwhile they provide useful markers for monitoring the process. FT-IR and PY-GC-ITD require less manipulation than HPLC, but absolute quantitations are more difficult with the former techniques than with the latter.

More detailed information on the nature of the material is obtainable by PY-GC-ITD, although the energy involved in the pyrolysis process might be too high to discriminate between small changes in the matrices due to mild fermentation processes.

HPLC is a more selective and specific technique, but sample preparation is required prior to analysis.

REFERENCES

Buta, J.G. and Galletti, G.C. (1989). FT-IR investigations of lignin components in various agriculture lignocellulosic by-products. J. Sci. Food Agric. **49**, 37-43.

Buta, J.G., Zadrazil, F. and Galletti, G.C. (1989). FT-IR Details of lignin degradation in wheat straw by the white-rot fungus *Stropharia rugosoannulata* with differing oxygen concentrations. J. Agr. Food Chem. **37**, 1382-84

Chiavari, G., Concialini, V. and Galletti, G.C. (1988). Electrochemical detection in the high-performance liquid chromatographic analysis of plant phenolics. Analyst **113**, 91-94.

Chiavari, G., Francioso, O., Galletti, G.C., Piccaglia, R. and Zadrazil, F. (1989). Characterization by pyrolysis-gas chromatography of wheat straw fermented with white-rot fungus *Stropharia rugosoannulata.* J. Anal. Appl. Pyrolysis, **15**, 129-136.

Faix, O. (1986). Investigations of lignin polymer models (DHP's) by FT-IR spectroscopy. Holzforschung, **40**, 273-280.

Faix, O., Meier, D. and Grobe, I. (1987). Studies on isolated lignins in woody materials by pyrolysis-gc-ms and off-line pyrolysis-pyrolysis-gc with flame ionization detection. J. Anal. Appl. Pyrolysis, **11**, 403-417.

Galletti, G.C., Piccaglia, R., Chiavari, G., Concialini, V. and Buta, J.G. (1988). HPLC characterization of phenolics in lignocellulosic materials. Chromatographia, **26**, 191-196.

Galletti, G.C., Piccaglia, R., Chiavari, G. and Concialini, V.(1989). HPLC/electrochemical detection of lignin phenolics from wheat straw by direct injection of nitrobenzene hydrolysates. J. Agric. Food Chem. **37**, 985-987.

Hartley, R.D. and Buchan, H. (1984). High-performance liquid chromatography of phenolic acids and aldehydes derived from plants or from the decomposition of organic matter in soil. J. Chromatogr. **180**, 139-143.

Hergert, H.L. (1971). Infrared spectra. In *Lignins: occurrence, formation, structure and reactions,* eds. K.V. Sarkanen and C.H. Ludwig, Wiley Interscience, New York, pp. 267-297.

Jung, H.G. and Himmelsbach, D.S. (1989). Isolation and characterization of wheat straw lignin. J. Agric. Food Chem. **37**, 81-87.

Pouwels, A.D., Eijkel, G.B. and Boon, J.J. (1989). Curie-point pyrolysis-capillary gas chromatography-high-resolution mass spectrometry of microcrystalline cellulose. J. Anal. Appl. Pyrolysis **14**, 237-280.

Reeves, J.B., III (1986). Use of nitrobenzene oxidation for study of lignin composition with an improved method for product extraction. J. Dairy Sci. **69**, 71-76.

Zadrazil, F. and Reiniger, P. (eds.) (1988). *Treatment of lignocellulosics with white-rot fungi,* Elsevier Applied Science, London.

Zadrazil, F. and Kamra, D.N. (1989). Influence of air and oxygen supplies on lignin degradation and its relation with in vitro digestibility of wheat straw fermented with *Stropharia rugosoannulata, Pleurotus eryngii* and *Pleurotus sajor caju.* Mush. J. Tropics, **9**, 79-88.

ULTRASTRUCTURAL ALTERATIONS OF WOOD CELL WALLS DURING DEGRADATION BY FUNGI

K. Ruel

Centre de Recherches sur les Macromolécules Végétales
(CERMAV), - CNRS - B.P. 53 X, 38041 Grenoble, Cedex (France)

Because it is made up of a compact polymer network, the wood cell wall is not easily degraded by wood-rotting fungi. Electron microscopic investigation of the ultrastructural modifications caused by the white-rot species, *Phanerochaete chrysosporium,* clearly shows that attack on wood by an individual microorganism may have different micromorphogical patterns. This variability in the behaviour of the fungus is related to the physiological changes that it undergoes during growth. As a consequence, the lytic enzymes are differentially located intracellularly or may diffuse through the fungus wall. Immunocytochemical observation of the extracellular enzymes demonstrates that they do not penetrate deeply in the wood cell wall. This limited penetration often starts in an already pre-degraded zone. Thus, participation of non-enzymic agents in the process is suggested.

INTRODUCTION

Lignin is the most important factor limiting wood cell wall biodegradation. Several microorganisms can partially degrade lignocellulosic material. But, only certain fungi and some bacteria have the capability of degrading, albeit to different extents, all of the plant cell wall polymers, viz. cellulose, hemicelluloses and lignin (Ruel and Barnoud, 1985). Fungi, in particular, can be efficient with respect to delignification. Among the best wood-degrading fungi, a distinction must be made between the soft-rot, the brown-rot and the white-rot fungi. The soft-rots are characterized by a functional cellulase system that can attack crystalline cellulose *in vitro.* However, because the enzymes show a limited ability to diffuse through wood cell walls, their actions give rise to the formation of small cavities in the wood (Liese, 1970). Degradation is never extensive since these types of fungi are particularly lacking in ligninolytic activities (Kirk and Cowling, 1984). Similarly, the brown-rot fungi primarily metabolize the carbohydrate components of the wood and have only a weak action on lignin (Kirk and Highley, 1973; Messner and

Stachelberger, 1984). Here again, hemicelluloses and cellulose are the preferred substrates for the fungal enzymes. The hemicellulosic component of the middle lamella is first removed before the secondary wall is attacked (Blanchette and Abad, 1988). Degradation results in swelling of the wood cell wall in which a great part of lignin remains.

The most efficient wood-degrading fungi are found amongst the white-rots. These species are categorized according to the micromorphological aspects of the degradation that they create in the wood. Some degrade all cell wall components simultaneously and some show a selectivity towards one or more constituents (Blanchette *et al.*, 1987). *Phanerochaete chrysosporium,* is considered to be the most typical of white-rot fungi. It's well-described arsenal of enzymes includes all of those required for degradation of the different wall polymers present in wood. However, due to variations in their relative activities, wood cell wall degradation is not always total. As a result, *P. chrysosporium* may show more than one pattern of degradation. This was clearly observed in a scanning electron microscopy study by Otjen and Blanchette (1986).

In this paper we present the results of transmission electron microscopy studies on several aspects of the degradation caused by *P. chrysosporium.* The ability of the fungus to release the enzymes responsible for the degradation of hemicellulose, cellulose and lignin was examined by immunocytochemistry. The difficulty with which enzymes may diffuse into the wood cell wall was demonstrated.

MATERIALS AND METHODS

Plant material

Wood samples were taken from a 20-year-old aspen tree (*Populus tremula*) harvested in France. Wood wafers (4 x 20 x 50 mm) were degraded (at STFI, Stockholm, Sweden) by the wild-type strain K3 of the white-rot fungus, *P. chrysosporium,* or by its cellulase-deficient mutant (Johnsrud and Eriksson, 1985).

Electron Microscopy

Lignin staining

Wood attacked by fungi was fixed in a 2.5% $KMnO_4$ solution for visualization of lignin. Samples were then dehydrated using solutions of ethanol of increasing concentration. Embedding was performed either in a 1:1 (v/v) mixture of methyl-butyl methacrylate, with 1% (w/v) benzoyl peroxide (w/v) as catalyst, or in Epon.

Polysaccharide staining

Periodic acid-thiocarbohydrazide-silver proteinate (PATAg), prepared according to Ruel *et al.* (1977), was used for contrasting the polysaccharide part of the wood cell wall.

Immunocytochemical labelling

Preparation of antisera

Antisera directed against the crude enzyme mixture secreted by *P. chrysosporium*, during cultivation on cellulose, were raised in rabbits. Immunoglobulin G (IgG), purified at the Institute Pasteur (Lyon), were used for immunolabelling. Secondary goat anti-rabbit gold marker was obtained from Sigma. The anti-ligninase antiserum was raised in rabbit, using a purified ligninase preparation (a gift from Dr. E. Odier, INRA, Paris-Grignon, France) as antigen.

Fixation and embedding

Tissues were fixed in 2% (w/v) paraformaldehyde, 2.5% (w/v) glutaraldehyde in 0.1 M phosphate buffer, pH 7.4, containing 0.02% (w/v) picric acid. They were dehydrated in glycol methacrylate monomer and embedded in glycol methacrylate (GMA)(Spaur and Moriarty, 1977).

Labelling

Antibodies were used as post-embedding markers. Sections of decayed wood were first incubated on a drop of TBS (0.1 M Tris-phosphate buffer, pH 7.4, containing 0.15 M or 0.50 M NaCl), in 0.15 M glycine. After rinsing in TBS, they were floated on a drop of 1% TBS-BSA (bovine serum albumin) or (non-immune goat serum) before treating either with IgG anti-crude enzymes (19 μg/ml diluted in TBS-BSA) (or TBS/normal goat serum, i.e. TBS/NGS), or anti-ligninase antiserum diluted 1:250 in the same buffer, for 60 min at room temperature. The secondary antisera labelled with gold (10 nm in diameter) was a goat anti-rabbit antisera purchased from Janssen (Pharmaceutica, Beerse, Belgium). It was diluted 1:30 in TBS/NGS. The sections were examined on a Philips 400T electron microscope without any counterstaining.

Immunocytochemical controls

(a) Substitution of the primary antibody with pre-immune rabbit serum IgG fraction.

(b) Treatment of section with goat-antirabbit gold-labelled secondary antibody alone, omitting the primary antibody step.

(c) Labelling with antisera pre-adsorbed with their respective antigens. Equal volumes of anti-crude enzymes and of the enzymic extract, or anti-ligninase and pure ligninase, were incubated for 1h before use.

Enzyme-gold complexes

Purified mannanase and xylanase were complexed with colloidal gold particles of 5 nm diameter according to Ruel and Joseleau (1984).

RESULTS AND DISCUSSION

Patterns of attack exhibited by *Phanerochaete chrysosporium*

Ultrathin sections of wood from *Populus tremula* that had been incubated for various lengths of time with *P. chrysosporium* were examined using electron microscopy. Although a reported pattern of wood degradation by white-rot fungi was said to be typical, it is interesting to note in Fig. 1 that *P. chrysosporium* displayed several patterns of action. Morphological modifications of the wood cell walls took the form of bore holes, or of a general swelling of the wall due to the selective solubilization of one of the wall constituents, or even a progressive thinning of the wall layers. The latter is the more common pattern attributed to white-rots.

From the preceding observations it can be seen that *P. chrysosporium* does not always behave in the same way. This is to be correlated with the fact that the morphology of the hyphae themselves undergoes dramatic variations. These morphological changes are accompanied by changes in the physiological status of the hyphae and may, in turn, result in modification of the factors responsible for cell wall degradation. In particular, it has been demonstrated that the enzymes involved in the breakdown of the wood cell wall components are produced by *P. chrysosporium* in varying proportions depending on growth conditions (Ander and Eriksson, 1975). Their release depends on hyphal cell metabolism during growth (Kubicek, 1983). This is clearly depicted in Fig. 2. Hyphae with or without cellular content and hyphae surrounded by an abundant extracellular mucilage, may be seen.

Removal of different cell wall polymers

Biochemical studies of the enzymes involved in wood degradation showed that *P. chrysosporium* first attacks hemicelluloses and lignin and then attacks cellulose (Ander and Eriksson, 1975). In order to verify this point *in vivo,* specific markers of the 3 main wall components were used. With potassium permanganate staining of lignin and periodic acid-thiocarbohydrazide-silver staining of the hemicelluloses (PATAg), it can be seen that one of the first events to occur is an enhancement of lignin reactivity and a removal of hemicelluloses (Fig. 3). A more precise aspect of the progressive removal of

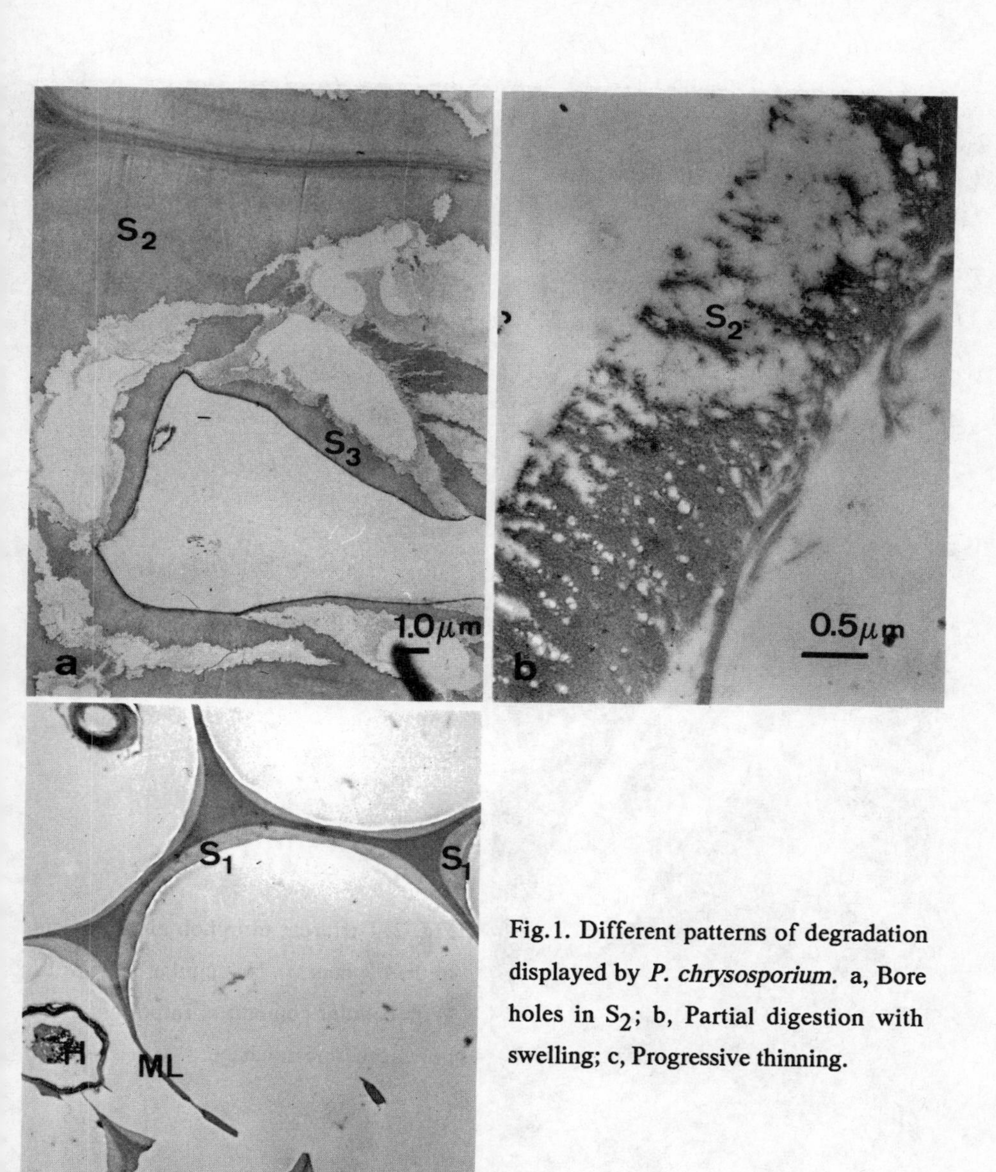

Fig. 1. Different patterns of degradation displayed by *P. chrysosporium*. a, Bore holes in S_2; b, Partial digestion with swelling; c, Progressive thinning.

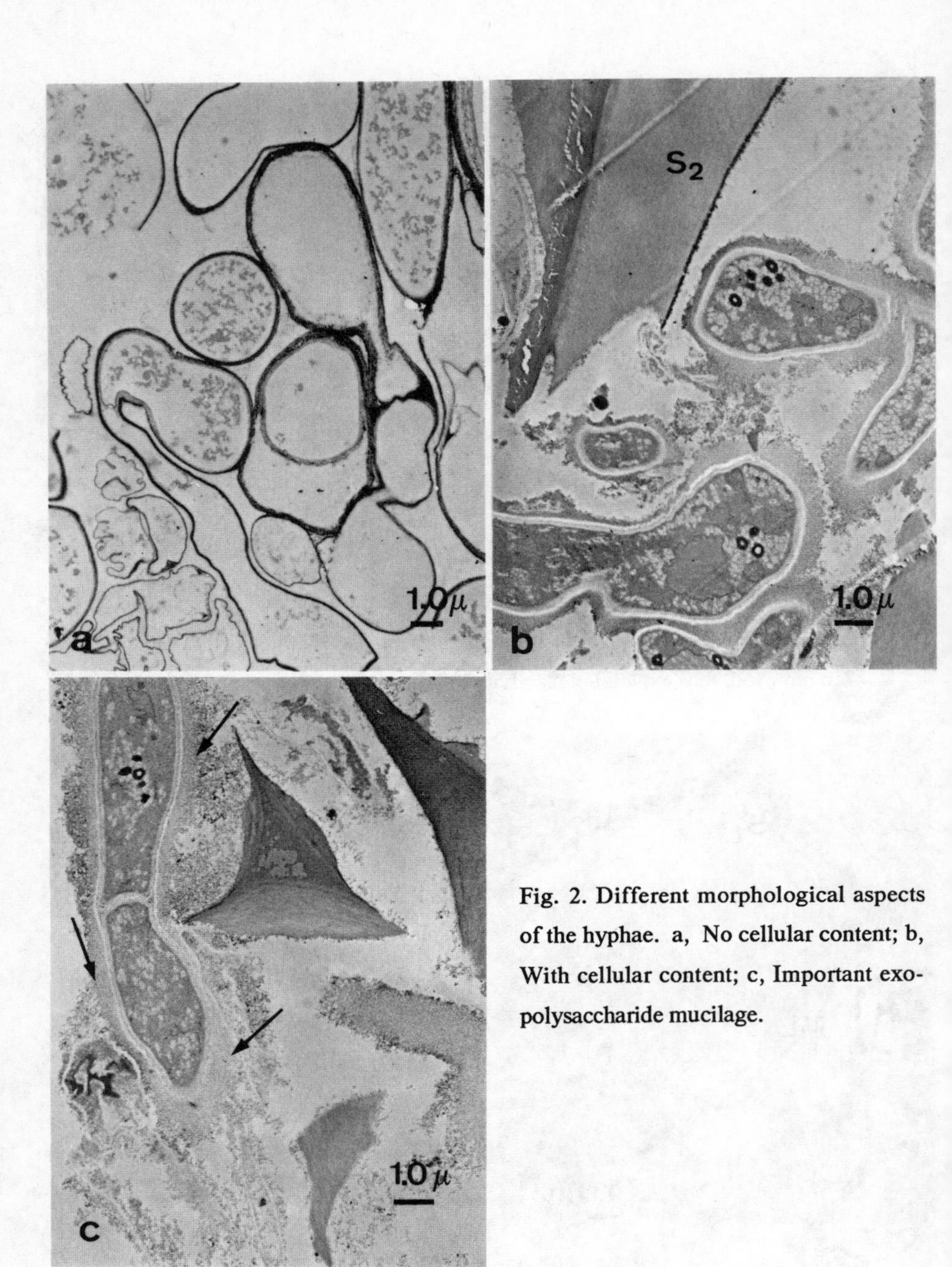

Fig. 2. Different morphological aspects of the hyphae. a, No cellular content; b, With cellular content; c, Important exopolysaccharide mucilage.

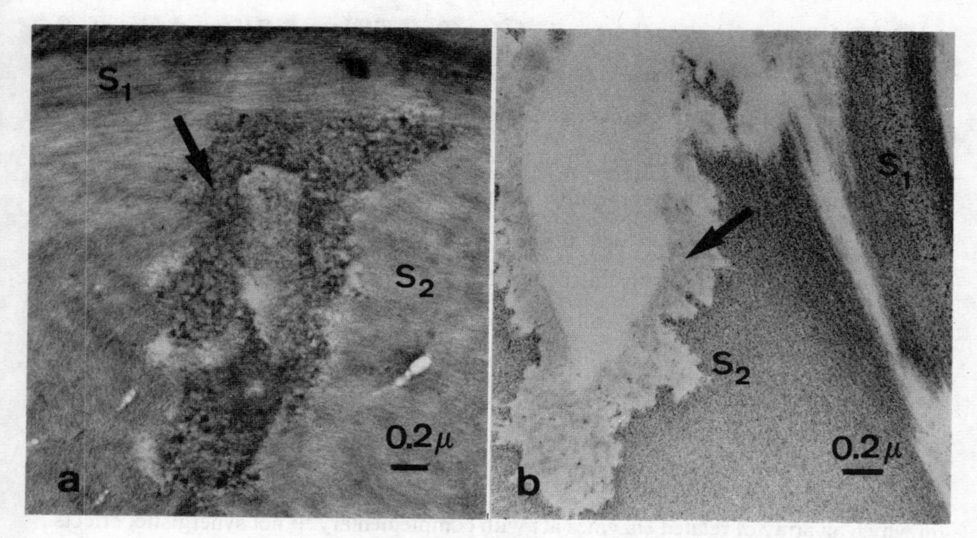

Fig. 3. Chronology in the attack. a, $KMnO_4$-staining showing enhanced reactivity of lignin; b, PATAg-staining showing that hemicelluloses have been removed, thus explaining reactivity in a.

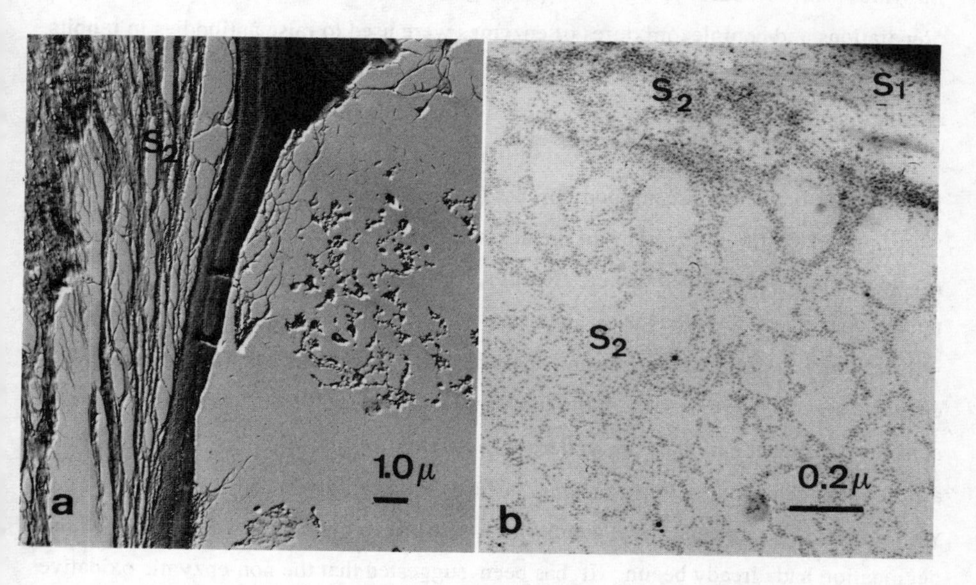

Fig. 4. Cellulase-deficient mutant unmasks the cellulose network in spruce wood. a, Shadowing; b, Mannanase-gold complex showing the presence of hemicelluloses associated with cellulose.

hemicelluloses was provided by the use of a cellulase-deficient strain of *P. chrysosporium* (Johnsrud and Eriksson, 1985). The action of this mutant allowed of the visualization of the ultrastructural organization that should be observed if the first steps of the degradation were limited to lignin and hemicelluloses.

In the case of spruce wood subjected to attack by the cellulase-deficient mutant, a typical image of the cellulose network was obtained but the removal of lignin was not complete (Fig. 4). This demonstrates that, due to interactions at the ultrastructural level between cellulose and the other wall constituents, all of the lignin is not equally accessible to enzymic removal. The same is true for hemicelluloses. These substances, and xylans in particular, have been shown to establish strong molecular associations with cellulose (Mora *et al.*, 1986). Thus, it is obvious that in order to achieve a complete degradation of the wood constituents, lignolytic fungi must possess a highly diversified enzyme system in which an array of related enzymes act with complementary, if not synergistic, effects.

Visualization of the lytic enzyme system of fungi

Enzymes can be observed directly by using immunocytochemical methods. Pure preparations and complex mixtures of enzymes were used to raise antibodies in rabbits. Thus, a purified lignin-peroxidase fraction (Buswell *et al.*, 1985) provided an anti-ligninase (anti-L) antibody. A culture filtrate obtained during primary growth of *P. chrysosporium,* and, therefore, devoid of lignin-degrading activities, provided the bases for cellulase-hemicellulase antibody probes (anti-polysaccharidases). These antibodies were applied to ultrathin sections of the degraded wood and the enzymes were indirectly revealed via a secondary marker. Results, both with anti-ligninase and with anti-polysaccharidases, revealed that the enzymes may be located within the hyphal cells (i.e. intracellular), within the periplasm, or be bound to the fungal wall. However, the enzymes are able to pass through the fungal wall (Fig. 5). This diffusion is particularly active whenever the hyphae are undergoing autohydrolysis. But, the most important aspect of the behaviour of these enzymes is their limited ability to penetrate the wood cell wall. Moreover, this superficial attack *in vivo* seems to be restricted to areas in which degradation had already begun. It has been suggested that the non-enzymic oxidative processes initiated by the fungus could induce the degradation process, thereby providing a susceptible site for attack by the hydrolytic enzymes (Kirk and Farrell, 1987).

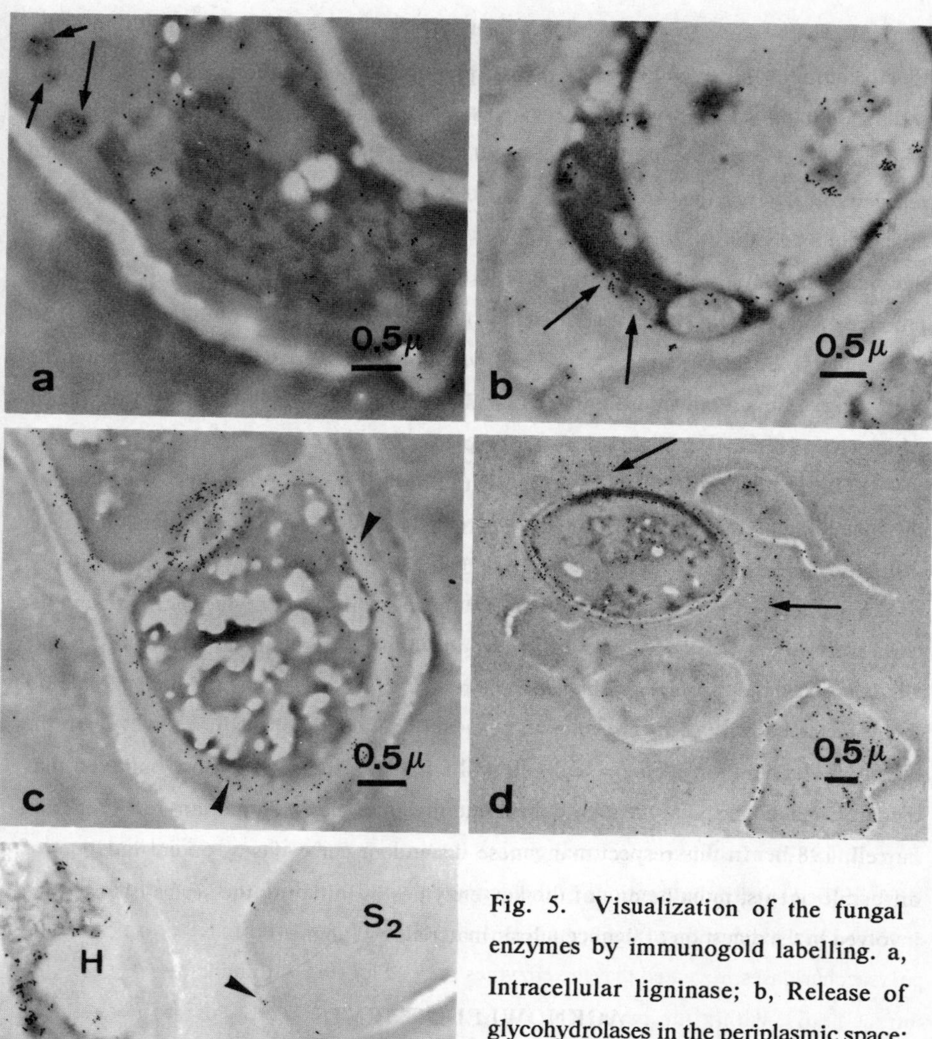

Fig. 5. Visualization of the fungal enzymes by immunogold labelling. a, Intracellular ligninase; b, Release of glycohydrolases in the periplasmic space; c. Wall-bound ligninase; d. Excretion; e. Limited penetration of the wood cell wall.

CONCLUSION

Electron microsopic examination of the morphological modifications created in the cell walls of wood inoculated with *P. chrysosporium* demonstrates that the subdivision of the white-rot fungi into several subclasses, based upon a typical image of decay, is somewhat restrictive. The various patterns of attack by *P. chrysosporium* noted here can be explained by the physiological variations of the fungus under the influence of external factors that modify growth conditions. These factors can affect the balance of enzymes produced and, thereby, affect the efficiency of degradation of some of the wood cell wall components. Specific markers of the cell wall polymers were useful for following the chronology of the removal of lignin and hemicelluloses by the cellulases-less mutants (in which case, degradation could not go to completion).

The lignin- and polysaccharide-degrading enzymes of *P. chrysosporium,* as revealed by immunocytochemistry, can be intracellular, wall-bound or extracellular. This diversity of localization is in agreement with the diversity of physiological states exhibited by the hyphae. Here again one can correlate the variability in enzyme secretion and the variability of growth conditions in the wood with the diversity in modes of attack. Obviously, a major problem in the application of *P. chrysosporium* to biological pulping is the limited ability of its enzymes to penetrate the wood cell wall. However, in the complete degradation of the wood cell walls by the fungus, it is now believed that oxidative reactions and aryl cation radical intermediates play an important role (Kirk and Farrell, 1987). In this respect manganese-dependent peroxidases, potent inducers of diffusible agents, must be one of the key enzymes in initiating the series of reactions involved in the digestion of lignocellulosic materials by fungi.

ACKNOWLEDGEMENTS

Thanks are expressed to Dr. K.-E. Eriksson (STFI, Stockholm) for providing samples of decayed wood, to Professor R. Guinet (Institut Pasteur, Lyon) for preparing anti-glycohydrolases, to Dr. Odier (INAPG - Paris) for the gift of anti-ligninase and to Professor J.P. Joseleau for his help in the preparation of this manuscript.

REFERENCES

Ander, P. and Eriksson, K.-E. (1975). Influence of carbohydrates on lignin degradation by the white-rot fungus *Sporotrichum pulverulentum*. Svensk Papperstidn. **78**, 643-652.

Blanchette, R.A. and Abad, A.R. (1988). Ultrastructural localization of hemicellulose in birch wood (*Betula papirifera*). decayed by brown- and white-rot fungi. Holzforschung, **42**, 393-398.

Blanchette, R.A., Otjen, L. and Carlson, M.C. (1987). Lignin distribution in cell walls of birch wood decayed by white-rot Basidiomycetes. Phytopathol. **77**, 684-690.

Buswell, J.A., Mollet, B. and Odier, E. (1985). Ligninolytic enzyme production by *Phanerochaete chrysosporium* under conditions of nitrogen sufficience. FEMS Microbiol. Lett. **25**, 295-299.

Johnsrud, S.C. and Eriksson, K.-E. (1985). Cross-breeding of selected and mutated homokariotic strains of *Phanerochaete chrysosporium* K3: New cellulase-deficient strains with increased ability to degrade lignin. Appl. Microbiol. Biotechnol. **21**, 320-327.

Kirk, T.K. and Cowling, E.B. (1984). Biological decomposition of solid wood. Adv. Chem. Ser. **207**, 455-487.

Kirk, T.K. and Farrell, R.L. (1987). Enzymatic combustion: the microbial degradation of lignin. Ann. Rev. Microbiol. **41**, 465-505.

Kirk, T.K. and Highley, T.L. (1973). Quantitative changes in structural components of conifer woods during decay by white and brown-rot fungi. Phytopathol. **63**, 1338-1342.

Kubicek, C.P. (1983). β-glucosidase excretion in *Trichoderma* strains with different cell wall bound β-1,3-glucanase activities. Can. J. Microbiol. **29**, 163-169.

Liese, W. (1970). Ultrastructural aspects of woody tissue disintegration. Ann. Rev. Phytopathol. **8**, 231-258.

Messner, K. and Stachelberger, H. (1984). Transmission electron microscope observations of brown-rot caused by Fomitopsis pinicola with respect to osmiophilic particles. Trans. Br. Mycol. Soc. **83**, 113-130.

Mora, F., Ruel, K., Comtat, J. and Joseleau, J.P. (1986). Aspect of native and redeposited xylans at the surface of cellulose microfibrils. Holzforschung, **40**, 85-91.

Otjen, L. and Blanchette, R.A. (1986). A discussion of microstructural changes in wood during decomposition by white-rot basidiomycetes. Can. J. Bot. **64**, 905-911.

Ruel, K. and Barnoud, F. (1985). Degradation of wood by microorganisms. In *Biosynthesis and Biodegradation of Wood Components,* ed. T. Higuchi,

Academic Press, New York, pp. 441-464.

Ruel, K. and Joseleau, J.P. (1984). Use of enzyme-gold complexes for the ultrastructural localization of hemicelluloses in the plant cell wall. Histochem. **81**, 573-580.

Ruel, K., Comtat, J. et Barnoud, F. (1977). Localisation histologique et ultrastructurale des xylanes dans les parois primaires des tissus d'Arundo donax. C.R. Acad. Sci. **284**, 1421-1424.

Spaur, C.R. and Moriarty, G.C. (1977). Improvements of glycol methacrylate. I. Its use as an embedding medium for electron microscopic studies. J. Histochem. Cytochem. **25**(3), 163-174.

FUNGAL TRANSFORMATION OF LIGNOCELLULOSICS AS REVEALED BY CHEMICAL AND ULTRASTRUCTURAL ANALYSES

A.T. Martínez[1], J.M. BARRASA[2], G. Almendros[3] and A.E. González[1]

[1] CIB, CSIC, Velázquez 144, 28006 Madrid, Spain

[2] Dept. Botany, Univ. Alcalá de Henares, Madrid, Spain

[3] IEBV, CSIC, Serrano 115 dpdo, 28006 Madrid, Spain

Ultrastructural and chemical aspects of lignocellulose transformation by fungi during natural decay and solid-state fermentation are presented. An extensive fungal transformation of wood by *Ganoderma australe*, showing both selective and simultaneous degradation patterns, was found in Chilean rain forest, and studied under *in vitro* conditions. After a preferential decay of lignin and xylans, this fungus produces a partially delignified and highly digestible material, 'huempe', that is used as cattle feed. Lignin alteration during wood decay was analyzed by spectroscopic and degradative techniques. The S/G ratios from CuO alkaline degradation of wood were only slightly greater that those obtained from CPMAS ^{13}C-NMR. Very high values were obtained in some of the Chilean woods studied. The composition of the hemicellulose fraction of the latter was also unusual. The CuO degradation gave information on the alteration of lignin by fungi. This was followed by the study of the non-condensed fraction by acidolysis and thioacidolysis. A decrease in the S/G ratio during fungal degradation was found in both condensed and non-condensed fractions of lignin and, compared with that by other fungal species, an extensive oxidation of lignin was observed during wood decay by *G. australe.*

INTRODUCTION

The study of fungal degradation of lignin has shown a very significant development during the last decade. Such progress is related to the increasing interest in the rational utilization of plant biomass and has been made possible by recently available analytical methods for the study of complex polymers. In this paper we present the results of studies on

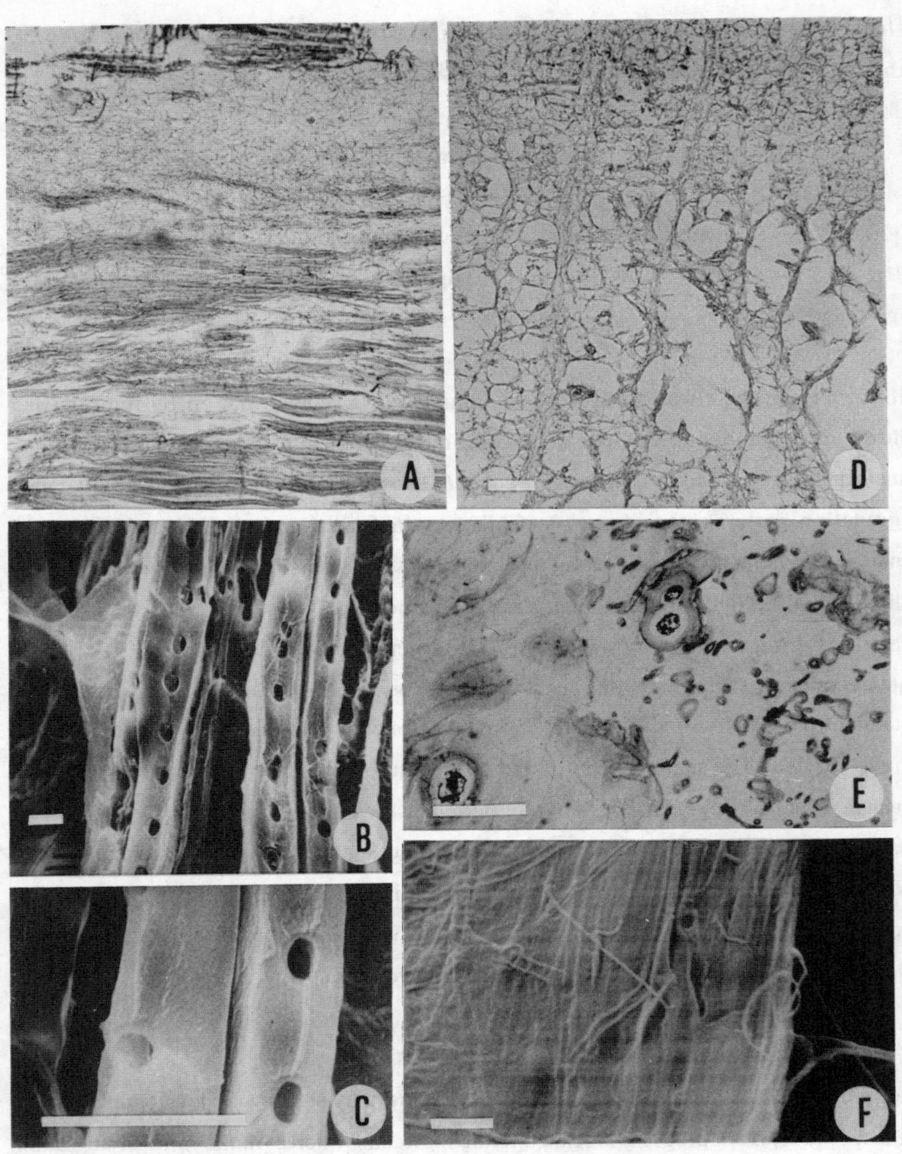

Fig. 1. Natural decay of *E. cordifolia* by *G. australe* (A-C) and of *L. philippiana* by *P. chrysocrea* (D-F). A, Zones of selective and simultaneous degradation; B-C, Fibre separation; D, Network formed by mycelium and decayed walls; F-G, Hyphae in aggregates of decayed walls. The bars indicate 50 μm in A-D and 10 μm in E-F.

different aspects of the natural decay of lignocellulose, the possibilities for the utilization of ligninolytic fungi in biotransformation of plant materials, and an overview of different techniques used in these studies.

Our first investigations on fungal degradation of lignin were based on those of González (1980), Zadrazil *et al.* (1982) and González *et al.* (1986) who had studied the natural decay of wood in Chilean rain forest. This decay produces a highly digestible delignified material, huempe or palo podrido, that is consumed by cattle. As such it represents a suitable model for biotechnological processes of wood delignification (Kirk, 1983). Several aspects of this natural process had not been previously studied and the results obtained during the present investigations have been recently presented by Barrasa *et al.* (1987), Barrasa and Martínez (1989) and González *et al.* (1987, 1989).

The development of biotechnological processes for the production of delignified materials of potential interest to the paper industry or to animal feeding has been also investigated. These processes were based on the solid-state fermentation (SSF) of forest and agricultural lignocellulosic materials with ligninolytic fungi selected for the delignification of these substrates (González *et al.*, 1988; Valmaseda *et al.*, 1987; Valmaseda, 1989).

ULTRASTRUCTURAL ASPECTS OF NATURAL DECAY OF WOOD

Analysis of the ultrastructural aspects of fungal transformation of wood is an important study in the development of biological processes for selective delignification (Blanchette *et al.*, 1987). The latter is characterized by the degradation of the middle lamella, the separation of fibres and the production of a material with potential use in the production of paper pulp.

Wood delignification by *Ganoderma australe* and *Phlebia chrysocrea* was studied under natural and *in vitro* degradation conditions. Extensive fungal growth was characteristic of this natural process of wood delignification. In the zones of active fungal growth, polysaccharides were consumed as a carbon source and an indiscriminate destruction of cell walls took place (Fig. 1A,D). Such a degradation pattern corresponded to that described as "simultaneous degradation" (Blanchette *et al.*, 1987). However, in the case of wood of *Eucryphia cordifolia* decayed by *G. australe,* the separation of fibres

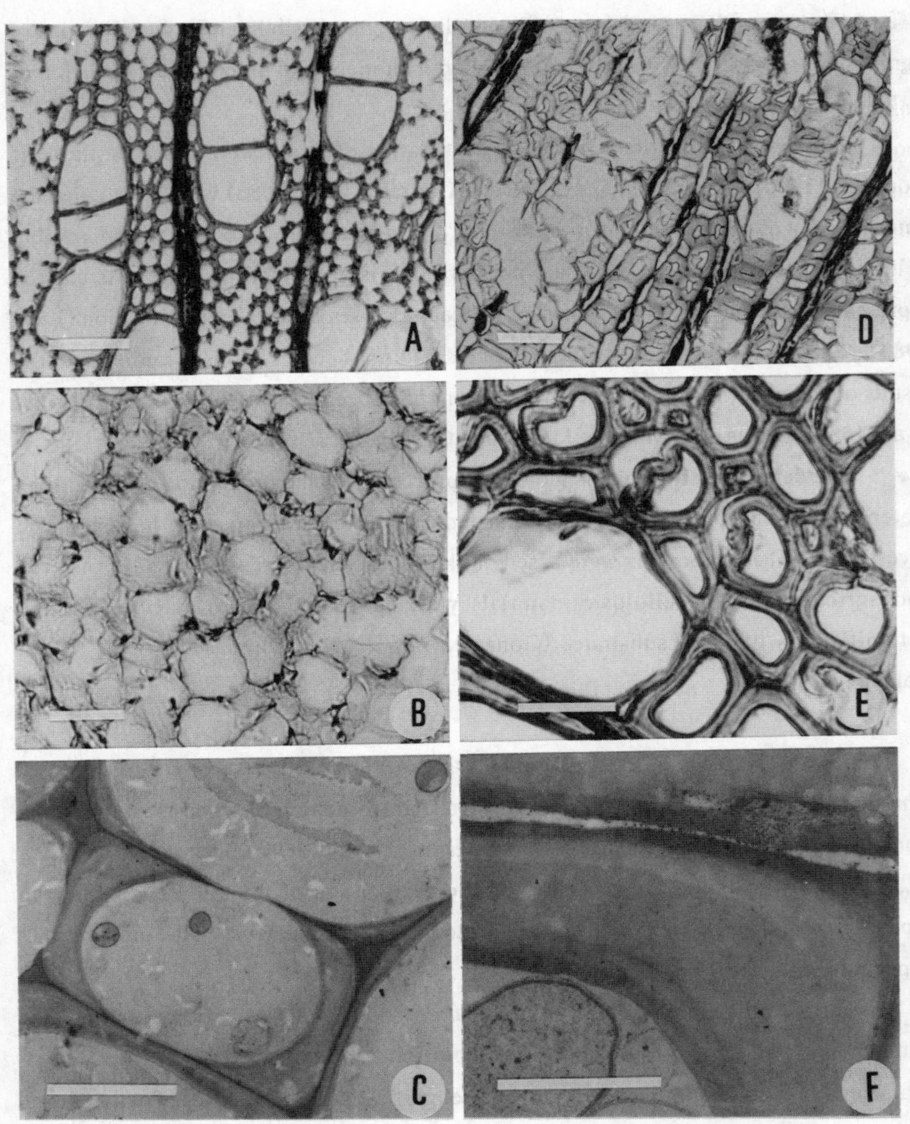

Fig. 2. *In vitro* decay of wood by *G. australe* (A,B,D) and *P. chrysocrea* (C,E,F). A, Degradation of *N. dombeyii* ; B, Network formed by decayed walls and corners of *N. dombeyii* ; C, Thinning of walls of *N. dombeyii* ; D, Bundles of thick-walled fibres of *E. cordifolia* ; E, Detachment of secondary wall of *L. philippiana* ; F, Degradation of middle lamella of *L. philippiana*. The bars indicate 50 µm in A,D; 25 µm in B,E; 10 µm in C,F.

after degradation of middle lamella (reported by Dill & Kraepelin, 1986) was also observed in those zones in which mycelia were scarcely present (Fig. 1A-C). The diffusion of enzymes or intermediate compounds in lignin biodegradation (as oxidized manganese) from zones of active fungal growth to the surrounding wood (Fig. 1A) was probably implicated in this type of cell-wall degradation. As opposed to other types of wood decay, this 'selective degradation' was not restricted to 'white-pockets' in wood.

A noticeable aspect of this natural process was the extent of wood biodegradation at the final stages of decay. The chemical characteristics of the jelly-like material produced after fungal decay will be described below. Its ultrastructural aspect is shown in Fig. 1D-F. Amorphous aggregates of mycelium and highly decayed cell-walls were observed and cellulose microfibrils became visible after degradation of hemicellulose and lignin.

A network-like structure, roughly conserving the tissue arrangement, was found in wood of *Laurelia philippiana* decayed by *P. chrysocrea* (Fig. 1D). This decay pattern (also found during *in vitro* decay, Fig. 2C) was produced by a progressive thinning of cell-walls from cell lumen to middle lamella. The material produced at advanced stages of this type of decay was of a higher consistency than the soft material formed during decay of *E. cordifolia* by *G. australe* which showed selective degradation.

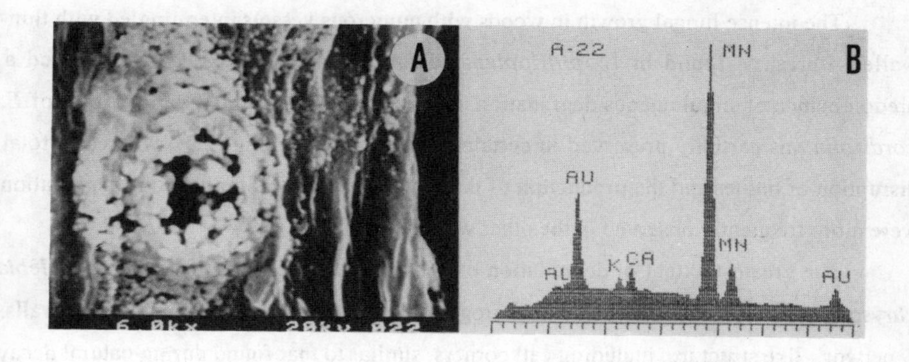

Fig. 3. Deposits of manganese oxide (A) in wood of *Eucryphia cordifolia* degraded by *Ganoderma australe,* analyzed by SEM-microanalysis (B). The bar indicates 5 μm.

Black spots and flecks in some samples of wood from *E. cordifolia* decayed by *G. australe* had a very high content of manganese. Granular deposits of manganese oxide were observed and analyzed by scanning electron microscopy and X-ray microanalysis (Fig. 3).

ULTRASTRUCTURAL STUDIES DURING *IN VITRO* DEGRADATION

The agar-block technique, with a total incubation of 12 months, was used in studies of *in vitro* degradation of the wood of three Chilean trees (viz. *E. cordifolia, L. philippiana* and *Nothofagus dombeyii*) by two basidiomycetes (viz. *G. australe* and *P. chrysocrea*) responsible for this natural process of wood decay .

Different decay patterns were found during *in vitro* degradation of wood from the three species. Changes in water and air diffusion, and in the resistance to fungal colonization, seemed to be related to the variation in the size and distribution of vascular elements in the different wood types. The wood of *E. cordifolia* had the lowest volume of vessels, thicker cell-walls in fibres (Fig. 2D) and the greater bulk density (0.64 g/cm^3). The two other woods had thinner fibre cell-walls and a lower density, but the great number of vessels (Fig. 2A). The highest lignin content (25.6 %) was found in *E. cordifolia* and the lowest (21.6%) in *L. philippiana.* This contributed to the higher biodegradability of the latter wood.

The intense fungal growth in woods with numerous vessels intermingled with thin-walled fibres, as found in *L. philippiana* and *N. dombeyii* (Fig. 2A), produced a predominance of simultaneous degradation. The integrity of the thick-walled fibres of *E. cordifolia* was partially preserved in certain regions of wood (Fig. 2D), whereas a total disruption of tissues and the production of the typical pattern of simultaneous degradation were more frequently observed in the other woods (Figs. 2A).

The greatest extent of degradation of the woods studied was effected by *Phlebia chrysocrea.* This was characterized by a progressive destruction and thinning of cell-walls. A network-like structure, including cell corners, similar to that found during natural decay of wood by this fungal species, was occasionally found during *in vitro* degradation (Fig. 2B).

At the initial stages of cell-wall degradation the detachment of secondary wall, after a general delignification of wood and partial degradation of the middle lamella, was occasionally observed (Fig. 2E,F). These pictures suggested an initial step of selective degradation but, during *in vitro* decay, the detached secondary wall was progressively degraded by the fungus.

DEGRADATION OF CELLULOSE, HEMICELLULOSES AND LIGNIN

Differential degradation of cellulose, hemicelluloses and lignin has been observed during natural degradation and SSF of lignocellulosic materials by ligninolytic fungi. A preferential degradation of lignin and pentosans, and the conservation of glucans was found at the different stages of huempe formation (Fig. 4). This was also found during SSF of wood by *G. australe.* A significant degradation of cellulose and lignin was also effected by the white-rot fungus *Trametes versicolor.* This fungus brought about simultaneous degradation of the different constituents of wood.

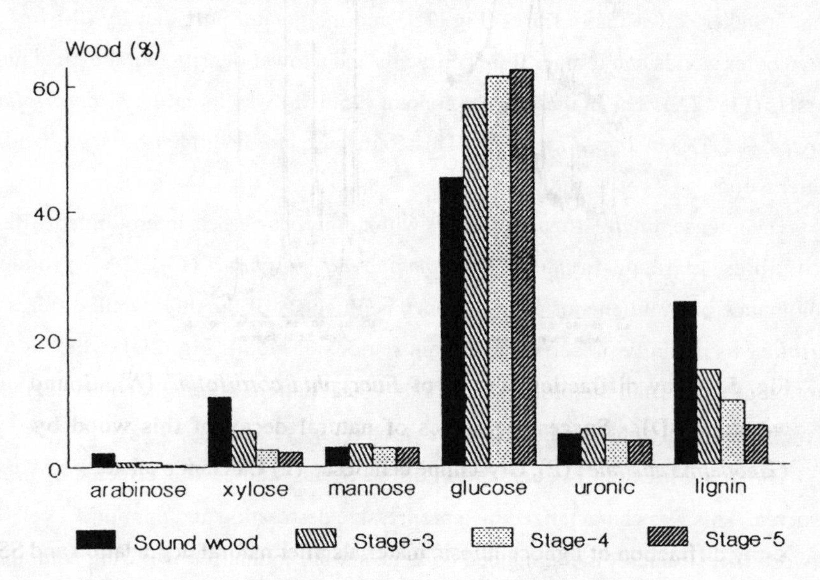

Fig. 4. Lignin content and polysaccharide composition of wood at the different stages of natural decay of *Eucryphia cordifolia* by *Ganoderma australe.*

The preferential degradation of lignin and pentosans during natural decay of Chilean woods by *G. australe* could be related to the unique composition of their lignin (see above) and hemicellulose fractions. The latter have a low content of pentoses (13% in *E. cordifolia* and *L. philippiana*) and a relatively high content of mannose (nearly 6% in *L. philippiana*), when compared with those of other hardwoods.

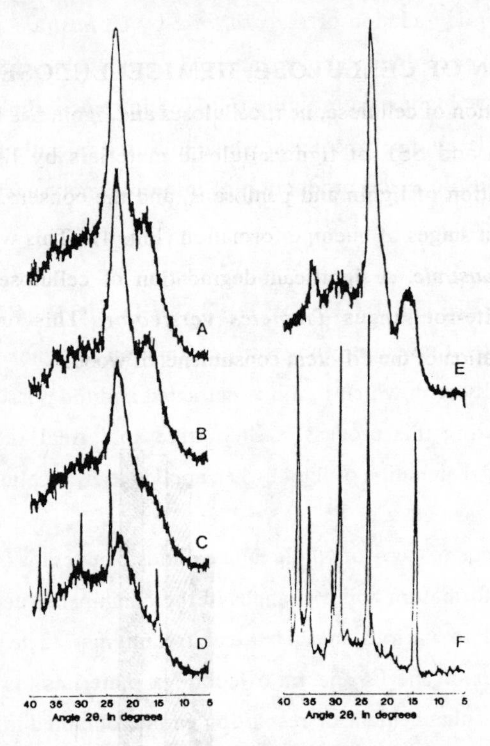

Fig. 5. X-ray diffraction spectra of *Eucryphia cordifolia.* (A), Sound wood; (B-D), Successive stages of natural decay of this wood by *Ganoderma australe*; (E), Crystalline cellulose; (F) Calcium oxalate.

X-ray diffraction of lignocellulosic materials after natural degradation and SSF by fungi generally showed a decrease in the crystallinity of cellulose. A marked reduction in crystallinity was found at the advanced stages of natural decay of *E. cordifolia* by *G. australe* (Fig. 5C-D) when compared with sound wood (Fig. 5A). However, a high crystallinity value and a diffraction pattern similar to that of crystalline cellulose was found at the intermediate stage, consisting of a cellulose-enriched material (Fig. 5C). Numerous crystals were found on hyphae at advanced stages of natural decay, and the diffraction pattern of this material revealed the presence of calcium oxalate produced by fungal metabolism (Fig. 5D-E).

During natural degradation of *E. cordifolia* by *G. australe* the losses of weight (from bulk density of freeze-dried samples), attained a mean value of 85% at the final stage of decay. However, the highest digestibility (60-70% in some samples) was found at the intermediate stage of decay, i.e. the stage at which it is used as cattle feed. This suggested that degradation of cellulose and production of microbial polysaccharides occur at the advanced stages of decay.

SPECTROSCOPIC STUDIES OF DECAYED LIGNOCELLULOSES

The study of lignin alteration during fungal transformation of lignocellulosic materials is limited by the difficulty with which lignin is extracted and the possibility that chemical alterations occur during this process. Alternative analytical techniques, yielding information about fungal alteration of lignin when applied to lignocellulosic materials, have been investigated.

The infrared spectroscopy of whole lignocellulosic materials (4000-300 cm^{-1}) can provide additional information about changes in the chemical structure of lignin if the spectra are subjected to various mathematical treatments. The analysis of bands characteristic of lignin in the IR spectra of complex materials (as wood) is difficult. However, significant enhancement of resolution can be obtained after subtraction of a positive multiple of its Laplacian. Changes in lignin bands at 1510, 1460 and 1425 cm-1 (and also in those corresponding to carbohydrates, protein, and carboxyl groups) were evident in transformed IR spectra of beech wood decayed by ligninolytic fungi. The most intense alteration of lignin was effected by *Ganoderma* species, whereas *T. versicolor* gave rise to the simultaneous degradation of wood constituents.

Nuclear magnetic resonance (NMR) spectroscopy constituted one of the most useful tools for the elucidation of lignin chemical structure. The development of cross-

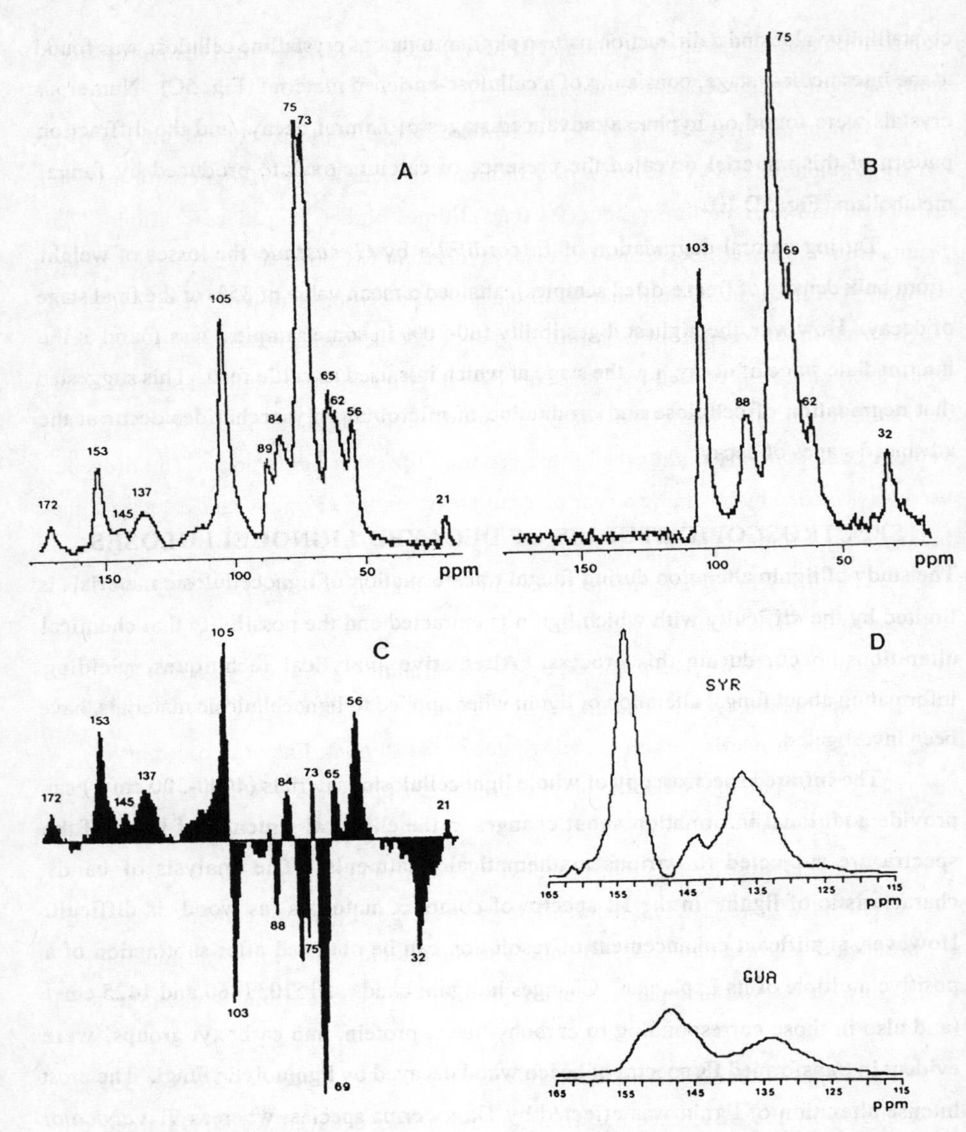

Fig. 6. CPMAS ^{13}C-NMR spectra of wood from *Eucryphia cordifolia*. A, Sound wood. B, Final stage of wood decay by *Ganoderma australe*. C, Difference spectrum (sound - decayed wood). D, Syringyl and guaiacyl components of the 165-115 ppm region of the wood spectrum (A).

polarization and magic angle spinning (CPMAS) ^{13}C-NMR introduced the possibility of studying lignin in lignocellulosic materials.

One application of CPMAS ^{13}C-NMR of wood is the 'direct' calculation of the syringyl/guaiacyl ratio (S/G) of wood lignin. The traditional S/G calculation (lignin extraction and degradation) produces values 3 times higher than in 'real' lignin. This results from the lower liberation of G units involved in C-C links at C_4. The quantitative estimation of the S and G components in the aromatic carbon region of the CPMAS ^{13}C-NMR spectrum of wood from *E. cordifolia* (Fig. 6D) was carried out by subtraction of the cedar spectrum, as proposed by Manders (1987). After spectral scaling and obtaining a zero subtraction value in the 146-148 ppm region, an S/G ratio of 1.2 was obtained. This value is comparatively higher than those reported for other hardwoods. The procedure used was verified by obtaining a near unit ratio between the 137 ppm and 153 ppm signals in the syringyl component of *E. cordifolia*).

The different signals in the spectrum of wood of *E. cordifolia* (Fig. 6A) were assigned to the main wood constituents:

- cellulose (C_1 = 105 ppm; C_4 = 89 and 84 ppm; C_2, C_3, C_5 = 75 and 73 ppm; and C_6 = 65 and 62 ppm)
- lignin aromatic carbon (unsubstituted = c. 120 ppm; linked to oxygen = 153 ppm, and linked to alkyl chains = 137 ppm) and methoxyl carbon (56 ppm)
- hemicellulose acetyl groups (methyl = 21 ppm; and carbonyl = 172 ppm)

The signals of C_2 and C_6 from hemicelluloses overlapped those of cellulose, and that of C_1 produces a slight shoulder at 103 ppm. The shoulder at 148 ppm, corresponding to C_3 and C_4 in G units, was also partially overlapped by the signal attributed to tannins at 145 ppm.

Changes in the CPMAS ^{13}C-NMR spectrum during wood degradation (Fig. 6C) were shown by the difference spectrum (sound minus decayed wood) obtained after subtraction of spectra (the signals at 173, 153, 145, 137, 120, 105, 84, 73, 65, 56 and 21 ppm diminished during fungal decay, and those at 103, 88, 75, 69, and 32 ppm increased). In transformed wood (Fig. 6B) the signals corresponding to lignin and hemicelluloses have dissappeared, indicating a preferential degradation of these wood constituents by *G. australe.* Several cellulose signals (84 and 65 ppm) were missing, and a unique signal at 75-73 ppm was observed. New signals were shown at 69 and 32 ppm and that of C_1 was displaced to 103-104 ppm. The first signal corresponded to free-C_4 (absent in cellulose) and revealed the existence of additional polysaccharides of microbial origin. The polymethylene signal at 32 ppm suggested the accumulation of alkyl compounds during microbial degradation of wood.

ISOLATION AND CHARACTERIZATION OF ALTERED LIGNIN

Isolation of lignin is time-consuming and problematic but the study of the polymer obtained gives information about the chemical alterations produced by fungi. Björkman lignin (MWL) is the reference material in studies of lignin chemical structure. However, analysis of this polymeric fraction alone is not adequate for the study of fungal degradation of lignin. In studies of lignin alteration during transformation of beech wood and wheat straw, the extractive-free transformed material was treated with 0.2 M HCl in dioxane-water and three fractions of different molecular size (dioxane-lignin, oligomeric fraction, and low-molecular weight acidolysis products) were studied separately.

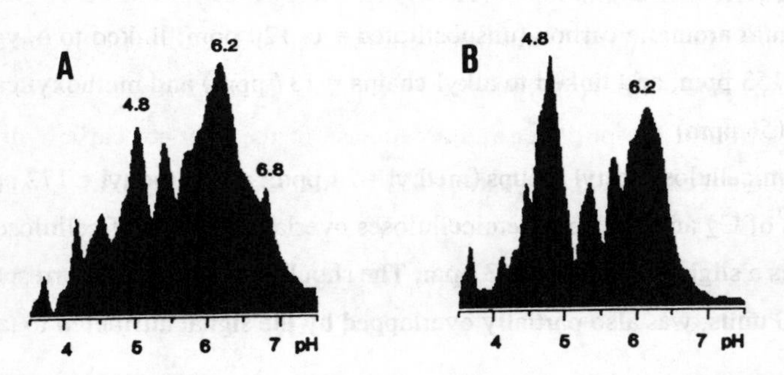

Fig. 7. Isoelectric focusing of lignin from beech wood. A, Sound wood. B, Wood decayed by *Ganoderma australe.*

The polymer fraction was of a polydisperse nature and changes in molecular size and isoelectric point distribution during fungal degradation were studied by gel permeation and isoelectric focusing, respectively. A decrease in high molecular weight fractions and the increase in oligomer and monomer products were produced during degradation of wood by *G. australe.* However, repolymerization of the degradation products was frequently produced and the molecular size distribution varied after lignin degradation by different fungi.

The isoelectric pattern (pH 3.5-7.5) of dioxane-lignin from sound wood of beech showed several major bands, the most important being centered at pH 6.2 (Fig. 7). As a consequence of the oxidative degradation of lignin and the production of carboxyl groups during fungal growth on wood, lignin isoelectric points were shifted to lower pH values.

Because of their polydisperse nature, lignin preparations cannot be characterized by most of the analytical techniques used for the study of simple chemical compounds. However, ^{13}C-NMR spectroscopy can be used for the study of chemical alterations of polymeric lignin during fungal degradation. Changes produced by fungi were observed in the subtracted spectra (control lignin minus altered lignin). The transformed lignins present a relative increase in the amount of alkyl (46-0 ppm) and carboxyl (200-160 ppm) carbons, suggesting their oxidative alteration. The lignins altered by the white-rot fungi, *P. chrysosporium* and *G. australe*, presented decreased amounts of aryl carbons, and the former species caused additional decreases in the resonances produced by the side-chain of phenylpropane units and the lignin-hemicellulose links. The cellulolytic fungi studied (i.e. *Trichoderma* and *Chaetomium* spp.) produced a significant carboxylation and demethoxylation of the polymer and a decrease of the residual carbohydrate linked to lignin, but did not cause changes in the 160-110 ppm aromatic region.

FUNGAL ALTERATION OF LIGNIN STUDIED BY CHEMICAL DEGRADATION

Non-destructive techniques (IR, NMR, etc) may be complemented with the results obtained from the identification of simple compounds released during chemical degradation of lignin. The 'degradative' techniques are limited by the possibility of chemical alteration of the lignin units and the production of artifacts. However, they permit precise chemical identification of the products obtained (e.g. by the use of mass spectrometry). In the studies on microbial transformation of lignin, the degradation products released during the depolymerization of the whole transformed substrate were recovered and identified by gas chromatography-mass spectrometry (GC-MS). The results of the total degradation of lignin by CuO alkaline oxidation, yielding simple products, and the acidolysis procedure, producing a mild depolymerization of non-condensed lignin, are discussed below.

The low molecular weight acidolysis products from beech wood decayed by *G. australe, T. versicolor, F. fomentarius* and *G. lucidum* were studied. They amounted to almost 6% of the total sample and included several minor alkyl compounds (n- and b-alkanoic mono- and diacids, ketoacids and alkanes) and were more frequent in decayed samples.

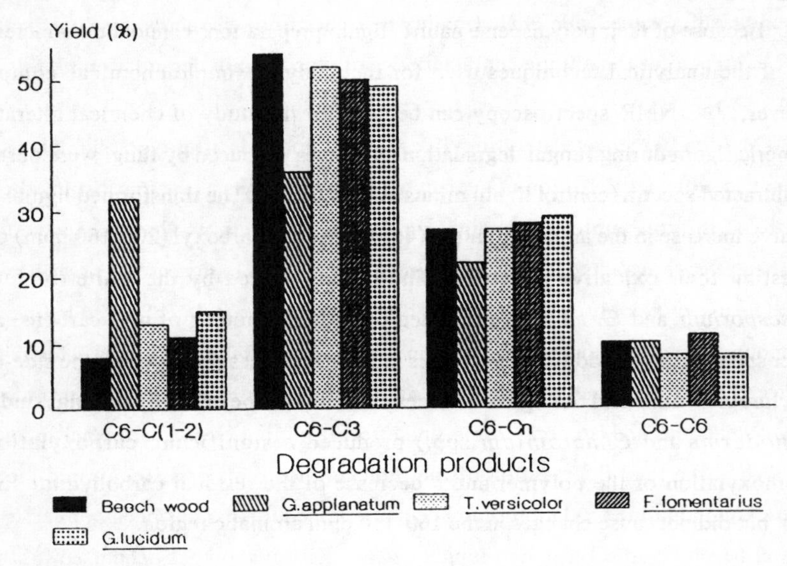

Fig. 8. Principal groups of products obtained by acidolysis of beech wood decayed by ligninolytic fungi.

Acidolysis destroys ether links between lignin units. Thus, compounds of the aryl-propanone and aryl-OH-propanone types were the major products obtained from beech wood. Several compounds with higher molecular weights were included in these fractions but their definitive identification by GC-MS was not always possible. Compounds of the aryl-propane type (C_6-C_3) amounted to nearly 50% of the products obtained. Those with a shorter side-chain accounted for 8%, while complex products, including some dimers (C_6-C_6), amounted to 36% of the compounds obtained (Fig. 8).

After fungal degradation of wood, increased yields of benzenecarboxylic acids and a decrease in aryl-OH-propanones were obtained by acidolysis, as reported by Pometto & Crawford (1985) after alteration of lignin by actinomycetes. However, in the case of *G. australe*, these changes were very important, the benzenecarboxylic acids amounting to 20% of the aromatic products while the aryl-OH-propanones were reduced to 3.8%. The general reduction in compounds of the aryl-propane type and the marked increase in compounds with reduced side-chain (mainly benzenecarboxylic acids) by *G. australe* (Fig. 8) were evidence for the unique characteristics of wood lignin degradation by this fungal species.

The high percentage of syringyl-derived compounds after wood acidolysis was a consequence of the participation of guaiacyl units in C-C links at the unsubstituted C_5. The S/G ratio of the low molecular weight acidolysis products was near 3 in wood, and diminished after fungal degradation, attaining a value of 1.8 in the case of *G. australe*.

The use of acidolysis in the study of lignin biodegradation is limited by the variety of compounds obtained, their complexity in some cases and the low yields. Thioacidolysis (Lapierre *et al.*, 1985) overcomes several of these inconveniences, viz. the production of secondary reactions and isomeric forms, and effects a similar cleavage of ether linkages.

When the total degradation of the lignocellulosic material by a rapid method is required, CuO alkaline oxidation is an adequate procedure and has been applied to wheat straw and beech wood subjected to fungal decay. This method has been used for the total degradation of lignin (e.g. for estimation of S/G ratios) in lignocellulosic or lignin-containing materials (such as sediments or humic substances). Lignin is depolymerized during oxidation but a subsequent oxidative degradation of the side-chains and production of aromatic acids, aldehydes and ketones takes place (Fig. 9).

The S/G ratio of lignin from different woods was estimated after alkaline degradation by CuO and compared with those obtained from CPMAS ^{13}C-NMR, acidolysis and thioacidolysis. As a rule, the S/G ratios estimated from the two latter techniques were 2-3 times higher than those from CuO degradation or solid-state NMR.

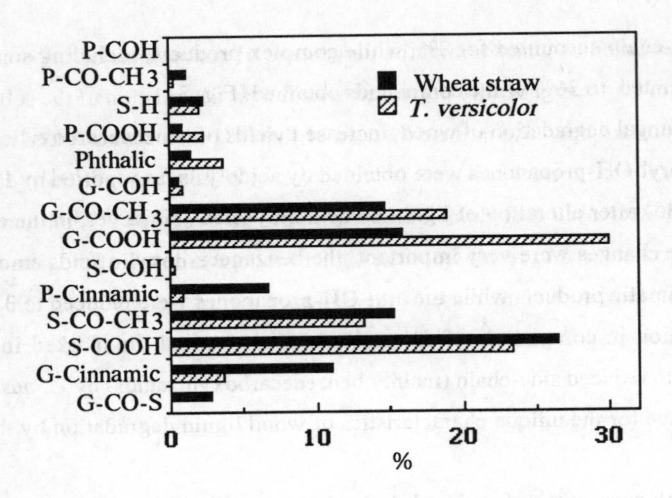

Fig. 9. Aromatic compounds found after CuO alkaline degradation of wheat straw decayed by *Trametes versicolor* (P = 4-OH-phenyl; G = 3-methoxy, 4-OH-phenyl; S = 3,5-dimethoxy, 4-OH-phenyl).

This reflected the predominance of S units in the non-condensed (ether-linked) lignin. The S/G ratio, i.e. 1.6, from *E. cordifolia,* obtained after CuO degradation, was only slightly higher than that calculated from the CPMAS ^{13}C-NMR spectrum (Fig. 6). This value is high when compared with those of other hardwoods analyzed by Manders (1987). The S/G ratio, i.e. 1.0, of lignin from *L. philippiana* was rather low, but an extremely high ratio, i.e. 2.1, was found in *N. dombeyii.* The two Chilean woods with a high S/G ratio also showed, after CuO degradation, a significant proportion of p-OH phenols, viz. p-OH-benzaldehyde and p-OH-benzoic acid. These have not previously been found in hardwoods but are generally found in non-woody angiosperms after degradation of p-OH-phenyl lignin units. This unique characteristic will be verified in extracted lignin (MWL) from these woods and its effects on delignification by fungi is being investigated.

Decreased yields of cinnamic acids and increased amounts of benzenecarboxylic acids were observed after fungal decay of wheat straw (Fig. 9). Cinnamic acids are relatively common in wheat straw and were degraded by the different fungi studied. After oxidative degradation of wheat straw, the percentage of compounds of the p-OH-phenyl type was low (c. 9%) and generally decreased after fungal degradation of the substrate.

An important decrease in the S/G ratio of lignin (Fig. 10) was found after wheat straw decay by fungal species, such as *T. versicolor,* producing a significant degradation of lignin. The decreased S/G ratio found during fungal decay of other lignocellulosic materials was considered to result from the higher biodegradability of the non-condensed (S-rich) lignin. However, a decrease of the S/G ratio (from 2.5-3.0 to 1.8-2.0) was also found after acidolysis and thioacidolysis of decayed beech wood (degraded by *G. australe*).

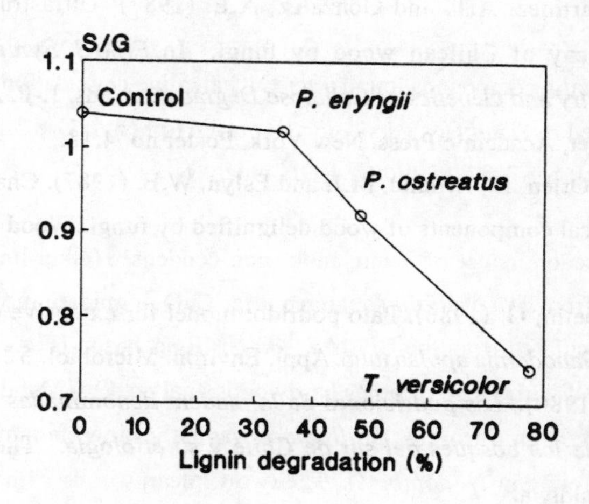

Fig. 10. S/G ratio and lignin degradation after wheat straw decay by ligninolytic fungi.

These results showed that the higher biodegradability of the ether linkages (more abundant in S-rich lignin) was not the sole explanation for the decrease in the S/G ratio brought about by fungi. Topological differences in the distribution of lignin types through the cell-wall (high content of S-G lignin in the highly-lignified middle lamella and of S lignin in the secondary wall) could be also implicated in the decrease in the S/G ratio during fungal decay.

ACKNOWLEDGEMENTS

The authors are indebted to R. Bechtold, E. Gómez, F.J. González-Vila, A. Prieto, G. Jurzitza, M.J. Martínez and M. Valmaseda for provision of data and discussion of results. The work presented was supported by the Spanish Biotechnology Program (Grants PBT 85-10 and BIO 88-0185) and by the Volkswagenverk Foundation.

REFERENCES

Barrasa, J.M., Martínez, A.T. and González, A.E. (1987). Ultrastructural aspects of natural decay of Chilean wood by fungi. In *FEMS Symposium No. 43, Biochemistry and Genetics of Cellulose Degradation,* eds. J.-P. Aubert, P. Béguin and J. Millet, Academic Press, New York, Poster no. 4.18.

Blanchette, R.A., Otjen, L., Effland, M.J. and Eslyn, W.E. (1987). Changes in structural and chemical components of wood delignified by fungi. Wood Sci. Technol. **19**, 35-46.

Dill, I. and Kraepelin, G. (1986). Palo podrido: model for extensive delignification of wood by *Ganoderma applanatum.* Appl. Environ. Microbiol. **52**, 1305-1312.

González, A.E. (1980). *Las pudriciones de la madera denominadas 'Huempe' o 'Palo podrido' de los bosques del sur de Chile y su etiología.* Thesis, Universidad Austral, Valdivia.

González, A.E., Barrasa, J.M., Martínez, A.T. and Almendros, G. (1988). Basic investigations for biotechnological processes of wood delignification with *Ganoderma applanatum.* In *Biotec-88, Abs. 2nd Spanish Conf. on Biotechnology,* Barcelona, p 379.

González, A.E., Martínez, A.T., Almendros, G. and Grinbergs, J. (1989). A study of yeasts during the natural delignification and fungal transformation of wood into cattle feed in Chilean rain forest. Antonie van Leeuwenhoek, **55**, 221-236.

González, A.E., Grinbergs, J. and Griva, E. (1986). Biologische Umwandlung von Holz in Rinderfutter - 'Palo podrido'. Zentralbl. Mikrobiol. **141**, 181-186.

González, A.E., Martínez, A.T. and Almendros, G. (1987). Chemical characterization of wood at different stages of fungal degradation in Chilean forest. In *FEMS Symposium No. 43, Biochemistry and Genetics of Cellulose Degradation,* eds. J.-P. Aubert, P. Béguin and J. Millet, Academic Press, New York, Poster no. 4.17.

Kirk, T.K. (1983). Degradation and conversion of lignocelluloses. In *The Filamentous Fungi, Vol. IV. Fungal Technology,* eds. J.E. Smith, B.R. Berry and B. Kristiansen, Arnold, London, pp 266-295.

Lapierre, C., Monties, B. and Rolando, C. (1985). Thioacidolysis of lignin: comparison with acidolysis. J. Wood Sci. Technol. 5, 277-292.

Martínez, A.T. and Barrasa, J.M. (1989). Cell-wall transformation by ligninolytic fungi. In *Proceedings OECD Workshop on Cell walls: Structure, Biodegradation and Utilization,* ed. J. Delort-Laval, J. Anim. Food Sci. in press.

Manders, W.F. (1987). Solid-state ^{13}C-NMR determination of syringyl/guaiacyl ratio in hardwoods. Holzforschung, **41**, 13-18.

Pometto, A.L. III and Crawford, D.L (1985). Simplified procedure for recovery of lignin acidolysis products for determining the lignin-degrading abilities of microorganisms. Appl. Environ. Microbiol. **49**, 879-881

Valmaseda, M. (1989). *Fermentación en estado sólido por hongos ligninolíticos.* Thesis, Universidad Complutense, Madrid.

Valmaseda, M., Almendros, G. and Martínez, A.T. (1988). Multivariate analysis on chemical transformation of lignocellulosic materials by ligninolytic basidiomycetes. In *Biotec-88, Abs. 2nd Spanish Conf. on Biotechnology,* Barcelona, p 398.

Zadrazil, F., Grinbergs, J. and González, A.E. (1982). 'Palo podrido' - decomposed wood used as feed. Eur. J. Appl. Microbiol. Biotechnol. **15**, 167-171.

CHAIRMAN'S REPORT ON SESSION II

Changes in lignocellulosic materials during biological treatment

The discussion on the analytical aspects of untreated and treated lignocellulosic feedstuffs ranged from traditional wet chemistry to advanced spectroscopy (including near and mid-infrared, cross-polarization/magic angle spinning ^{13}C-nuclear magnetic resonance and x-ray diffraction), chromatographic techniques (including pyrolysis-gas chromatography and high performance liquid chromatography) and electron microscopy combined with immunocytochemical labelling.

The effectiveness of rapid, non-chemical treatments became clear, especially when large numbers of samples have to be characterized. However, the limitations of some of the methods became equally obvious: for example, infrared spectra need extensive mathematical treatments. Calibration of the instrument is of tremendous importance. However, this technique seems to be the method of choice, for routine analysis of feedstuffs of similar composition.

Wet chemistry still remains the basis for the interpretation of spectra and pyrograms. Therefore, further development of wet chemical methods with regard to reproducibility and specificity was generally recommended.

Information (on substrates) obtained using only one method is of limited use. The discussants were of one voice in recommending that various methods be used to provide the necessary information on the structure and composition of lignocellulosic material at the molecular level.

The use of cross-polarization/magic angle spinning ^{13}C-NMR spectroscopy allows of the application of NMR to insoluble substrates. The usefulness of this method in the context of modification of lignin as a result of microbial attack was demonstrated. The same was true for pyrolysis-gas chromatography. However, minor, but important changes, could not be monitored. For these purposes high performance liquid chromatography, and immunocytochemical labelling in particular, seemed to be the methods of choice.

Electron microscopy also reveals some of the ultrastructural modifications caused by microbial attack. Staining of lignin and of polysaccharides allows of the visualization of these components and some conclusions on the sequence of attack can be made. Immunocytochemical methods lead to elegant and direct observations on enzyme action within the woody tissue.

The applicability of a broad array of methods currently available for studies on fungal attack on lignocellulosic feedstuffs were introduced and examples were discussed. It became obvious that each method can contribute to our overall knowledge on lignocellulose biodegradation. However, some procedures appear to be so sophisticated that their use cannot be taken on board by any one laboratory. Thus, collaboration between microbial-orientated laboratories and those having the different analytical facilities and the personnel with the necessary expertise is of the utmost importance if we are to achieve progress in our understanding of microbial degradation of lignocellulosic materials.

Jürgen Puls

SESSION III

Bioconversion of lignocellulosic materials in submerged and solid-state cultivation and in reactors

SOLID-STATE VERSUS LIQUID CULTIVATION OF *TALAROMYCES EMERSONII* ON STRAWS AND PULPS: ENZYME PRODUCTIVITY

Maria G. Tuohy, Tara L. Coughlan and Michael P. Coughlan

Department of Biochemistry, University College, Galway, Ireland

Thermostable extracellular polysaccharide-degrading enzyme systems are produced by the aerobic thermophilic fungus, *Talaromyces emersonii,* when grown on appropriate substrates. In this paper we compare the production of such enzymes by two strains of the organism, CBS814.70 and UCG208, when grown on a variety of agricultural residues by submerged and solid-state cultivation procedures.

INTRODUCTION

The large quantities of lignocellulosic wastes and residues that are generated annually represent a largely untapped source of food, fuel and chemical feedstocks. Cellulose, hemicellulose and, to a lesser extent, pectin comprise the bulk of the dry weight of such materials. Consequently, exploitation of surplus biomass of agricultural, forestry, domestic and industrial origin requires that the enzyme systems for hydrolysis of these polysaccharides be available. Indeed, it has been calculated that the cost of producing the necessary cellulolytic enzymes is the major factor limiting the fullest exploitation of these most abundant of renewable resources (Mandels, 1985; Pourquié and Desmarquest, 1989).

There is considerable evidence in the literature that different enzyme systems interact synergistically to effect the saccharification of biomass. For example, hydrolysis of pectin affords cellulases and hemicellulases greater access to their respective substrates thereby allowing of greater conversion of the appropriate types of biomass (Voragen *et al.*, 1980; Beldman *et al.*, 1984; Moloney *et al.*,1984; Coughlan *et al.*, 1985). Similarly, hydrolysis of the hemicellulose fraction of straws and woods has been shown to increase considerably the extent of cellulose degradation in these materials (Wood, 1981; Schwald *et al.*, 1988; Macris and Kekos, 1989). The retting of flax straw is also facilitated by the use of enzyme preparations having hemicellulase as well as pectin-degrading activity (Gillespie *et al.*, 1989). Thus, the cocktail of enzymes to be used for saccharification is more important than the activity of any individual enzyme or enzyme system. For other applications, the absence of a particular enzyme activity would be essential. Thus, a hemicellulase preparation devoid of cellulase activity would be appropriate to the production of cellulose fibres in the paper industry (Biely, 1985; Linko *et al.*, 1989).

The operational stability of the enzymes used is also an important consideration. The ability to work at high temperatures is frequently found to give greater conversion and to obviate or minimize viscosity and contamination problems. The extracellular enzymes produced by *Talaromyces emersonii* have been shown to have good thermostability (McHale and Coughlan, 1981a; Moloney *et al.*, 1985; Tuohy and Coughlan, 1989) and as such are promising candidates for various applications. In this paper we report on the production of enzyme systems relevant to lignocellulose utilization by two strains of this organism, viz. CBS814.70 and UCG208, when grown on various straws and pulps by liquid and solid-state cultivation procedures. Solid-state cultivation (reviewed by Cannel and Moo-Young, 1980) has several advantages over liquid cultivation (Weiland, 1988; Considine and Coughlan, 1989). Among these is the possibility that it may be a more economical way of utilizing agricultural residues. This is of considerable importance since the cost of the inducing substrate accounts for 31% of total cost of enzyme production by liquid fermentation (Pourquié and Desmarquest, 1989).

MATERIALS AND METHODS

Substrates for growth, enzyme induction and assay

Beet pulp (average size was 0.2 x 0.8 cm), wheat bran, spring barley straw (var. Klaxon), winter barley straw (var. Panda), spring oat straw (var. Leanda) and winter wheat straw (var. Norman) were obtained locally. Before use the straws were chopped using a blender into pieces of about 0.5 cm long. Glucose, lactose, oat spelts xylan, arabinan, p-nitrophenyl-ß-D-glucoside, p-nitrophenyl-ß-D-galactoside, p-nitrophenyl -ß-D-xyloside and p-nitrophenyl-α-L-arabinoside were obtained from Sigma Chemical Co. Ltd. (U.K.). Arabinan was also isolated from beet pulp. Solka floc (BW.40; purified ball-milled Spruce cellulose) was obtained from Brown and Co., Berlin, NH, USA.

Organisms and culture conditions

Talaromyces emersonii strain CBS814.70 (obtained from Centraal Bureau voor Schimmelcultures, Baarn, The Netherlands) and a mutant, UCG208, derived from CBS814.70 (Moloney *et al.*, 1983a) are routinely subcultured on Sabouraud dextrose agar medium. Liquid cultures (final volume 170 ml) in 250 ml Erlenmeyer flasks shaken at 250 rpm were grown at 45°C, pH 4.5, in the mineral salts/inducing substrate (2%

w/v)/corn steep liquor (0.5% w/v)/ammonium sulphate (1.5% w/v)/yeast extract (0.1% w/v) medium described earlier (Moloney *et al.*, 1983a). Solid-state cultivation (approx. 30 ml final volume) was carried out under static conditions in 250 ml Erlenmeyer flasks at 45°C in the same medium but at an inducing substrate concentration of 33% (w/w) (Tuohy *et al.*, 1989).

Enzyme extraction

After suitable periods of growth the solid-state cultures were mixed with approx. 4 volumes of 25 mM sodium acetate buffer, pH 5, containing 0.01% (v/v) Tween 80. The mixture was blended for 20 sec in a homogeniser, then incubated with shakingat 140 rpm at room temperature for 2 h, and finally centrifuged for 1 h at 1,300 x g. The supernatant was filtered through glass wool and used for assay of enzyme activity and extracellular protein concentration. The liquid cultures were centrifuged, filtered and assayed as above.

Assay of enzyme activity and protein concentration

For logistical convenience, and in the absence of information on temperature optima for many of the enzymes studied, all activities were carried out at 50°C out in 0.1 M sodium acetate buffer, pH 5. This underestimates the true activity at least in the case of ß-glucosidase, total cellulase, endoglucanase and xylanase which have temperature optima of 65-80°C under assay conditions (Folan and Coughlan, 1978; McHale and Coughlan, 1981a; Moloney *et al.*, 1985; Tuohy and Coughlan, 1989). For the purpose of comparing solid-state with liquid cultivations, enzyme yields are expressed as IU/g inducing substrate used in growth media. Total activity in liquid cultures can be calculated by multiplying the values given as IU/g inducing substrate (Tables 1to 9) by 3.4. Those in the directly comparable solid-state cultures can be calculated by mutiplying the values given in Table 10 by a factor of 8 to 10 (the differences between 8 and 10 reflect differences in dry weights).

Filter paper-degrading activity in extracts was determined by the DNS method (Miller, 1959) and the procedure described by Mandels *et al.* (1976). The hydrolysis of low viscosity CM-cellulose (final conc. 4%), barley ß-glucan, beet pulp arabinan and oat spelts xylan (all at a final conc. of 0.67%, w/v) was in each case measured by the DNS method (using appropriate standards) following incubation with neat or diluted enzyme for 10-30 min as appropriate. ß-glucosidase, ß-galactosidase, ß-xylosidase and α-arabinofuranosidase activities were measured by noting the increase in A_{410} following 10-15 min incubation of p-nitrophenyl-ß-D-glucoside (final conc. 0.9 mM), p-nitrophenyl-ß-D-galactoside (final conc. 2.5 mM), p-nitrophenyl-ß-D-xyloside (final conc. 0.5 mM) and p-nitrophenyl-α-L-arabinofuranoside (final conc. 0.5 mM), respectively, with an aliquot of neat or diluted enzyme (the reaction was stopped by the addition of 0.5-1.0 volume of 1 M sodium carbonate). Trichloroacetate-precipitable protein concentration in solid-state extracts or liquid culture filtrates was determined by the method of Lowry *et al.* (1951) and is expressed as mg/g inducing substrate in growth medium.

RESULTS AND DISCUSSION

Comparison of strains CBS814.70 and UCG208

The results presented in the following pages show that *Talaromyces emersonii* strains CBS814.70 and UCG208 produce a range of enzyme activities when grown on glucose, lactose or a range of lignocellulosic substrates by liquid-state (Tables 1 to 9) or solid-state cultivation procedures (Table 10, Figs. 1 A and B). One is reminded that the activities given, while suitable for comparative purposes, are underestimated because the assays were carried out at suboptimal temperatures (see Methods section). In the liquid cultures generally, the mutant strain UCG208 yielded more FPase and CMCase activities than did the parent strain CBS814.70 regardless of substrate. This was to be expected since UCG208 had been isolated for its improved ability to produce cellulases (Moloney *et al.*, 1983a). Of interest is the fact that, unlike the parent strain, UCG208 yielded substantial amounts of these enzymes even when grown on glucose or lactose. This would suggest that synthesis of endo- and exoglucanases is at least partially derepressed in the mutant. CBS814.70 was at least as good, if not better than UCG208, with respect to production of ß-glucosidases whatever the inducing substrate. However, there was little to choose between parent and mutant strains with respect to xylanase, ß-xylosidase and ß-galactosidase. In fact, little ß-galactosidase was produced by either strain in liquid cultures regardless of the growth medium used (Tables 1 to 9). Surprisingly, solid-state cultures of strain CBS814.70 harvested at 5 days yielded more of each enzyme activity on all but one substrate than did UCG208 harvested at the same time (Table 10). The only exception was the wheat bran/winter wheat straw (1:9) medium on which UCG208 performed better. This apparent reversal of the superiority of strain UCG208 over CBS814.70 in going from liquid- to solid-state cultivation appears to be related to the time of harvesting. Thus, in the solid-state time course studies (Figs. 1 A and 1 B) it can be seen that, regardless of inducing substrate, the mutant strain UCG208 yields more of all activities than does the parent strain CBS814.70 when account is taken of the growth

Table 1. Enzyme production (IU/g inducing substrate) by liquid cultures of Talaromyces emersonii strains CBS814.70 and UCG208.

Inducing substrate	Growth time (h)	FPase IU/g		CMCase IU/g		ß-Glucosidase IU/g		Xylanase IU/g		ß-Xylosidase IU/g		ß-Galactosidase IU/g		Ext. protein mg/g	
		CBS	UCG	CBS	UCG	CBS	UCG	CBS	UCG	CBS	UCG	CBS	UCG	CBS	UCG
Glucose	24	0	4.9	45.5	118.8	0.4	4.1	126.1	353.0	17.4	4.6	-	-	48.3	74.8
	36	0	7.4	95.0	185.0	1.9	13.8	351.5	626.5	13.3	3.0	6.2	1.2	90.5	80.3
	48	0	9.1	-	311.9	1.2	2.5	0	354.0	2.7	2.8	5.8	1.5	128.1	80.8
	60	0.2	11.4	-	-	1.6	7.0	0	321.0	48.0	10.2	-	-	142.1	92.3
	72	0	-	86.5	188.3	0.7	4.0	0	237.5	0.0	0.0	-	-	115.9	64.5
	84	0	18.2	80.5	162.6	1.2	4.3	0	221.5	9.0	16.4	7.5	3.0	100.4	56.7
	96	0	18.5	-	-	6.8	30.0	0	233.5	0.0	1.9	6.5	2.1	70.7	52.5
	108	0	19.8	74.0	185.7	23.8	39.7	0	332.5	13.9	0.0	-	-	54.6	52.6
	120	0	14.2	71.5	208.6	29.8	8.9	0	426.5	14.7	10.4	-	-	51.5	62.5
	132	0	6.4	53.5	176.0	37.6	11.6	0	550.0	22.0	20.3	-	-	40.5	94.4
	144	0	-	35.5	174.2	30.3	14.2	0	428.0	43.4	45.6	7.5	3.6	49.6	79.5
Lactose	24	0.08	1.1	183.0	60.5	6.5	1.2	285.8	347.5	18.2	8.2	-	-	73.4	81.9
	36	0.14	16.9	-	24.2	11.3	33.0	290.7	53.5	25.8	14.6	0.9	0.3	66.6	109.5
	48	0.19	17.3	-	43.9	6.1	6.3	12.1	12.0	13.0	16.1	0.0	0.4	66.4	60
	60	0.32	-	-	72.5	2.9	5.7	71.2	30.5	14.2	79.3	-	-	99.9	80.4
	72	0.36	31.9	163.8	111.4	3.3	6.4	61.4	130.5	8.7	10.1	-	-	78.3	62.5
	84	0.48	34.5	246.0	61.4	3.3	5.2	6.9	26.5	23.1	28.0	1.1	0.6	57.8	50
	96	0.44	89.5	254.0	133.9	27.8	41.1	4.7	0.0	4.7	26.7	-	-	65.7	42.5
	108	0.4	78.4	280.0	282.1	23.3	10.6	0	54.0	9.5	0.0	-	1.1	45.8	64.5
	120	0.36	34.8	168.9	282.1	33.2	18.5	0	97.0	17.0	23.0	1.4	1.1	46.5	61.5
	132	0.21	14.3	128.5	218.6	39.1	16.2	-	464.5	21.4	25.7	1.4	1.3	45.5	61.2
	144	0.08	-	68.5	385.9	38.3	19.3	72.7	414.0	29.7	83.0	1.5	1.3	72.3	82.7

Table 2. Enzyme production (IU/g inducing substrate) by liquid cultures of Talaromyces emersonii strains CBS814.70 and UCG208.

Inducing substrate	Growth time (h)	FPase IU/g		CMCase IU/g		ß-Glucosidase IU/g		Xylanase IU/g		ß-Xylosidase IU/g		ß-Galactosidase IU/g		Ext. protein mg/g	
		CBS	UCG	CBS	UCG	CBS	UCG	CBS	UCG	CBS	UCG	CBS	UCG	CBS	UCG
Solka floc	24	0.7	7.7	61.9	101.6	1.8	1.5	334.2	722.0	11.6	17.6	-	-	53.7	89.6
	36	0.6	11.6	117.6	260.1	5.8	4.4	1668.0	1211.5	30.5	60.7	-	-	63.2	85.8
	48	2.6	16.2	-	616.1	1.7	5.0	1233.5	404.0	23.6	18.9	-	-	59.0	78.8
	60	12.0	18.2	-	366.1	1.7	5.9	1470.5	1475.0	8.3	0.0	0.6	1.0	90.3	83.4
	72	-	-	593.1	435.1	1.8	7.1	1444.0	947.5	51.9	83.7	0.4	1.3	64.6	68.9
	84	16.6	29.6	507.6	517.3	0.9	37.3	1147.0	488.5	93.6	62.3	1.8	2.4	50.6	81.0
	96	25.8	51.6	848.5	661.9	23.9	41.2	1188.5	521.5	73.6	59.7	-	-	79.0	65.8
	108	33.8	62.1	857.6	978.2	36.7	-	1105.8	570.0	191.4	57.4	-	-	85.4	82.8
	120	28.6	57.9	1026.1	1273.5	35.8	19.1	1035.6	643.0	126.6	60.7	-	-	79.7	68.9
	132	17.8	21.3	1333.1	1595.3	40.3	13.1	2801.8	1925.5	174.9	66.0	-	-	120.5	69.4
	144	-	-	1116.9	826.8	38.0	17.8	2693.0	1161.5	99.3	101.0	-	-	109.0	89.1
Xylan (oat spelts)	24	3.0	0.0	78.3	108.1	2.5	11.6	240.5	470.5	8.9	6.0	-	-	78.6	87.9
	36	3.7	0.3	109.5	127.7	6.0	38.9	596.7	1274.0	13.9	31.9	-	-	79.5	80.1
	48	4.4	2.1	-	147.0	1.4	2.9	291.0	383.0	9.2	17.3	-	-	87.6	60.4
	60	5.2	4.6	-	160.2	3.3	4.5	359.1	564.0	6.1	16.2	1.1	0.6	55.3	59.3
	72	-	-	142.9	170.1	2.2	5.9	267.6	539.5	32.1	47.6	-	4.2	65.1	81.1
	84	7.1	15.8	-	216.9	2.4	4.1	174.5	703.0	36.4	38.4	-	-	69.4	61.8
	96	10.0	22.9	140.5	377.7	21.7	40.5	7.3	896.5	18.6	100.3	-	-	77.7	57.5
	108	13.3	34.5	115.7	455.0	28.0	21.4	354.9	1081.5	34.1	77.1	1.2	-	53.3	75.7
	120	11.2	19.7	80.4	464.4	32.6	27.9	443.9	929.0	32.1	106.1	-	-	42.8	89.0
	132	7.9	7.0	261.9	342.3	29.3	28.5	1186.5	1563.0	64.1	114.7	-	-	84.1	86.5
	144	-	-	208.4	173.9	28.0	20.4	1295.6	1962.0	96.1	132.8	-	7.4	70.4	90.1

Table 3. Enzyme production (IU/g inducing substrate) by liquid cultures of Talaromyces emersonii strains CBS814.70 and UCG208.

Inducing substrate	Growth time (h)	FPase IU/g		CMCase IU/g		ß-Glucosidase IU/g		Xylanase IU/g		ß-Xylosidase IU/g		ß-Galactosidase IU/g		Ext. protein mg/g	
		CBS	UCG	CBS	UCG	CBS	UCG	CBS	UCG	CBS	UCG			CBS	UCG
Beet pulp	24	2.9	0.0	95.4	141.7	6.1	9.2	242.8	321.0	10.7	7.1	-	-	82.3	97.8
	36	3.9	2.6	80.8	165.1	4.4	13.7	1602.0	814.0	13.0	33.5	-	-	82.4	95.6
	48	4.6	3.5	-	194.8	2.7	4.8	373.0	251.5	10.0	23.5	0.6	-	83.7	94.1
	60	5.8	13.2	-	217.0	3.6	5.5	730.0	743.5	17.4	59.8	0.4	-	78.9	108.4
	72	6.8	-	205.5	375.0	2.3	9.0	388.0	705.5	28.6	3.4	-	-	83.8	98.6
	84	9.7	22.7	224.0	460.8	2.6	6.4	69.0	223.0	54.0	20.6	0.9	-	59.8	93.6
	96	9.9	44.2	274.9	523.3	22.8	30.7	179.0	540.5	30.0	79.4	-	-	54.3	84.1
	108	9.2	59.7	223.3	604.3	20.0	30.2	326.0	613.0	26.8	46.4	1.1	-	60.5	111.0
	120	7.2	23.3	295.8	750.7	25.3	15.0	358.0	647.5	39.3	53.7	0.8	-	74.9	122.1
	132	5.4	12.3	199.9	671.7	23.3	17.7	708.0	787.5	44.5	53.2	-	-	74.6	115.3
	144	3.5	-	183.4	667.9	34.8	20.7	1081.0	1404.5	55.2	73.4	-	-	91.0	161.0
Wheat bran	24	0.1	0.6	-	189.3	0.8	13.7	464.5	1013.5	7.9	17.5	-	-	37.6	136.3
	36	2.2	1.6	75.8	142.9	2.8	16.9	766.0	1465.0	14.1	25.0	-	2.4	115.1	188.0
	48	2.7	2.4	-	-	2.5	7.4	400.0	554.5	12.3	34.3	2.8	3.0	115.0	140.1
	60	3.1	7.8	-	-	8.7	9.1	613.0	387.6	24.5	62.4	-	-	91.9	122.0
	72	4.6	-	82.0	291.5	3.9	16.7	608.5	-	23.3	73.3	3.1	-	87.9	118.7
	84	6.2	15.6	79.0	304.7	4.5	17.6	197.7	147.5	27.7	52.4	4.7	4.6	81.6	104.6
	96	7.6	23.2	80.0	322.6	-	39.6	100.7	717.5	24.7	81.3	6.3	-	85.8	87.2
	108	7.1	36.9	133.0	456.5	62.5	23.4	326.5	843.5	22.4	34.5	3.3	-	70.4	99.4
	120	5.5	8.3	110.3	253.8	60.3	29.5	439.0	311.5	33.7	65.3	14.9	3.7	77.6	102.8
	132	1.7	4.5	51.2	148.8	60.5	20.3	614.0	1069.0	36.8	51.5	-	3.1	116.5	108.3
	144	-	-	63.1	148.8	28.0	11.7	556.7	1633.0	28.4	72.4	10.8	-	129.4	109.4

Table 4. Enzyme production (IU/g inducing substrate) by liquid cultures of Talaromyces emersonii strains CBS814.70 and UCG208.

Inducing substrate	Growth time (h)	FPase IU/g		CMCase IU/g		ß-Glucosidase IU/g		Xylanase IU/g		ß-Xylosidase IU/g		ß-Galactosidase IU/g		Ext. protein mg/g	
		CBS	UCG	CBS	UCG	CBS	UCG	CBS	UCG	CBS	UCG	CBS	UCG	CBS	UCG
S. barley straw	24	0.00	0.0	7.2	41.9	3.3	8.4	53.2	443.0	16.0	12.0	-	-	85.9	55.5
	36	0.03	2.3	-	-	17.3	16.9	649.0	754.5	38.0	18.6	1.8	1.2	63.5	53.5
	48	0.09	2.1	-	-	2.8	9.0	276.0	419.5	23.8	27.5	2.9	-	66.3	47.6
	60	0.12	6.3	-	-	4.4	8.3	270.0	409.5	35.8	41.0	5.0	4.1	70.3	56.4
	72	-	-	38.8	122.9	4.4	11.2	231.0	0.0	31.0	38.0	3.0	7.1	72.5	66.6
	84	0.22	12.8	120.0	587.5	2.8	16.6	170.5	180.5	40.0	53.9	3.1	2.8	37.7	65.6
	96	0.33	42.4	145.3	-	32.4	38.6	317.5	269.5	14.7	84.1	3.5	3.5	53.6	58.2
	108	0.31	37.1	259.5	587.2	62.8	16.0	402.5	443.0	18.2	27.1	3.5	-	71.4	68.0
	120	0.26	25.4	309.7	596.5	28.0	14.6	492.5	458.0	41.0	36.9	5.8	3.5	57.0	71.9
	132	0.13	5.2	339.9	-	67.5	7.6	811.5	1208.5	55.4	62.2	5.0	5.7	66.6	66.8
	144	-	-	282.5	769.0	93.9	8.7	613.5	1270.0	42.2	123.1	5.4	3.5	79.3	-
Spring oat straw	24	0.04	1.1	23.8	44.4	3.9	6.7	115.0	300.0	19.5	11.5	-	-	66.9	72.7
	36	0.07	4.0	71.5	-	16.4	15.6	346.5	474.5	23.0	6.5	1.6	1.0	65.8	84.1
	48	0.10	3.7	-	-	3.3	7.0	122.5	120.5	18.6	14.4	2.5	2.4	56.9	72.6
	60	0.14	4.3	-	-	-	5.9	310.0	410.5	9.2	-	2.3	1.9	64.0	86.7
	72	-	-	101.1	187.5	4.3	9.8	258.0	193.5	15.3	41.7	1.7	7.8	56.5	69.7
	84	0.23	22.5	132.2	195.0	4.4	10.9	76.5	257.0	22.8	28.0	2.2	2.4	93.4	78.2
	96	0.22	32.1	-	298.3	30.6	33.5	169.0	351.5	21.5	43.7	2.6	2.4	56.9	67.4
	108	0.21	30.7	165.8	320.6	34.4	25.7	207.5	221.5	-	9.5	-	2.6	-	98.5
	120	0.15	14.5	213.2	437.2	60.3	23.0	276.5	361.5	24.7	31.4	2.6	2.9	66.4	78.8
	132	0.08	8.1	207.0	-	47.5	-	537.0	434.0	47.7	41.5	3.3	3.0	80.9	76.1
	144	-	-	112.9	558.5	55.3	12.5	467.0	594.5	64.7	152.0	2.8	3.6	78.5	78.0

Table 5. Enzyme production (IU/g inducing substrate) by liquid cultures of Talaromyces emersonii strains CBS814.70 and UCG208.

Inducing substrate	Growth time (h)	FPase IU/g		CMCase IU/g		ß-Glucosidase IU/g		Xylanase IU/g		ß-Xylosidase IU/g		ß-Galactosidase IU/g		Ext. protein mg/g	
		CBS	UCG	CBS	UCG	CBS	UCG	CBS	UCG	CBS	UCG	CBS	UCG	CBS	UCG
W. barley straw	24	0.0	0.0	0.0	97.6	2.8	8.3	0.0	129.0	16.2	6.8	-	-	87.9	83.7
	36	1.9	0.1	20.3	-	6.0	11.2	542.5	363.5	19.6	3.4	1.8	1.7	74.3	69.5
	48	4.8	0.2	-	-	3.4	6.3	178.5	256.5	18.4	13.0	3.1	1.5	68.7	68.3
	60	6.2	2.6	-	-	2.0	8.6	412.7	243.5	17.7	15.3	1.6	1.5	49.8	64.6
	72	-	-	172.1	186.0	4.2	1.3	209.0	215.5	8.7	14.0	1.5	4.3	57.4	54.9
	84	13.2	11.3	147.6	210.0	3.6	7.7	211.0	107.0	20.0	24.0	4.4	1.6	57.3	54.9
	96	10.2	19.9	167.3	219.1	20.1	21.8	229.0	0.0	17.6	38.3	2.7	2.4	48.8	51.9
	108	8.6	20.8	213.0	296.5	26.0	24.8	255.2	350.5	12.7	5.9	3.1	-	60.0	64.8
	120	8.9	20.0	213.7	323.6	21.8	24.8	275.5	379.5	21.2	36.9	3.4	2.9	55.0	74.3
	132	5.1	10.4	141.9	-	39.0	9.5	612.5	438.5	27.7	66.7	-	2.6	85.5	71.3
	144	-	-	156.2	-	31.5	8.3	721.5	823.5	78.3	96.2	10.2	3.3	76.3	64.7
W. wheat straw	24	0.0	0.0	0.0	0.0	3.8	4.7	0.0	158.0	15.4	8.0	-	-	67.3	69.3
	36	0.0	0.0	22.6	-	11.9	14.2	314.3	618.0	22.8	15.2	5.1	2.0	74.4	79.4
	48	0.0	2.8	-	-	3.0	7.4	43.2	188.0	20.0	23.8	7.8	-	71.6	84.3
	60	0.0	2.7	-	-	7.2	6.4	80.6	203.5	16.4	48.4	6.0	2.1	55.0	72.5
	72	0.0	-	56.8	183.8	3.3	9.4	104.6	208.0	17.3	51.9	5.1	4.4	70.8	68.4
	84	2.3	10.3	116.1	244.5	4.8	10.4	71.5	0.0	27.8	40.6	5.0	3.5	46.7	61.3
	96	3.6	22.9	-	262.1	39.2	36.6	151.4	225.5	38.0	42.3	5.3	3.6	47.4	56.4
	108	4.7	18.5	150.8	279.6	58.8	13.0	180.1	230.0	25.8	15.3	5.8	3.8	47.6	60.2
	120	4.2	12.9	183.6	358.2	65.0	7.7	318.0	344.0	29.4	41.8	-	4.3	50.0	69.1
	132	3.6	4.3	179.8	-	72.5	11.5	479.6	393.5	64.0	60.8	6.8	3.9	68.2	59.6
	144	-	-	114.2	351.1	-	-	436.5	489.5	-	61.5	6.4	4.1	-	60.5

Table 6. Enzyme production (IU/g inducing substrate) by liquid cultures of Talaromyces emersonii strains CBS814.70 and UCG208.

Inducing substrate	Growth time (h)	FPase IU/g		CMCase IU/g		ß-Glucosidase IU/g		Xylanase IU/g		ß-Xylosidase IU/g		ß-Galactosidase IU/g		Ext. protein mg/g	
		CBS	UCG	CBS	UCG	CBS	UCG	CBS	UCG	CBS	UCG	CBS	UCG	CBS	UCG
WB/SBS (1 to 1)	24	0.9	2.9	80.5	126.2	2.1	11.0	191.2	1224.0	18.7	24.2	-	-	67.0	79.0
	36	1.2	3.2	76.5	-	17.4	29.5	563.7	1567.5	33.8	29.4	0.9	2.2	66.6	98.7
	48	3.2	5.6	-	-	0.7	7.9	500.2	781.0	118.0	29.0	0.8	4.3	85.3	90.0
	60	5.3	11.5	-	-	0.4	15.0	212.2	773.0	58.6	41.2	0.9	-	116.7	94.1
	72	-	-	168.2	340.7	1.1	10.7	681.2	525.5	49.1	54.6	-	-	88.2	73.4
	84	5.0	14.1	197.4	443.3	0.6	13.5	463.7	184.5	52.2	45.2	0.4	10.3	80.8	76.7
	96	3.2	27.3	232.4	410.4	0.0	13.0	392.7	335.0	46.0	64.7	0.4	-	84.6	63.9
	108	2.5	17.1	275.3	446.9	0.0	11.7	565.2	443.0	23.3	15.2	0.0	-	67.6	79.3
	120	1.2	16.1	346.4	582.5	0.0	42.1	722.2	514.0	61.5	40.8	-	5.8	97.3	97.3
	132	0.6	9.8	294.7	-	0.0	17.2	1061.2	1089.5	65.3	57.6	0.5	1.3	120.0	97.0
	144	-	-	328.2	676.9	0.0	22.2	808.7	1551.5	81.0	87.7	11.3	-	133.9	86.2
WB/SBS (1 to 9)	24	0.0	-	0.0	-	4.2	-	58.7	297.5	11.1	8.9	-	-	59.0	65.4
	36	0.0	-	11.2	-	11.5	-	723.7	597.5	15.5	5.0	2.1	0.5	124.4	93.4
	48	0.9	-	37.7	-	2.2	-	98.7	-	14.7	5.8	-	5.3	52.3	87.7
	60	2.5	-	-	-	2.4	-	163.2	-	18.0	18.4	3.9	1.7	62.3	98.5
	72	-	-	267.0	-	3.2	-	196.7	-	24.1	0.0	1.4	2.3	52.1	83.0
	84	2.6	-	134.3	-	2.3	-	109.2	-	33.1	13.1	2.1	0.5	47.8	81.6
	96	2.3	-	174.4	-	20.0	-	115.7	-	21.4	10.7	5.9	0.4	46.9	77.5
	108	2.0	-	212.7	-	-	-	227.7	-	19.5	0.0	6.0	2.3	61.7	82.9
	120	1.3	-	264.5	-	-	-	440.7	-	-	5.8	-	0.7	68.0	89.2
	132	0.0	-	199.3	-	-	-	1345.2	-	23.6	7.5	-	-	71.9	71.8
	144	0.0	-	215.7	-	-	-	698.7	-	-	68.7	-	-	66.8	78.0

WB, wheat bran; SBS, spring barley straw

Table 7. Enzyme production (IU/g inducing substrate) by liquid cultures of Talaromyces emersonii strains CBS814.70 and UCG208.

Inducing substrate	Growth time (h)	FPase IU/g		CMCase IU/g		ß-Glucosidase IU/g		Xylanase IU/g		ß-Xylosidase IU/g		ß-Galactosidase IU/g		Ext. protein mg/g	
		CBS	UCG	CBS	UCG	CBS	UCG	CBS	UCG	CBS	UCG	CBS	UCG	CBS	UCG
WB/SOS (1 to 1)	24	5.8	0.8	59.1	62.4	3.1	9.3	241.0	435.6	19.0	19.0	-	-	79.0	87.7
	36	6.9	0.6	166.2	-	5.6	16.8	1681.0	440.5	27.3	22.1	1.2	3.8	97.6	93.3
	48	10.4	2.6	-	-	2.5	7.6	639.0	123.0	33.1	30.4	1.4	1.7	62.3	86.5
	60	11.8	5.1	-	-	3.3	7.4	608.0	92.0	36.9	59.0	1.1	2.3	61.3	85.6
	72	-	-	221.1	86.6	4.5	5.7	653.0	48.5	44.7	26.0	1.4	2.1	79.7	83.8
	84	14.0	6.7	246.8	110.9	2.9	7.5	678.0	0.0	48.8	36.6	1.8	4.4	55.6	82.0
	96	10.3	22.1	225.7	103.5	18.4	11.3	194.5	110.0	44.4	82.2	5.5	3.1	81.1	73.1
	108	11.6	20.7	342.9	200.1	35.0	23.1	354.5	138.5	23.6	35.6	1.8	8.9	62.0	73.6
	120	9.6	5.0	461.1	188.0	35.0	5.9	496.0	428.5	52.1	34.0	1.3	5.6	81.9	76.5
	132	9.4	0.5	295.2	-	21.5	5.5	1803.0	647.5	37.8	46.5	1.3	6.5	85.3	83.6
	144	-	-	264.5	212.1	45.0	4.4	662.0	576.0	30.0	130.6	1.4	5.0	108.0	97.3
WB/SOS (1 to 9)	24	1.4	1.1	78.3	229.3	2.9	11.5	290.5	693.0	16.9	17.4	-		61.4	73.4
	36	3.8	5.5	157.2	-	11.4	16.5	978.0	1098.5	41.5	31.7	0.7	-	76.9	98.3
	48	4.5	7.1	-	-	3.3	7.6	350.5	497.5	39.4	45.7	4.3	1.2	55.9	74.3
	60	9.0	8.8	-	-	3.4	8.3	477.0	595.5	40.7	51.0	1.4	0.9	83.1	72.2
	72	12.4	-	229.5	418.7	4.7	7.1	621.5	546.0	49.0	38.9	2.1	3.4	52.1	71.4
	84	13.8	12.8	273.2	569.8	4.0	12.2	304.0	408.0	35.1	43.5	2.6	2.6	58.6	69.7
	96	5.3	23.3	-	686.6	22.8	7.9	200.5	584.0	27.0	85.3	3.0	3.0	55.7	56.6
	108	4.4	21.5	332.9	753.0	22.5	9.8	200.7	626.5	21.8	28.8	2.7	4.2	44.0	74.4
	120	3.9	16.7	417.5	474.4	16.7	10.6	299.0	701.5	28.2	39.2	2.8	5.1	54.0	72.4
	132	3.2	6.7	296.2	-	17.3	11.1	865.0	1289.5	44.8	61.7	2.6	5.5	62.6	70.0
	144	-	-	306.9	581.1	13.9	11.1	743.5	1040.5	63.9	78.5	2.7	-	72.7	74.8
													-		

WB, wheat bran; SOS, spring oat straw

Table 8. Enzyme production (IU/g inducing substrate) by liquid cultures of Talaromyces emersonii strains CBS814.70 and UCG208.

Inducing substrate	Growth time (h)	FPase IU/g		CMCase IU/g		ß-Glucosidase IU/g		Xylanase IU/g		ß-Xylosidase IU/g		ß-Galactosidase IU/g		Ext. protein mg/g	
		CBS	UCG	CBS	UCG	CBS	UCG	CBS	UCG	CBS	UCG	CBS	UCG	CBS	UCG
WB/WBS (1 to 1)	24	0.2	3.0	24.3	78.4	4.9	10.6	650.0	907.5	19.9	22.0	-	-	73.6	101.8
	36	2.4	5.3	78.6	-	15.4	27.7	860.5	1599.5	36.6	29.1	2.3	1.1	89.7	109.4
	48	6.3	9.5	-	-	6.0	10.8	900.0	990.5	35.9	15.4	2.3	1.1	90.8	107.8
	60	12.6	11.5	-	-	2.1	15.5	865.0	1380.5	49.8	-	1.8	-	76.5	133.1
	72	-	-	162.2	323.7	3.7	12.2	783.0	1092.5	51.6	59.4	2.6	1.1	77.2	87.7
	84	12.6	9.3	225.3	420.2	4.0	14.8	565.0	505.5	52.4	86.8	3.2	1.9	59.8	87.5
	96	12.8	28.4	248.0	351.1	35.4	17.9	335.5	662.0	32.9	90.6	6.8	1.5	52.8	85.9
	108	12.6	15.6	270.9	599.1	43.1	17.7	652.0	699.0	42.6	35.6	3.0	-	64.7	69.9
	120	11.5	13.9	279.8	527.7	45.5	12.4	795.0	754.5	78.6	44.2	3.4	1.8	68.0	84.4
	132	7.6	12.4	-	-	60.0	10.2	1394.0	953.5	103.5	66.5	3.8	2.1	69.8	-
	144	-	-	200.0	349.0	9.6	8.8	1132.5	1120.0	95.9	107.7	3.5	1.9	85.0	97.0
WB/WBS (1 to 9)	24	0.0	0.0	47.7	93.1	0.0	9.3	112.0	307.5	15.0	8.0	-	-	66.8	67.1
	36	0.0	0.0	94.0	-	9.8	15.8	323.0	768.0	19.7	15.8	7.1	0.9	75.1	92.1
	48	0.6	1.6	-	-	3.3	10.1	167.0	550.5	16.7	13.2	8.1	1.5	61.0	77.8
	60	0.7	9.4	-	-	1.3	15.5	350.5	498.5	31.0	-	1.3	1.4	59.9	55.1
	72	-	-	233.4	419.5	1.6	9.6	477.5	451.5	34.0	40.2	1.4	1.1	67.6	70.7
	84	5.7	7.5	228.0	514.8	3.5	12.7	540.5	262.5	34.8	38.8	9.6	2.1	66.1	67.9
	96	6.7	29.0	237.0	-	28.2	9.8	93.5	99.0	24.6	67.1	2.9	-	46.1	74.5
	108	6.6	14.9	260.1	-	35.1	10.3	297.5	277.5	36.7	18.8	1.2	3.3	36.6	72.6
	120	4.9	9.1	324.1	533.4	37.3	8.4	330.0	335.5	37.5	37.8	7.6	-	59.6	75.2
	132	3.3	3.1	217.2	-	34.6	11.5	1873.5	875.5	62.0	72.0	2.1	2.0	59.8	-
	144	-	-	222.1	462.2	32.4	13.8	1430.5	118.3	112.0	126.6	3.7	3.1	52.2	72.1

WB, wheat bran; WBS, winter barley straw.

Table 9. Enzyme production (IU/g inducing substrate) by liquid cultures of Talaromyces emersonii strains CBS814.70 and UCG208.

Inducing substrate	Growth time (h)	FPase IU/g		CMCase IU/g		ß-Glucosidase IU/g		Xylanase IU/g		ß-Xylosidase IU/g		ß-Galactosidase IU/g		Ext. protein mg/g	
		CBS	UCG	CBS	UCG	CBS	UCG	CBS	UCG	CBS	UCG	CBS	UCG	CBS	UCG
WB/WWS	24.0	0.0	0.0	0.0	211.9	2.0	8.5	604.5	1645.5	22.2	26.4	-	-	91.0	80.5
(1 to 1)	36.0	4.5	5.0	78.0	-	7.7	27.5	1767.5	1813.5	42.7	37.0	2.7	-	102.1	108.3
	48.0	5.9	5.8	-	-	4.8	10.4	766.0	773.5	34.6	54.7	2.4	1.0	58.5	95.6
	60.0	6.1	9.2	-	-	1.7	18.3	652.5	778.0	43.1	57.0	3.1	2.1	110.5	68.6
	72.0	-	-	170.0	340.6	5.1	9.5	837.0	583.5	47.0	48.0	3.5	2.1	96.9	80.3
	84.0	7.9	12.1	175.0	373.8	3.9	15.7	521.0	412.5	47.0	42.8	3.5	2.9	73.7	75.7
	96.0	10.0	25.8	205.4	367.1	24.6	10.4	471.5	575.0	44.4	117.7	4.0	2.8	70.8	68.7
	108.0	8.9	22.5	288.1	727.5	33.0	20.3	736.0	823.0	32.4	29.8	5.9	2.4	67.5	87.5
	120.0	8.6	22.0	289.0	602.6	35.7	17.6	794.5	875.5	57.0	47.1	4.0	2.6	76.7	88.0
	132.0	7.9	12.8	249.8	-	36.9	14.2	826.0	1465.0	83.0	60.4	3.8	2.6	68.9	84.9
	144.0	-	-	260.9	594.8	39.0	19.0	1418.7	1433.5	101.0	112.7	3.7	3.0	85.3	85.6
WB/WWS	24.0	0.0	0.0	21.9	177.7	2.3	8.6	199.0	770.0	16.0	16.8	-	-	79.8	87.5
(1 to 9)	36.0	0.7	0.0	76.2	-	4.8	16.5	566.5	1125.5	28.8	22.1	2.0	-	93.9	101.1
	48.0	3.7	5.7	-	-	4.1	12.3	267.5	543.0	28.2	33.7	2.8	5.1	55.6	83.1
	60.0	5.3	6.3	-	-	5.1	21.7	403.5	477.5	12.0	43.8	-	3.4	62.8	75.1
	72.0	6.8	-	137.2	305.2	6.0	7.5	318.5	450.5	28.6	44.8	4.5	2.9	94.4	61.0
	84.0	9.5	7.3	146.5	343.7	3.7	8.4	207.0	90.0	37.0	40.7	4.4	3.1	62.6	76.9
	96.0	9.3	18.8	157.1	-	7.9	24.8	133.0	261.0	15.4	49.4	4.6	3.5	56.2	72.5
	108.0	6.4	12.9	203.6	398.7	17.6	56.4	236.5	507.0	11.6	20.1	5.4	3.4	45.0	75.1
	120.0	5.7	10.7	164.2	554.8	61.0	28.8	252.5	723.0	16.8	40.0	5.8	3.8	56.9	80.0
	132.0	3.2	7.8	133.8	-	49.5	27.0	350.0	779.0	24.3	43.5	5.9	3.7	67.8	81.1
	144.0	-	-	124.6	679.6	65.5	31.0	389.0	744.0	27.5	120.5	5.9	3.9	62.5	80.7

WB, wheat bran; WWS, winter wheat straw.

Table 10. Enzyme activities in extracts of solid-state cultures of Talaromyces emersonii strains CBS814.70 and UCG208

Strain	Inducing substrate	FPase	CMCase	ß-Glucosidase	Xylanase	ß-Xylosidase	Arabinofuranosidase	Arabinanase	Pectinase	Polygalacturonase	ß-Galactosidase	Extracellular protein
		IU/g	IU/g	IU/g	IU/g	IU/g	IU/g	IU/g	IU/g	IU/g	IU/g	mg/g
CBS814.70	Wheat bran	1.79	83.4	5.91	64.2	4.64	21.02	2.38	43.1	69.0	2.12	10.89
UCG208	Wheat bran	1.95	67.6	3.37	88.5	3.29	4.06	2.48	47.8	82.7	1.18	9.5
CBS814.70	Beet pulp	2.58	81.9	5.62	42.4	1.92	30.15	1.76	27.4	42.7	13.6	6.49
CBS814.70	WB/BP (1:1)	5.02	157.2	10.86	74.5	-	-	6.66	51.7	88.4	-	14.03
UCG208	WB/BP (1:1)	1.44	47.1	2.83	40.5	-	-	2.88	61.4	113.4	-	7.61
CBS814.70	WB/WWS (1:1)	1.40	67.7	1.92	32.3	1.39	11.22	1.62	28.6	56.0	1.16	6.27
UCG208	WB/WWS (1:1)	0.42	54.0	0.42	32.7	0.10	1.65	1.69	49.1	85.3	0.27	10.22
CBS814.70	WB/SOS (1:1)	1.63	63.3	1.84	43.0	0.14	1.42	1.29	30.6	44.8	0.16	3.47
UCG208	WB/SOS (1:1)	1.31	41.6	1.48	30.3	0.08	1.24	1.48	33.7	55.2	0.16	5.11
CBS814.70	WB/SBS (1:1)	0.45	6.9	0.41	21.6	0.08	1.57	1.66	37.2	59.7	0.25	9.28
UCG208	WB/SBS (1:1)	2.04	74.3	1.47	70.8	0.39	3.56	2.65	47.9	73.0	1.82	11.6
CBS814.70	WB/WBS (1:9)	2.60	108.5	2.35	88.3	2.29	3.49	2.91	52.3	80.0	0.97	8.66
UCG208	WB/WBS (1:9)	0.96	47.7	1.80	33.3	0.10	0.85	1.53	31.2	51.9	0.27	6.43
CBS814.70	WB/WBS (1:1)	1.10	56.6	2.53	40.5	0.43	7.83	1.40	27.3	47.3	0.44	5.83

WB, wheat bran; BP, beet pulp; SBS, spring barley straw; SOS, spring oat straw; WBS, winter barley straw; WWS, winter wheat straw. Arabinanase activity was measured using arabinan isolated from beet pulp. All cultures were harvested at 5 days.

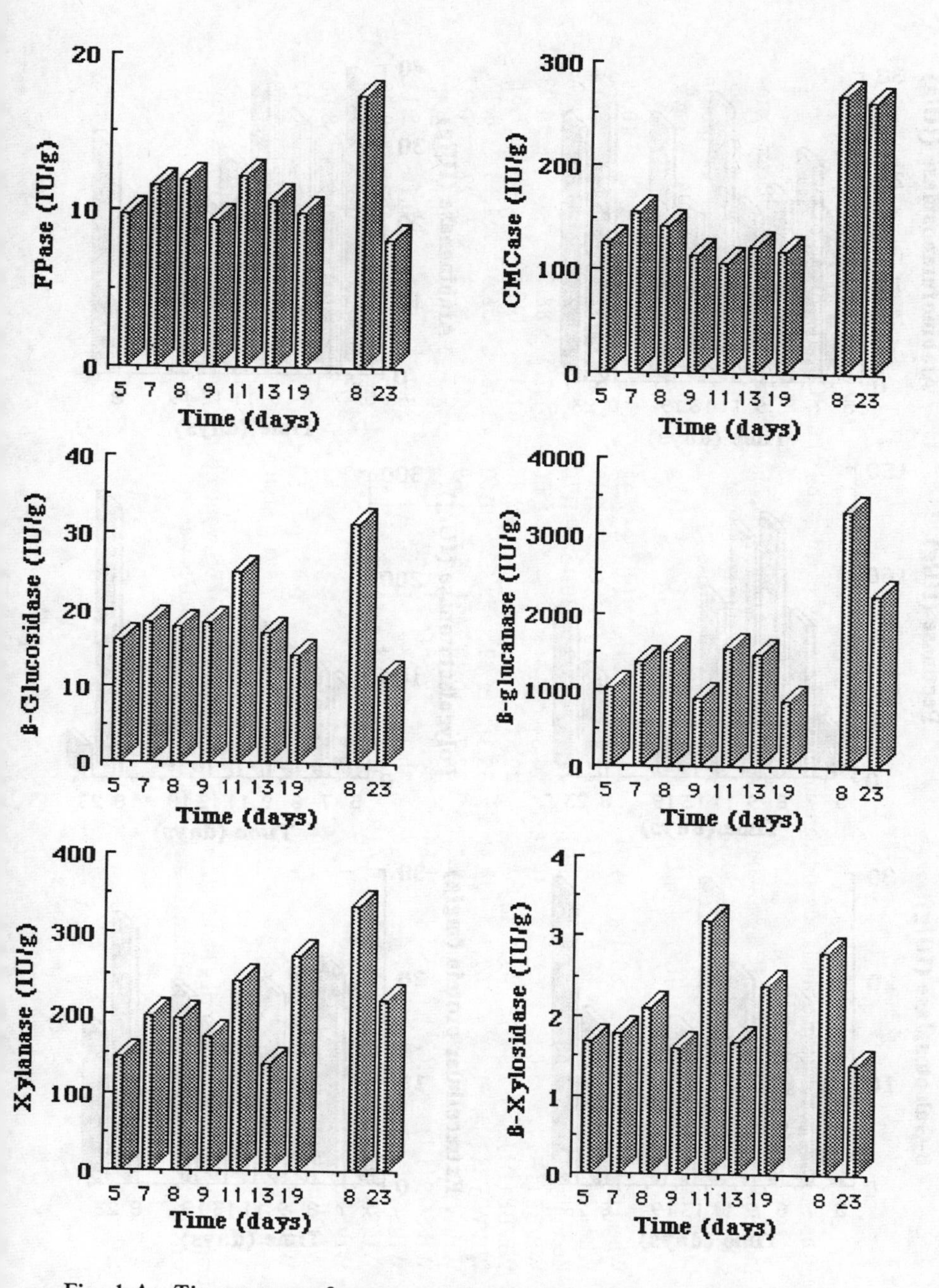

Fig. 1 A. Time course of enzyme production by solid-state cultures of *T. emersonii* CBS814.70 (5 to 19 days) and UCG208 (8 and 23 days) grown on unbuffered wheat bran/beet pulp (1:1) medium.

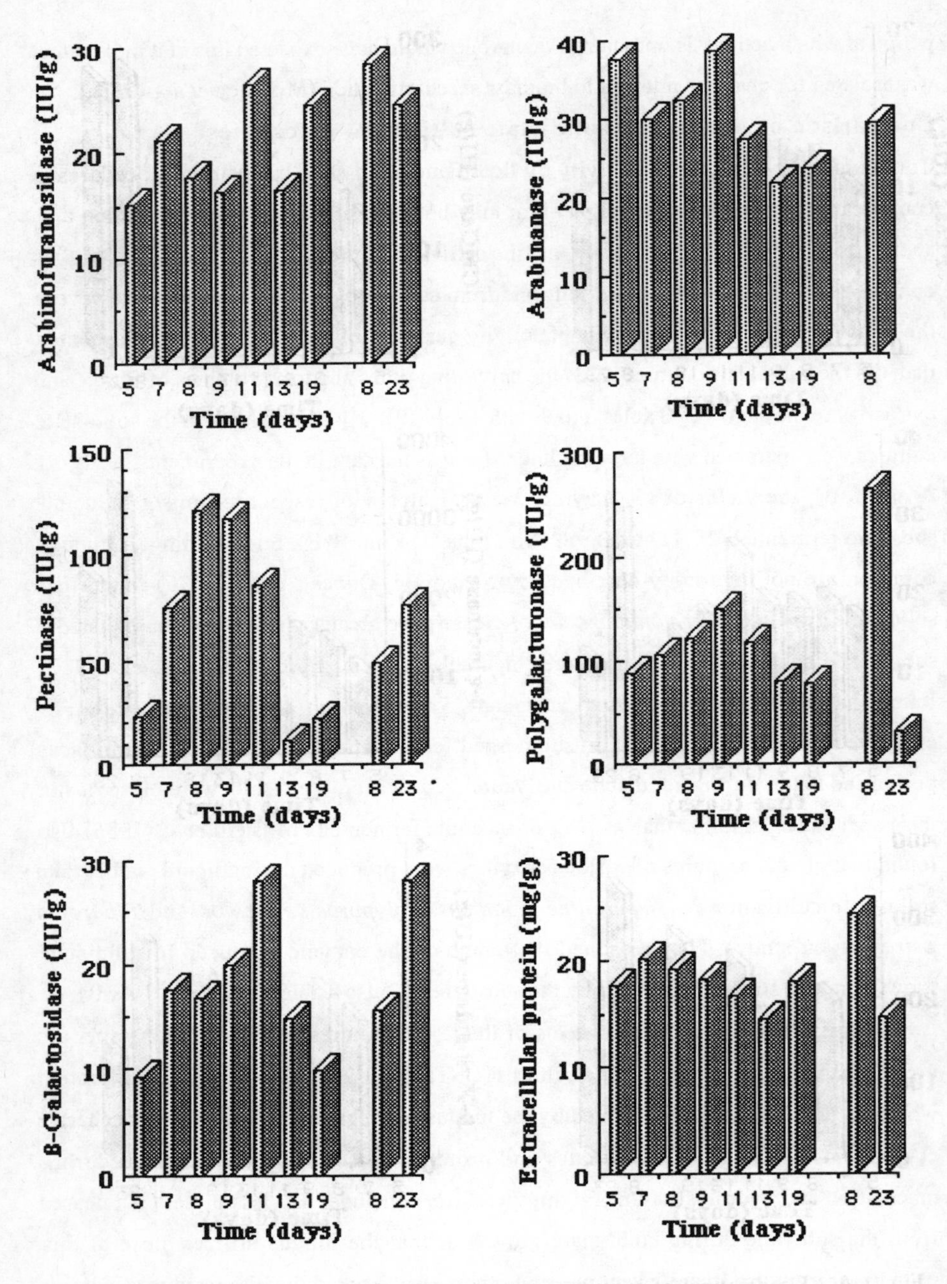

Fig. 1 B. Time course of enzyme production by solid-state cultures of *T. emersonii* CBS814.70 (5 to 19 days) and UCG208 (8 and 23 days) on wheat bran/beet pulp (1:1) medium. Arabinanase measured using Sigma substrate.

period at which activity is optimal. Again, one would have expected this of a mutant that was isolated for qualities relevant to biomass saccharification (Moloney *et al.*, 1983a).

Comparison of liquid- and solid-state cultivation procedures

Comparison of enzyme productivity by liquid and solid-state cultivation procedures is complicated by the fact that one medium may be more suited to liquid-cultivation and another to solid-state and by the fact that the cultivation time at which a specific activity is optimal varies from enzyme to enzyme and from culture to culture. In general, however, the liquid cultures yield more units of activity per gram of inducing substrate fermented than did the solid-state cultures when the harvesting times and total volumes of filtrate and extract were the same (cf. Tables 3 to 9 with Table 10). By contrast, when the solid-state cultures were extracted with larger volumes (as was the case in the experiments of Figs. 1 A and 1 B), the yields of all enzymes were greater with respect to units/g inducing substrate fermented (cf. Tables 3 to 9 with Figs. 1 A and 1 B). Such findings, although common are not invariably obtained. For example, Dubeau *et al.* (1986) found that solid-state growth of *Chaetomium cellulolyticum* gave greater yields of xylanase than did submerged cultivation albeit at longer incubation times. By contrast, Grajek (1987) found that while *Humicola languinosa* and *Sporotrichum thermophile* were more effective at producing xylanase in submerged fermentation, *Thermoascus aurantiacus* performed better during solid-state cultivation. However, the results were presented in this paper as IU/ml rather than as IU/g of substrate fermented. Milstein *et al.* (1986) also found that greater amounts of various enzymes were produced during liquid- rather than solid-state cultivation of *Aspergillus japonicus, Polyporus versicolor* and *Pleurotus ostreatus* on straw. They reasoned that much of the enzyme produced in solid-state growth may be strongly adsorbed by the substrate (cf. also Table 4 and Figs. 1 A, B).

The extracellular protein contents of the liquid cultures reported in this paper are greater than those of the solid-state cultures (cf. Tables 3-9 with Table 10). Such protein would include the enzymes secreted by the fungus for degradation of the polysaccharide components of the growth substrates and protein released from the substrate during maceration and hydrolysis. One might, therefore, conclude that less protein is released from the substrate during solid-state growth or that the fungus utilizes more of this protein for growth. In any event we, and others, have noted that the residue remaining after solid-state cultivation is enriched (3 to 4-fold) with protein by comparison with the starting solid substrate (Pandey *et al.*, 1988; Considine *et al.*, 1989; Tuohy *et al.*, 1989). Nevertheless, when account is taken of the weight loss of the solid substrate during cultivation, few cultures yield an actual increase in protein (see e.g. Considine *et al.*, 1989; Tuohy *et al.*, 1989).

Comparison of inducing substrates

Comparison of enzyme productivity on various substrates is also complicated in that no single growth substrate is best for all activities whatever the cultivation procedure, and in that the various activities do not always reach optimum values at the same cultivation time on any one growth medium. However, if one disregards the complications of harvesting time completely, one may note that the media containing wheat bran and straw were the best of the liquid cultures used (Tables 3 to 9) while wheat bran/beet pulp was consistently the best for solid-state cultivation (Table 10).

Periodicity in enzyme production during cultivation

Examination of the data in Tables 1 to 9 shows that many of the enzyme activities undergo a characteristic peaking and troughing during liquid cultivation. The peaking can be explained, at least in part, by the fact that different forms of a particular enzyme are secreted at different times during cultivation. For example, during growth on Solka floc, ß-glucosidase III, which is secreted cotemperaneously with cellulases, reaches maximum values at 36 h and then disappears completely by 48 h (McHale and Coughlan, 1981b). Subsequently, a second electrophoretically- and otherwise different form, ß-glucosidase I, appears and reaches maximum values at about 80 h.

There are several possible reasons for the observed troughing in activity. These include inactivation of the polysaccharidases by proteases and their inhibition or repression by phenolic compounds (see e.g. Sharma *et al.*, 1985; Martin and Akin, 1988) or sugar products released during digestion of the growth substrate. However, it may be noted that the growth media used in the above experiments were not buffered. Thus, the pH of the culture fluid during growth on glucose or Solka floc dropped sharply to a value of 3-3.5 and then slowly increased again, whereas on beet pulp and wheat bran, for example, it rose steadily to about 7.0 from the initial value of 4.5 (Tuohy and Coughlan, 1989). Similar alterations in pH have been found to occur during growth of other fungi in unbuffered media (Mandels *et al.*, 1975; Dekker, 1983; Dubeau *et al.*, 1987).

ß-glucosidase III disappeared from the Solka floc medium (see above) between 36 and 48 h because of its instability at the low pH values that developed due to the metabolism of cellulose at that time (McHale and Coughlan, 1981b). By contrast, when the growth media used were buffered against change in pH, ß-glucosidase III still plateaued at 36 h but persisted in the medium even when ß-glucosidase I accumulated.

The fact that pH values dropped to values close to 3 or increased to about 7 during growth on particular substrates also explains, at least in part, the observed troughing of activities other than ß-glucosidase since it was not observed when buffered media were used (not shown; but see also Tuohy and Coughlan, 1989). Mandels *et al.* (1975) have also noted the disappearance of preformed ß-glucosidase and, to a lesser extent, cellulase from the medium as a result of the low pH that develops during growth of *Trichoderma reesei* on soluble carbohydrates or cellulose. Sharp decreases in enzyme activity during early stages of cultivation were much less marked in the case of the unbuffered solid-state cultures of *T. emersonii* (Figs. 1 A and 1B) perhaps because of the protective effect of the greater concentration of substrate to which the enzymes could bind. Dubeau *et al.* (1987) have reported that xylanase production during growth of *C. cellulolyticum* on various nitrogen-supplemented solid-state media was greater when the pH was maintained at pH 7 than when pH was not maintained. By contrast, Brown *et al.* (1987) found, as we do, that production of some enzyme activities is greatest when the pH of the medium is allowed to fluctuate naturally.

Thermostability of the enzymes produced

The extracellular enzyme systems produced by *Talaromyces emersonii* during liquid cultivation are characterized by high temperature optima and thermal stability at pH values between 4 and 5 (Folan and Coughlan, 1978; McHale and Coughlan, 1981a; Moloney *et al.*, 1983a, 1985; Tuohy and Coughlan, 1989) as are those from other *Talaromyces sp.* (Nishio *et al.*, 1981; Clanet and Durand, 1987). The enzymes produced by solid-state cultivation are also thermostable. For example, xylanase in crude extracts lost no activity during incubation at 60°C, pH 5 over a period of a week. At 80°C, pH 5, the average half-life of xylanase in crude extracts of solid-state cultures of CBS814.70 and UCG208, grown on different substrates and harvested at various times, was about 20 min (Tuohy and Coughlan, 1989). Surprisingly, the average half-life of the enzyme from various liquid cultures under these conditions was considerably greater, i.e. 140 min. By contrast, Deschamps and Huet (1984) found the ß-glucosidase produced during solid-state growth of *Aspergillus phoenicis* to be more thermostable than those obtained from liquid cultures.

Economic considerations dictate that agricultural residues or surpluses, which are of course heterogeneous, rather than purified substrates be used in growth media for the large-scale production of the enzymes in question. However, which residue or surplus should be used would depend on several factors: (i) the availability of relevant lignocellulosic materials, (ii) the enzymes that one wishes to obtain, (iii) the use to which the enzymes are to be put. In general, the arsenal of enzymes produced will reflect the composition of the inducing substrate used. Thus, a specific lignocellulosic material comprised of cellulose, hemicellulose and pectin would induce the synthesis of each of the enzyme systems required for degradation of each of these polysaccharides. Another, lignocellulosic material may have a low content of pectin, for example. Consequently, the level of pectin-degrading activity produced during growth on that substrate would be low (see e.g. Considine and Coughlan, 1989). One can also generalize about the use to which enzyme preparations are to be put. Thus, if saccharification of a specific lignocellulosic material is desired, one frequently, though not always, finds that the most suitable mixture of enzymes for the saccharification are those produced using that material as the growth/inducing substrate (see. e.g. Moloney *et al.*, 1983b).

ACKNOWLEDGEMENT

M.G.T. thanks the Dept. of Education Ireland and University College, Galway for graduate studies maintenance awards. M.P.C. acknowledges receipt of EEC grant, Contract No. MA1D-0017-IRL.

REFERENCES

Beldman, G., Rombouts, F.M., Voragen, A.G.J. and Pilnik, W. (1984). Application of cellulase and pectinase from fungal origin for the liquefaction and saccharification of biomass. Enzyme Microb. Technol. **6**, 503-507.

Biely, P. (1985). Microbial xylanolytic systems. Trends Biotechnol. **3**, 286-290.

Brown, J.A., Collin, S.A. and Wood, T.M. (1987). Enhanced enzyme production by the cellulolytic fungus *Penicillium pinophilum,* mutant strain NTGIII/6. Enzyme Microb. Technol. **9**, 176-180.

Cannel, E and Moo-Young, M. (1980). Solid-state fermentation systems. Process Biochem. June/July, pp. 2-7 and August/September, pp. 24-28.

Clanet, M. and Durand, H. (1987). Study of polysaccharolytic enzymes produced by a new thermophilic fungus, *Talaromyces sp.* CL240. In *Biomass for Energy and Industry,* eds. G. Grassi, B. Delmon, J.-F. Molle and H. Zibetta, Elsevier Applied Science, London, pp. 722-726.

Considine, P.J. and Coughlan, M.P. (1989). Production of carbohydrate hydrolyzing enzyme blends by solid-state fermentation. In *Enzyme Systems for Lignocellulose*

Degradation, ed. M.P. Coughlan, Elsevier Applied Science, London, pp. 273-281.

Considine, P.J., Buckley, R.J., Griffin, T.O., Tuohy, M.G. and Coughlan, M.P. (1989). A simple and inexpensive method of solid-state cultivation. Biotechnol. Tech. **4**, 85-90.

Coughlan, M.P., Mehra, R.K., Considine, P.J., O'Rorke, A. and Puls, J. (1985). Saccharification of agricultural residues by combined cellulolytic and pectinolytic enzyme systems. Biotechnol. Bioeng. Symp. **15**, 447-458.

Dekker, R.F.H. (1983). Bioconversion of Hemicellulose: Aspects of hemicellulase production by *Trichoderma reesei* QM9414 and enzymic saccharification of hemicellulose. Biotechnol. Bioeng. **25**, 1127-1146.

Deschamps, F. and Huet, M.C. (1984). ß-Glucosidase production in agitated solid-state fermentation. Study of its properties. Biotechnol. Lett. **7**, 451-456.

Dubeau, H., Chahal, D.S. and Ishaque, M. (1986). Production of xylanases by *Chaetomium cellulolyticum* during growth on lignocelluloses. Biotechnol. Lett. **8**, 445-448.

Dubeau, H., Chahal, D.S. and Ishaque, M. (1987). Xylanase of *Chaetomium cellulolyticum* : Its nature of production and hydrolytic potential. Biotechnol. Lett. **9**, 275-280.

Folan, M.A. and Coughlan, M.P. (1978). The cellulase complex in the culture filtrate of the thermophilic fungus, *Talaromyces emersonii.* Int. J. Biochem. **9**, 717-722.

Gillespie, A.-M., Keane, D., Griffin, T.O., Tuohy, M.G., Donaghy, J., Haylock, R.W. and Coughlan, M.P. (1989). The application of fungal enzymes in flax retting and the properties of an extracellular polygalacturonase from *Penicillium capsulatum.* In *Fourth Int. Conf. Biotechnology in the Pulp and Paper Industry,* eds.T.K. Kirk and H.-M. Chang, Raleigh, NC, USA, May 16-19, in press.

Grajek, W. (1987). Production of D-xylanases by thermophilic fungi using different methods of culture. Biotechnol. Lett. **9**, 353-356.

Linko, M., Poutanen, K. and Viikari, L. (1989) New developments in the application of enzymes for biomass processing. In *Enzyme Systems for Lignocellulose Degradation,* ed. M.P. Coughlan, Elsevier Applied Science, London, pp. 331-346.

Lowry, O.H., Rosebrough, N.J., Farr, A.L. and Randall, R.J. (1951). Protein measurement with the Folin phenol reagent. J. Biol. Chem. **193**, 265-275.

McHale, A. and Coughlan, M.P. (1981a). The cellulolytic enzyme system of *Talaromyces emersonii*. Purification and characterization of the extracellular and intracellular ß-glucosidases. Biochim. Biophys. Acta, **662**, 152-159.

McHale, A. and Coughlan, M.P. (1981b). The cellulolytic enzyme system of *Talaromyces emersonii*. Identification of the various components produced during growth on cellulosic media. Biochim. Biophys. Acta, **662**, 145-151.

Macris, B.J. and Kekos, D. (1989). Enhanced cellulase activities using straw as growth substrate. In *Enzyme Systems for Lignocellulose Degradation,* ed. M.P. Coughlan, Elsevier Applied Science, London, pp. 261-271.

Mandels, M. (1985). Applications of cellulases. Biochem. Soc. Trans. **13**, 414-416.

Mandels, M., Sternberg, D. and Andreotti, R.E. (1975). In M. Bailey, T.-M. Enari and M. Linko (eds.), *Symposium on Enzymatic Hydrolysis of Cellulose,* SITRA, Helsinki, pp. 81-109.

Mandels, M., Andreotti, R.E. and Roche, C. (1976). Measurement of saccharifying cellulase. Biotechnol. Bioeng. Symp. **6**, 21-33.

Martin, S.A. and Akin, D.E. (1988). Effect of phenolic monomers on the growth and ß-glucosidase activity of *Bacteroides ruminicola* and on the carboxymethyl-cellulase, ß-glucosidase and xylanase activities of *Bacteroides succinogenes.* Appl. Environ. Microbiol. **54**, 3019-3022.

Miller, G.L. (1959). Use of dinitrosalicylic acid reagent for the determination of reducing sugars. Analyt. Chem. **31**, 426-428.

Milstein, O., Vered, Y., Sharma, A., Gressel, J. and Flowers, H.M. (1986). Heat and microbial treatments for nutritional upgrading of wheat straw. Biotechnol. Bioeng. **28**, 381-386.

Moloney, A.P., Hackett, T.J., Considine, P.J. and Coughlan, M.P. (1983a). Isolation of mutants of *Talaromyces emersonii* CBS814.70 with enhanced cellulase activity. Enzyme Microb. Technol. **5**, 260-264.

Moloney, A.P., Considine, P.J. and Coughlan, M.P. (1983b). Cellulose hydrolysis by the cellulases produced by *Talaromyces emersonii* when grown on different inducing substrates. Biotechnol. Bioeng. **25**, 1169-1173.

Moloney, A.P., O'Rorke, A., Considine, P.J. and Coughlan, M.P. (1984). Enzymic saccharification of sugar beet pulp. Biotechnol. Bioeng. **26**, 714-718.

Moloney, A.P., McCrae, S.I., Wood, T.M. and Coughlan, M.P. (1985). Isolation and characterization of the endoglucanases of *Talaromyces emersonii.* Biochem. J.

225, 365-374.

Nishio, N., Kurisu, H. and Nagai, S. (1981). Thermophilic cellulase production by *Talaromyces sp.* in solid-state cultivation. J. Ferment. Technol. **59**, 407-410.

Pandey, A., Nigam, P. and Vogel, M. (1988). Simultaneous saccharification and protein enrichment fermentation of sugar beet pulp. Biotechnol. Lett. **10**, 67-72.

Pourquié, J. and Desmarquest, J.-P. (1989). Scale up of cellulase production by *Trichoderma reesei.* In *Enzyme Systems for Lignocellulose Degradation,* ed. M.P. Coughlan, Elsevier Applied Science, London, pp. 283-292.

Schwald, W., Chan, M., Brownell, H.H. and Saddler, J.N. (1988). Influence of hemicellulose and lignin on the enzymatic hydrolysis of wood. In J.-P. Aubert, P. Béguin and J. Millet (eds.), *Biochemistry and Genetics of Cellulose Degradation,* Academic Press, New York, pp. 303-314.

Sharma, A., Milstein, O., Vered, Y. and Flowers, H.M. (1985). Effects of aromatic compounds on hemicellulose-degrading enzymes in *Aspergillus japonicus.* Biotechnol. Bioeng. **28**, 1095-1101.

Tuohy, M.G. and Coughlan, M.P. (1989). Production of thermostable xylan-degrading enzymes by *Talaromyces emersonii.* In *IEA Workshop: Biotechnology for Conversion of Lignocellulosics,* September 7-12, Lund, Sweden, in press.

Tuohy, M.G., Buckley, R.J., Griffin, T.O., Connelly, I.C., Shanley, N.A., Ximenes F. Filho, E., Hughes, M.M., Grogan, P. and Coughlan, M.P. (1989). Enzyme production by solid-state cultures of aerobic fungi on lignocellulosic substrates. In Enzyme Systems for Lignocellulose Degradation, ed. M.P. Coughlan, Elsevier Applied Science, London, pp. 293-312.

Voragen, A.G.J., Heutink, R. and Pilnik, W. (1980). Solubilization of apple cells with polysaccharide-degrading enzymes. J. Appl. Biochem. **2**, 452-468.

Weiland, P. (1988). Principles of solid-state fermentation. In Treatment of Lignocellulosics with White-rot Fungi, eds. F. Zadrazil and P. Reiniger, Elsevier Applied Science, London, pp. 64-76.

Wood, T.M. (1981). Enzyme interactions involved in fungal degradation of cellulosic materials. In *Proc. Int. Symp. Wood Pulping Chemistry,* Vol. **3**, SPCI, Stockholm, pp. 31-38.

PROTEIN ENRICHMENT OF SUNFLOWER SEED SHELL BY FERMENTATION WITH *TRICHOSPORON PENICILLATUM*

M.J. Fernandez, E. Roche, J. Pou, F. Garrido and D. Garrido

Institute of Industrial Fermentations

Dept. of Fermentation Technology and Bioengineering

P.O. Box 56, 28500 Arganda del Rey, Madrid, Spain

The aim of this work was to enrich sunflower seed shell with protein by way of solid-state fermentation (SSF) with yeast, and to use the fermented product (rich in fibre, protein and vitamins) in ruminant feed. *Trichosporon penicillatum* was chosen for solid-state fermentation because of the yield of protein achieved, the quantity of reducing equivalents utilized and the short period of cultivation required for maximum yield. Analysis showed that this species used all of the trace of glucose, 55% of the xylose and 46% of the fats present in the seed shell. It can be predicted that the use of a solid-state fermentation system that allows of appropriate oxygenation in the semi-solid culture virtually complete utilization of reducing and fatty matter in the shell could be obtained. These studies have provided sufficient data for proposing the development of the process with suitable fermentation facilities at the pilot-scale. This development is vital if one is to carry out the necessary economic assessment and establish the working conditions for future industrial application.

INTRODUCTION

The aim of this work was to treat sunflower seed shell with yeast by solid-state fermentation (SSF) with a view to using the fermented product (rich in fibre, protein and vitamins) in ruminant feed. The research forms part of the studies carried out to develop a simple, clean technological process that would optimize the nutritional and energy potential of the waste used as substrate.

For several reasons solid-state fermentation techniques are considered to be more suitable than those of traditional liquid state cultivation (liquid with aeration and agitation; LSF) for the exploitation of the lignocellulosic wastes. Such wastes are of little value *per se* and may differ considerably in composition. Thus, it is essential that procedures used to exploit their potential be simple and that facilities be adapted to deal with large volumes of material concentrated at one point, to the nature of the wastes in question and to the subsequent use of the endproduct. Above all one should note the considerable savings to be made due to the fact that the new biomass (i.e. the yeast) does not have to be harvested separately. Indeed, the yeasts and the hydrolyzed wastes and residues form a product that, once dry, may be used directly as animal feed. In LSF systems, by contrast, the biomass must be separated from the culture medium by costly filtration and centrifugal processes.

For this reason, many researchers have been working in recent years on the development of SSF techniques for the utilization of lignocellulosic wastes (Hesseltine, 1972; Bellamy, 1974; Srinivasan *et al.*, 1975; Grant *et al.*, 1978; Pamment *et al.*, 1978; Cannel and Moo-Young, 1980a, b; Viesturs *et al.*, 1981; Volfova *et al.*, 1978; Aidoo *et al.*, 1982; Moo-Young *et al.*, 1981, 1986; Laukevics *et al.*, 1984; Rolz, 1984).

We at IFI are widely experienced in the treatment of diverse lignocellulosic wastes (corn cobs, esparto, bagasse, fine-cut bagasse, shells, etc.) by LSF and by SSF. This experience has been accumulated since 1950 during investigations on the bioconversion of these materials (Garrido *et al.*, 1951; Marcilla *et al.*, 1951; Schnabel and Garrido, 1964; Garrido *et al.*, 1978; Pou *et al.*, 1985, 1987a, b; Garrido, 1986; Pou and Driguez, 1987).

The physical characteristics of the sunflower seed shell render it very suitable for SSF treatment. It does not have to be crushed prior to use and it allows the passage of air by forming a mesh in which the micellar yeasts, specifically selected for this purpose, can develop.

In this work it was seen to be more appropriate, with a view to industrial scale-up, to consider only the hydrolysis of the hemicellulose matter. Consequently the treatment of sunflower seed shells was studied to determine the optimal conditions for hydrolysis of the hemicellulose content. The physical characteristics of this waste were also examined with respect to the hydrolysis pretreatment and SSF.

MATERIALS AND METHODS

Analytical determinations

Protein was calculated from the total nitrogen content as measured by the Kjeldahl method, operating with an Auto 1030 Kjeltec analyzer (Tecator Co.), equipped with one digestor unit (mod. system 20 1015 Digestor) with the capacity to digest 20 samples simultaneously. Fibre and fat, ash and humidity were analysed, in keeping with Spanish standards (UNE 34-082/72, UNE 34-318/79 and UNE 34-078/76, respectively) for the analysis of animal feed. Total reducing sugars were determined by the DNS method (Miller, 1959) Considering that reducing compounds, other than sugar, are measured by this method, it would be more appropriate to speak of the total reducing equivalents (R.E.). Sugars were also quantitated by HPLC using the method of Gil de la Pena *et al.* (1983). Samples were pre-filtered using a Sartorius syringe filter (cellulose nitrate, 0.45 μ; SM-11306). Two types of HPLC column were used: a, Tracer sugar column at 90°C with double-distilled water as a mobile phase at a flow of 0.5ml/min; b, Tracer Spherisorb-NH_2 column at room temperature with acetonitrile/water (75:25 or 60:40, v/v) as a mobile phase and flow rates of 1 or 2 ml/min, respectively.

Acid hydrolysis

The shell was hydrolyzed by autoclaving at 1 bar pressure. Two solid:liquid ratios were tested (1:5 and 1:7) with different H_2SO_4 concentrations (0.5-3.0%, v/v) and different heating times (15 to 60 min). At the end of the treatment the solubilized compounds were eluted with H_2O (100 ml/g sunflower seed shell). The concentrations of sugars in the eluent were determined as described above.

Microorganisms

Three species of yeast, viz. *Hansenula anomala* (strain 926), *Geotrichum candidum* (strain 1471) and *Trichosporon penicillatum* (strain 1735), were utilized in the selection study, as was the yeast-like organism *Aureobasidium pullulans* (strain 1470). All were from the culture collection maintained at the Institute of Industrial Fermentations (IFI).

Culture conditions

Inocula were prepared by LSF using liquid hydrolysates (under the conditions described in the hydrolysis study) neutralized to pH 4.5 and diluted to an R.E. content 10-12 g/l. In the LSF tests the increase in dry biomass and the amount of R.E. used was monitored.

SSF tests were carried out in an oven at 30°C with hydrolyzed shell, neutralized to pH 4.5, as solid substrate. Given that it was not possible to measure cell growth directly, the increase in the protein content of the total dry material (hydrolyzed shell plus yeast biomass) was determined and the use of sugars and total R.E. was monitored.

The shell, being a natural product, contains metals, vitamins and other nutrients in amounts sufficient for the growth of the microorganisms. Thus, only mineral salts were added to the liquid hydrolysates used for both LSF and SSF. In the case of LSF, 1.37 g,

$MgSO_4.7H_2O$, 1.4 g $(NH_4)_2SO_4$ and 1.25 g of urea were added per litre of diluted hydrolysate. In the case of SSF, a concentrate of salts was prepared so that 2.6 KH_2PO_4, 0.74 g $MgSO_4 .7H_2O$, 2.8 g $(NH_4)_2SO_4$ and 2.5 g of urea were added per 100 g of hydrolyzed shell.

RESULTS AND DISCUSSION

Characterization of the substrate

The sunflower seed shell supplied by the company, KOIPE S.A., was used as a substrate. This lignocellulosic waste was chosen because it met two essential conditions for its use to be profitable: it is of low value and it is available in sufficient amounts at specific locations, i.e at sunflower seed oil factories. The shells, which are separated from the seed before the latter is pressed to extract the oil, are currently used only as fuel. In the separation process, the shells are splintered to a particle size appropriate to SSF. Moreover, they are dust-free. Thus, they may be utilized as they are accumulated, i.e. without pretreatment in the hydrolysis process.

The analysis of the sunflower seed shell according to Spanish Standards for Animal Feed is shown in Table 1.

Table 1. Analysis of sunflower seed shell (% of dry matter)

Protein	5.3
Fat	5.4
Fibre	38.4
Ash	3.2
Nitrogen-free extract	47.7

A notable feature of the analysis is the high fat content (5.4%) of the seed shell samples used. This was probably due to the presence of small particles of seed remaining after the industrial shelling process.

Hydrolysis of sunflower seed shells

Given that we are dealing with the use of wastes of little value and taking into account the techniques currently available, it seemed more appropriate, with a view to industrial scale-up, to consider only the hydrolysis of the hemicellulose content.

Acid hydrolysis has been and still is the most widely used treatment. This is especially true of industrial treatments of such wastes since the use of low concentrations of sulphuric acid for short periods at 100-120°C is sufficient to hydrolyze the hemicellulose.

Furthermore, since agro-industrial wastes are being used as the raw materials, it is essential that they be sterilized prior to cultivation of yeasts thereon. Thus, an acid pretreatment step would not only effect the hydrolysis of hemicellulose but would have the added advantage of serving to sterilize the substrate, and, since only one heating step would be required, would effect a considerable saving of energy. For these reasons, we pretreat the sunflower seed shell with dilute sulphuric acid at 1 bar pressure.

For the selection of the optimum conditions, it was necessary to study the influence of the solid:liquid ratio (1:5 or 1:7), acid concentration (0.5 to 3.0%) and reaction time (15, 30 and 45 min) on the acid hydrolysis process.

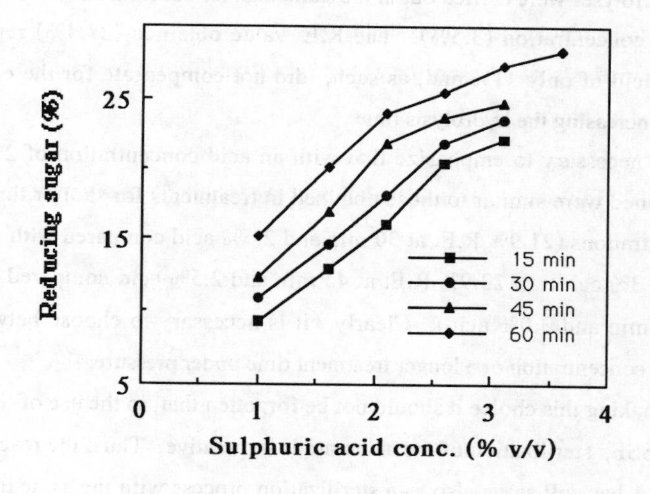

Fig. 1. Acid hydrolysis of sunflower seed shell. Comparison of the total R.E. obtained at a solid:liquid ratio of 1:7 at 1 atmosphere pressure as a function of and acid concentration and treatment temperature.

The results obtained with the solid:liquid ratios of 1:5 or 1:7 were much the same. However, the latter was chosen because mixing with acid was more homogeneous and the results more reproducible. As the concentration of acid and treatment time increase so does the amount of R.E. (i.e. total reducing equivalents) obtained (Fig. 1).

The differences between the R.E. obtained at 15 and 30 min were negligible. On the other hand, considerably greater amounts of R.E. were released by 45 min treatment and acid concentrations of up to 2%. The amounts released during the 45 min treatment tended to plateau at acid concentrations between 2.5 and 3%. The maximum yield of R.E. with a 45 min treatment and an acid concentration of 3% was 24.5%. The total R.E. values obtained (23-24.5%) were consistent with the hemicellulose content of this type of waste.

The R.E. value (22%) obtained with 3% acid even in the shortest time tested (15 min) was greater than that (20%) achieved by Eklun *et al.* (1976) who had treated this substrate under stronger conditions (3.5% acid and a 20 min treatment time at 1 bar).

In any case, to check whether the highest possible % yield of R.E. had been attained, hydrolyses were carried out in the same manner but for a longer time (1 h) and at higher acid concentration (3.5%). The R.E. value obtained (27.4%) represented an increased yield of only 11% and, as such, did not compensate for the extra expense incurred in increasing the hydrolysis time.

It is necessary to emphasize that with an acid concentration of 2%, the R.E. values obtained were similar to those obtained in treatments for shorter times at greater acid concentrations (21.9% R.E. at 30 min and 2.5% acid compared with 22% R.E. at 15 min and 3% acid; or 23.9% R.E. at 45 min and 2.5% acid compared with 23.7% R.E. at 30 min and 3.0% acid). Clearly, it is necessary to choose between using a greater acid concentration or a longer treatment time under pressure.

In making this choice it should not be forgotten that, in the use of lignocellulosic wastes for SSF, sterilization of the substrate is imperative. Thus, the reaction time at a pressure of 1 bar will serve also as a sterilization process with the same use of energy. However, taking into account the fact that long reaction times increase the degradation of pentoses to furfural, the following treatment was chosen: 30 min with 3% H_2SO_4 at a pressure of 1 bar. These conditions would also be sufficient for the sterilization of the substrate.

The sugars released on acid hydrolysis of the hemicellulose fraction under the chosen conditions were analyzed, qualitatively and quantitatively, by HPLC. The major components were xylose (10.6%) and arabinose (1.8%) whereas only a trace (0.6%) of glucose was found. From these results, one concluded that only the hemicellulose had been solubilized.

For comparative purposes, we also examined the use of a "steam explosion" hydrolysis system (Pou and Driguez, 1987; Pou *et al.*, 1987a, b, 1988a, b). The results obtained confirmed the fact that hydrolysis by "steam explosion" gave rise to lower R.E. levels than those obtained with acid hydrolysis. The R.E. value obtained by steam explosion under the most favourable conditions (treatment at 21 bar for 1 min) was only 7% compared with the value of 24.5% obtained with acid hydrolysis. Moreover, the smaller carbon-containing fractions (still incompletely identified) that were present in the hydrolysate from steam explosion treatment could not be assimilated by the yeasts selected for this kind of process. For this reason, treatment by steam explosion was rejected.

In view of the results above, it was decided to use acid hydrolysis for 30 min at a pressure of 1 bar, a solid:liquid ratio of 1:7 and a sulphuric acid concentration of 3% (v/v) as a pretreatment for the solid-state fermentation of sunflower seed shells.

Selection of microorganisms for LSF

The various microorganisms to be tested were grown on the hydrolysates, obtained under the chosen conditions, with a view to the selection of that species giving the greatest yield of protein, i.e. the greatest use of the sunflower seed shells. Such selection tests are essential when new substrates are used, since, as is well-known, even with total use of the carbon source, very different yields of biomass and protein may be found.

The species were selected both in liquid cultures with agitation and aeration (LSF) and in semi-solid cultures (SSF). Tests on liquid cultures were carried out, not only for comparison with SSF, but also because LSF was necessary to prepare the inocula for SSF. Four species were chosen for testing. All had been used with good results in previous work at the IFI and all showed mycelium-type growth under low aeration conditions. This quality is very important in SSF since filamentous microorganisms develop faster on the meshwork formed by the hydrolyzed waste.

The yeast, *Hansenula anomala,* had repeatedly been used in single-cell-protein production processes developed at the IFI. These studies included the production of yeasts in bisulphite lysates of *Eucalyptus globulus* and pre-hydrolysates of almond shell, and the production, for human consumption, of single-cell-protein from ethanol.

The yeasts, *Geotrichum candidum* and *Trichosporon penicillatum,* were chosen because they had been described in the literature as being cellulolytic and hemicellulolytic. Furthermore, in work carried out at IFI, these two yeasts have been used with good results in the treatment of fine-cut bagasse. In such treatment, *T. penicillatum* had an excellent protein transformation coefficient when used in SSF of the substrate in question.

The yeast-like microorganism, *Aureobasidium pullulans,* was also tested because its high productivity of extracellular ß-xylanases, ß-xylosidases, ß-glucosidases and CMCases had been demonstrated (Pou and Driguez, 1987; Pou *et al.*, 1988a). Such activities could favour the development of this species in SSF on hydrolyzed sunflower seed shell.

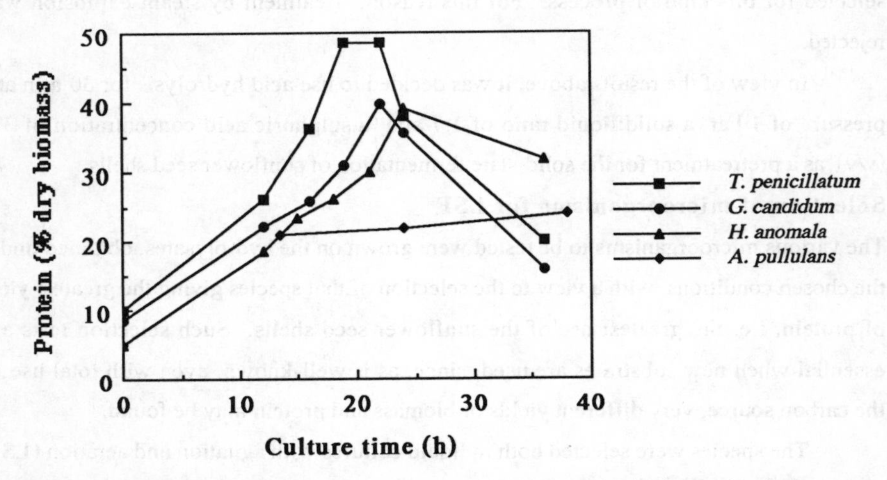

Fig. 2. Development of the species tested on liquid hydrolysate in shaken culture at 30°C. Initial R.E., 9.8 g/l.

The growth of all 4 species by LSF on liquid hydrolysate and by SSF on hydrolyzed shell were examined. The results of the former tests are shown in Fig. 2. The three species of yeast, viz. *H. anomala, G. candidum* and *T. penicillatum,* yielded maximum production of dry material after 24, 22 and 19 h, respectively. In the case of *A. pullulans,* strain 11740, yields of dry material did not reach a similar value (4.7 g/l) until 86 h. Accordingly, the use of the latter organism in an industrial process of this nature is not viable.

With respect to the production of protein (as a % of total dry matter), the best results were obtained with *T. penicillatum* , strain 1735 (Fig. 2). Not only did this strain give the greatest rate of protein production, it also yielded the greatest amount of protein as a percentage of dry matter (4.9 g/l compared with 4.0 g/l and 3.9 g/l in the case of *G. candidum* and *H. anomala,* respectively) and the highest yield of biomass with respect to the percentage of substrate R.E. used (69% compared with 64% and 57%, respectively).

In all cases, the yields obtained were far better than those (45-48% of substrate R.E. used) obtained by Eklun *et al.* (1976) who grew different *Candida* spp. on acid hydrolysates of the same substrate.

In view of the results obtained, it was decided that the liquid culture conditions for obtaining inocula should be 20 h in shaken culture at 30°C.

Selection of species for SSF

Once the conditions for the preparation of the inocula by LSF had been defined, the species of yeast that gave the best results were selected according to the percentage of protein obtained during growth, the quantity of R.E. utilized and the cultivation time needed to obtain maximum yield.

The SSF tests were carried out with hydrolyzed sunflower seed shell under the conditions chosen as best in the hydrolysis study. The hydrolysed shell was adjusted to pH 4.5 and the concentrated solution, containing the sources of phosphorus, nitrogen, magnesium and potassium, was added. As a result of this, the final solid:liquid ratio was 1:9. The samples were incubated in an oven at 30°C with adequate humidity and the increase in protein concentration and the amount of R.E. used were monitored.

The results obtained under these conditions are shown in Fig 3. As in the LSF tests, the strain of *A. pullulans* used grew at a slower rate than did any of the 3 yeasts tested (96 h to reach maximum values compared with 48 h). As a result it was rejected, because longer culture times involve higher risk of contamination and so invalidates its use in animal feed.

This fact underlines the need to carry out a selection of microorganisms for each new substrate. In general, yeasts are more appropriate for this type of process because of their faster growth rate. Pou and Driguez (1987, 1988) showed that the strain of *A. pullulans* tested secreted ß-xylanases, ß-xylosidases, ß-glucosidases and CMCases to the culture medium. Consequently, it appeared to have a perfect enzyme profile for use of lignocellulosic wastes. However, the tests above showed that it would be impractical to use it with the substrate in question.

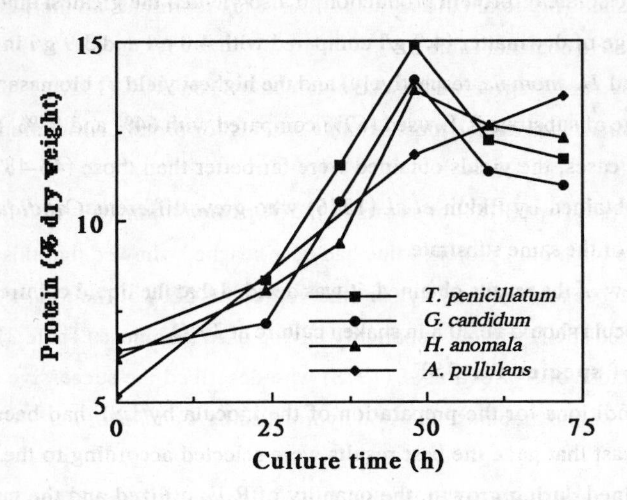

Fig. 3. Growth of yeasts during SSF at 30°C on hydrolyzed sunflower seed shell.

Table 2. Utilization in SSF of the sunflower seed shell hydrolysed by the species studied.

Yeast species	Protein (% DM)	% R.E. consumed	% Increase in [protein]	Protein coefficient *
H. anomala	13.6	8.3	7.3	88
G. candidum	13.8	8.1	7.4	91
T. penicillatum	14.9	8.8	8.1	92

*Protein transformation coefficient: g protein produced per 100 g of R.E. consumed. Initial R.E., 19.4%.

As was the case with LSF, *T. penicillatum* performed best on SSF yielding a product with 14.9% protein for an 8.8% use of R.E. The *G. candidum* and *H. anomala* strains gave protein yields of 13.8% and 13.6%, respectively. *T. penicillatum*, strain 1735, was chosen for SSF of sunflower seed shell because its use raised the protein content of the hydrolysed shell (i.e. total culture) to 14.9%. This represents an increase of 8.1% over the initial protein content (Table 2).

These results agree with those obtained by Pou *et al.* (1985, 1987a,b) who also chose *T. penicillatum* as the most suitable species for protein enrichment of bagasse by means of SSF.

Analyses (by HPLC) of sugars in the *T. penicillatum* cultures after 48 h growth, i.e. when the maximum protein value had been reached, showed that this strain had used traces of glucose and 5.8% of xylose (this represents 55% of the total xylose present in the hydrolysed shell) without apparently having used arabinose (Table 3). These results agree with those of Volfova *et al.* (1978) who described the successive assimilation of glucose, xylose and arabinose in wood hydrolyzed by *Candida utilis.*

Table 3. Utilization of sugars by *T. penicillatum* during SSF on hydrolyzed sunflower seed shell (initial R. E., 19.4%).

Culture time (h)	Xylose (%)	Arabinose (%)	Glucose (%)
0	10.6	1.8	0.6
48	4.8	1.8	0

All species tested showed high protein transformation coefficient values (Table 2). These high figures indicate that the yeast had used sources of carbon other than those included in the tabulated R.E., since, in the best of these cases, coefficients higher than 60% could not be obtained from sugars.

For this reason, an attempt was made to study whether the fat content of shell had also been used. As stated earlier, because of the presence of seed remains, this value is relatively high (5.4% of dry matter) in the actual substrate used. In the test, the chosen yeast, *T. penicillatum,* was grown under the conditions already described.

The results in Fig. 4 show that 2.5% fat was used at the time at which the protein production value (14.9%) was maximal. This value equals that found in previous tests, and the consumption of fat (which accounts for 46% of the total fat present in the hydrolysed shell) paralleled the consumption of reducing equivalents.

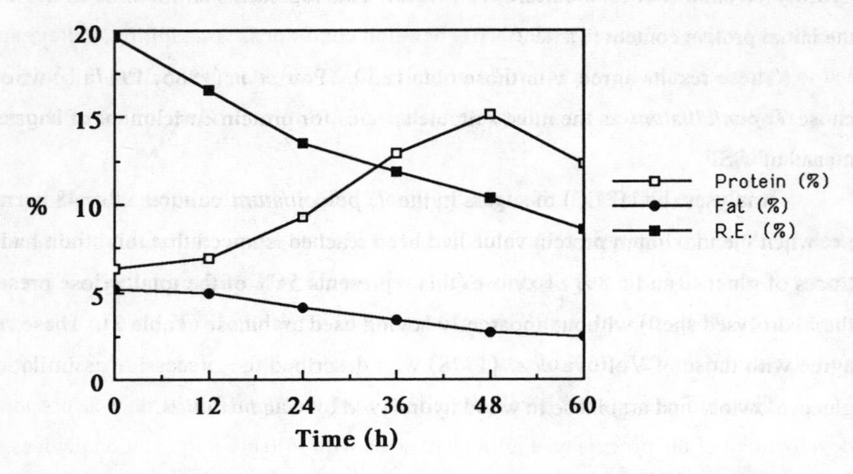

Fig. 4. Growth of *Trichosporon penicillatum* by SSF at 30°C on sunflower seed shell.

By taking account of the use of fat (2.5%), together with the use of the reducing equivalents (10.3%), one may calculate the "global" protein transformation coefficient (i.e. g protein produced/100 g R.E. + fat consumed) to be 68. This value is in keeping with the reality of the test.

Moreover, in all cases, including the *T. penicillatum* cultures, the use of R.E. was only partial. This was probably due to a lack of O_2, since the tests were carried out in ovens that may not have provided sufficient air. This lack of O_2 could be the cause of the continuing consumption of R.E. at late stages of cultivation at which time increases in protein concentration were not observed. This conclusion is logical if partial fermentation of sugar is taking place.

The final protein concentration attained, almost 15%, is 3-times the initial protein value of the shell. This is a very satisfactory result since a protein content of 15% is an acceptable value for a product to be used for animal feed. Thus, it may be predicted that with an SSF system that provides for appropriate oxygenation, an almost total use of the reducing and fatty matter present in the hydrolysed shell could be obtained. This supposition is supported by our experience in previous work (Garrido *et al.*, 1951; Garrido, 1986; Pou *et al.*, 1985, 1987a,b) in which we found that it is possible to use higher concentrations of carbon sources if the supply of O_2 is sufficient. For example, Pou *et al.* (1985, 1987), who investigated the growth of *Trichosporon penicillatum* on fine-cut bagasse, found that both the utilization of R.E. and the protein transformation coefficient value increased under better aeration conditions. In addition, better aeration led to a greater growth rate (Garrido *et al.*, 1978, 1986). Consequently, the cultivation time required to reach an adequate value of protein production was shortened considerably.

It may be concluded that, using lignocellulosic wastes, such as sunflower seed shell, with acid hydrolysis pretreatment and solid-state fermentation, and the species of yeasts specifically selected for this process, it would be possible to obtain a fermented product for use as animal feed. The product would be rich in fibre, protein and vitamins and could contain up to 24% of protein.

The studies reported here have already provided sufficient data to propose the development of the process on a pilot plant-scale with suitable fermentation devices. This intermediate scale-up is vital if one is to establish the working conditions for future industrial application and to carry out the necessary economic assessment.

The process could be of great economic interest to Spain since the volume of sunflower seed production is around 1,200,000 tonnes/year and the wastes (residues) accumulated in the oil factories amount to as much as 20% of this figure.

REFERENCES

Aidoo, K.E., Hendry, R. and Wood, B.J.B. (1982). Solid substrate fermentation. Adv. Appl. Microbiol. **28**, 201-237.

Bellamy, W.D. (1974). S.c.p. from cellulosic wastes. Biotechnol. Bioeng. **16**, 869-880.

Cannel, E. and Moo-Young, M. (1980a). Solid-state fermentation systems, Part I. Process Biochem. **15**, 2-7

Cannel, E. and Moo-Young, M. (1980b). Solid-state fermentation systems. Part II. Process Biochem. **15**, 24-28.

Eklun, E., Matakka, A., Mustranta, A. and Nybergh, P. (1976). Acid hydrolysis of sunflower seed husks for production of single cell protein. Eur. J. Appl. Microbiol. 2, 143-152.

Garrido, J. (1986). Proteinas en la alimentación humana. Quimica 2000, **4**, 45-50.

Garrido, J., Hidalgo, L. and Reus, A. (1951). Utilización de los prehidrolizados de esparto en la fabricación de levaduras alimenticias. Cienc. Aplic. **21**, 312-320.

Garrido, J., Gernandez, M.J., Amo, E., Tabera, J. and Garrido, F. (1978). Single cell protein from synthetic ethanol. Utilization of pure 0_2 in a new design fermentor. In *Proc. 5th Int. Congr. Food and Technology,* Kyoto (Japan), Abs. 8c-02, p. 266.

Gil de la Pena, M.L., Garrido, D., Moruno, E. and Garrido, J. (1983). Aplicación de la cromatografia líquida de alta eficacia (HPLC) al estudio de la evolución del mosto de cerveza. Cerveza y Malta XX, **78**, 20-28.

Grant, A.G., Han, Y.W. and Anderson, A.N. (1978). Pilot-scale semi-solid fermentation of straw. Appl. Environ. Microbiol. **35**, 549-553.

Hesseltine, C.W. (1972). Solid-state fermentation. Biotech. Bioeng. **14**, 517-532.

Laukevics, J.J., Apsite, A.F., Viesturs, U.E. and Tengerdy, R.D. (1984). Solid substrate fermentation of wheat straw to fungal protein. Biotechnol. Bioeng. **26**, 1465-1474.

Marcilla, J.., Feduchi, E., Hidalgo, L. and Garrido, J. (1951). El aprovechamiento industrial de los residuos agricolas. I. Estudio de la utilización de los prehidrolizados de carozos de maiz (mazorcas desgranadas) en la fabricación de las levaduras alimenticias. Cuadernos del Dep. de Fermentaciones Industriales (C.D.F.I.), Madrid, España, **1**, 3-50.

Miller, G.L. (1959). Use of dinitrosalicylic for determination of reducing sugar. Anal. Chem. **31**, 426-428.

Moo-Young, M., Lamtey, J. and Robinson, C.W. (1986). Pretreatment of lignocellulosics for bioconversion. In *Biotechnology and Renewable Energy,* eds. M. Moo-Young, S. Hasnain and J. Lamptey, Elsevier Applied Science, London , pp. 46-56.

Moo-Young, M., MacDonald, D.G. and Ling, A. (1981). Improved economics of Waterloo s.c.p. process by increased growth rates. Biotechnol. Lett. **3**, 154-157.

Pamment, N., Robinson, W.C. and Moo-Young, M. (1978). Solid-state cultivation of *Chaetomium cellulolyticum* on alkali-pretreated sawdust. Biotechnol. Bioeng. **20**, 1735-1744.

Pou, J. and Driguez, H. (1987). D-Xylose as inducer of the xylan-degrading enzyme system in *Pullularia pullulans.* Appl. Microbiol. Biotechnol. **27**, 134-138.

Pou, J., Fernandez, M.J. y Garrido, J. (1985). Obtención de proteinas a partir de bagacillo de cana. II. Estudio de la multiplicación de *Trichosporon penicillatum* en cultivo semi-sólido en bandeja. Microbiol. España, **38**, 89-95.

Pou, J., Roche, E.D., Garrido, D., Mulet, A. and Fernandez, M.J. (1987a). Sistemas de hidrólisis de residuos lignocelulosicos. In *Proc. XIV Congreso Nacional de Bioquimica (SEB),* Malaga, España, Abs. 17, **10**, 263.

Pou, J., Vittori, N., Fernandez, M.J. and Garrido, J. (1987b). Cultivo semi-sólido de *Trichosporon penicillatum* en jarras giratorias, para la obtención de proteinas a partir de bagacillo de caña. Interferón y Biotechnologia, **4** (2), 121-128.

Pou, J., Fernandez, M.J., Defaye, J. and Driguez, H. (1988a). Inducción de celulasas y xilanasas en *Aureobasidium pullulans.* Microbiología-SEM, **4**, 87-96.

Pou, J., Fernandez, M.J., Canellas, J., Mulet, A., Excoffier, G. and Vignon, M. (1988b). Obtención de hidrolizados por "steam explosion" para el enriquecimiento proteico de cascarilla de pipa de girasol. In *Proc. III Congreso Luso-Espanol de Bioquimica.* Sec. Biotechnologia, Santiago de Compostela, España, Abs. 16-26, 321.

Rolz, C. (1984). Microbial biomass from renewables: a second review of alternatives. Ann. Rep. Ferm. Proc. **7**, 213-356.

Schnabel, I. and Garrido, J. (1964). Producción de levaduras sobre lejías bisulfíticas de *Eucalyptus globulus* y pre-hidrolizados de cáscara de almendra. VI. Multiplicación de *Candida utilis* (1255) y *Hansenula anomala* (925), sobre prehidrolizados de cáscara de almendra. Rev. Ciencia Aplic. **101**, 498-513.

Srinivasan, V.R., Callihan, C. and Dunlap, C. (1975). *Final Report on Project no.APR-7303077,* National Science Foundation, Washington D.C.

Viesturs, V.E., Apsite, A.F., Laukevics, J.J., Ose, V.P., Baers, M.J. and Tangerdy, R.P. (1981). Solid-state fermentation of wheat straw with *Chaetomium cellulolyticum* and *Trichoderma lignorum.* Biotech. Bioeng. Symp. **1**, 359-369.

Volfova, O., Kyslikova, E., Sikyta, B. and Zalabak, V. (1978). Title of paper. Biol. Chem. Vet. **14**, 237-242.

BIOCONVERSION OF LIGNOCELLULOSIC RESIDUES BY MIXED CULTURES

M.T. Amaral Collaço, A. Avelino and H. Teixeira Avelino
Departamento de Technologia das Industrias Alimentares
DTIA-LNETI
Rua Vale Formoso, 1, 1900 Lisboa, Portugal

The ultrastructural and chemical changes in lignocellulosic materials following microbial degradation by liquid, semi-solid and solid-state fermentation were compared. Hydrolysis of the substrate matrices was brought about by *Trichoderma viride, Aureobasidium pullulans, Sporotrichum* sp., *Papulaspora* sp. and by the use of enzymes. The extent of degradation of cellulosic material and the yield of protein were measured in order to facilitate the transfer of a small-scale technology to a farm-scale process and to develop the necessary regional skills.

INTRODUCTION

Portugal, a country with few natural resources, dissipates much of its potential by failing to integrate the use of its natural and human capacities. One example of this is agro-industrial activity, the major producer of lignocellulosic residues. Such residues are potentially capable of being converted biologically to protein products. Realization of this potential would help Portugal reduce her dependence on foreign food and particularly on animal feed. Thus, for some years the creation of inter-institutional projects for the bioconversion of residues has been recommended so as to maximize the use of material and human resources. These are generally out of range of small teams.

Lignocellulosic residues are composed on average of lignin (15-25% of dry weight), cellulose (30-50%) and hemicellulose (20-40%). The exploitation of such materials is clearly of scientific as well as of economic interest. However, as might be expected of substances with such complex composition they are degraded only with difficulty. Moreover, there are significant variations in the extent of biodegradability of different materials even when their chemical composition appears to be similar.

Cellulosic residues from the forestry industries as well as grape bagasse and tomato pulp from other important national industries merit attention because their accumulation on site minimizes additional transport costs. Bioconversion of these residues has been assessed using different technological processes. The aim of such exercises is to integrate the use of the cellulosic and hemicellulosic fractions (Fig. 1) with the joint purpose of developing simple processes whose application will encourage the acquisition of regional techno-scientific capabilities.

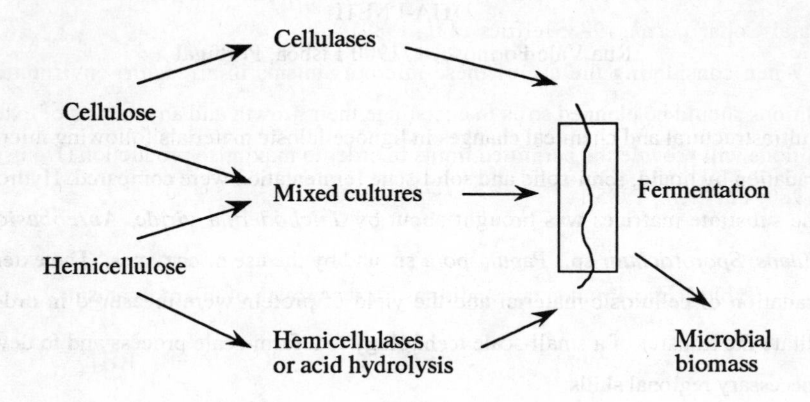

Fig. 1. Basic diagram for the use of fractions of lignocellulosic residues in microbial bioconversion.

THE MICROORGANISMS

The following microorganisms (hydrolytic starters) were assessed for their capacity to degrade cellulosic material: *Trichoderma viride* (5F), *Aureobasidium pullulans* (207F), *Sporotrichum* sp. (208F) and *Papulaspora* sp. (209F).

Table 1. Cellulolytic activities of microorganisms used.

	Fp activity IU/ml	Endoglucanase IU/ml
5F	0.23	0.26
207F	18.0	n.det.
208F	0.59	13.14
209F	0.44	20.46

Trichoderma viride (Gritzali & Brown, 1979), *Sporotrichum pulverulentum* (Eriksson, 1975) are known to produce exo-1,4-ß-D-glucanases, endo-1,4-ß-D -glucanases and ß-glucosidases and to have the ability to attack crystalline cellulose. The cellulolytic capacities of *Aureobasidium pullulans* and *Papulaspora* sp. were assessed by rapid tube test (Smith, 1977) and their enzymic activities were subsequently determined (Amaral-Collaço *et al.*, 1988; Jeffries *et al.*, 1989).

When considering the use of these microorganisms in a reactor, environmental conditions should be planned so as to encourage their growth and an analysis of external conditions will provide the permitted limits in order to maximise production (Domsch *et al.*, 1980; Pitt *et al.*, 1985).

Table 2. Environmental conditions for the growth of microorganisms.

	aw	pH	°C	RH%
5F	0.91	3.5-5.5	20-28	70-90
207F	0.90	3-7	25-30	70-90
208F	0.90	7	25-37	70-90
209F	0.90	4-7	20-35	70-90

Geotrichum candidum has been used in previous studies (Amaral Collaço, 1981, 1984; Amaral Collaço *et al.*, 1988) as a microorganism with a high protein value (52.3%) and with an amino acid content high in lysine. Its presence as a contaminant in dairy products leads to being included in the GRAS (Generally Recognised As Safe) list of food products. The optimal conditions for its growth are shown in Table 3.

Table 3. Optimal conditions for growth of *Geotrichum candidum.*

	aw	pH	°C	RH%
1F	0.90	3.7	25-30	70-90

THE REACTORS

The main function of a fermenter is to provide a controlled environment for the growth of a microorganism, or a defined mixture of microorganisms, so as to obtain the desired product. In designing and constructing a fermenter, parameters such as temperature control, available water (aw), relative humidity (RH), pH, O_2 transfer and viscosity should be considered, as should the cost of construction and production (Cannel *et al.*, 1980; Georgiou *et al.*, 1986).

For the processes of liquid and semi-solid fermentation using an STR type fermenter (Biolafite 50 l) for trials using SSF we prepared a tray reactor (48 cm wide, 7 cm high, 58 cm long; 0.1 cm diameter, lower hole; 0.5 cm diameter, upper hole).

FERMENTATION: LIQUID, SEMI-SOLID, SOLID STATE

Liquid and semi-solid

For this type of technological process the fermenter STR is used with the general scheme outlined in Fig. 2.

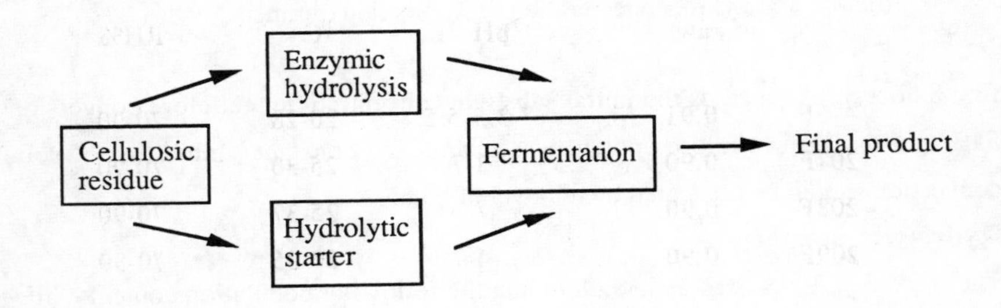

Fig. 2. General scheme for the optimization of enzymic hydrolysis.

Optimization of enzymic hydrolysis

In order to optimize enzymic hydrolysis the most appropriate concentrations of substrate and enzymes were determined. The best results were achieved at 16% (w/v) substrate concentration but for practical reasons in fermentation processes an initial substrate concentration of 12% (w/v) was chosen.

The optimal cellulase concentration of 1 IU/ml produced monomers of glucose and cellobiose that were removed by cellobiase at 0.04 IU/ml (about 90% conversion).

Conversion rates at the beginning of hydrolysis were 0.85 $g.\,l^{-1}.h^{-1}$ (first 15 h) and 0.37 $g.\,l^{-1}.h^{-1}$ for 30-40 h of hydrolysis (Fig. 3). The time chosen for syrup production was 24 h. Due to the solubility of the relevant components of the substrate, the fructose and xylose values are constant from the beginning of hydrolysis (after the

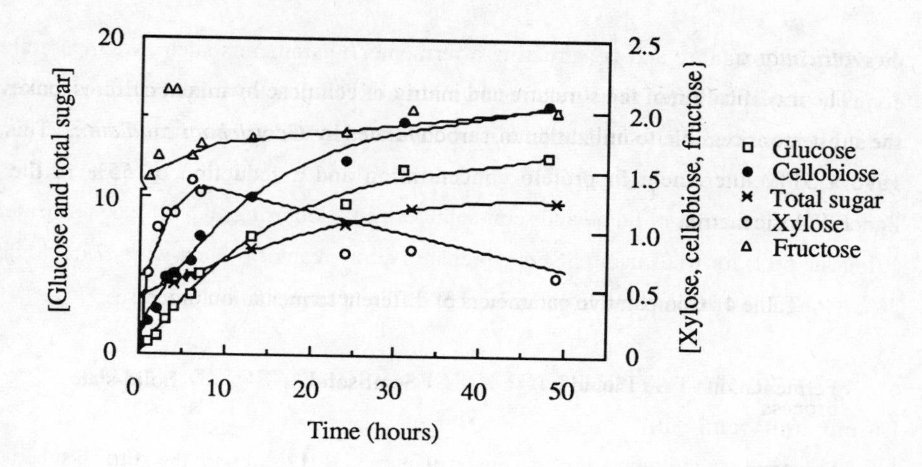

Fig. 3. Hydrolysis profile: Changes in total sugar, cellobiose, glucose, fructose and xylose concentrations as a function of time.

first 8 hours). As regards the initial substrate concentration, cellulose conversion was 55% and the hemicellulose content was maintained constant. Inhibition of cellulose conversion should be prevented by removal of glucose.

Hydrolytic starter

The mixed cultures used as inoculum had the following population contents: 5F - 10^7 colonies/ml; 207F - 10^9 colonies/ml; 208F - 3 x 10^7 colonies/ml; 209F - 3.7 x 10^4 colonies/ml. The value in dry matter of substrate was 9%, and temperature was maintained at 30°C for 5 days at 700 rpm.

A reduction of 34.2% in the lignocellulosic matrix was obtained.

Liquid and semi-solid fermentation with *Geotrichum candidum*

The hydrolysates from enzymic and microbial degradation were inoculated with 8 x 10^6 colonies/ml of *Geotrichum candidum,* supplemented with 0.5% ammonium sulphate and incubated at 30°C with regular mixing.

The liquors from enzymic hydrolyses did not reveal the presence of any inhibitory substances. *Geotrichum candidum* utilised every carbohydrate present, albeit only after the exhaustion of glucose, xylose and fructose. The same results were obtained with fermentations using syrup and syrup with solids.

Direct bioconversion using mixed cultures revealed a degradation of cellulose without production of carbohydrates. The best reduction, i.e 38%, was accomplished by

Sporotrichum sp.

The modification of the structure and matrix of cellulose by mixed cultures makes the substrate accessible to utilization of carbohydrates by *Geotrichum candidum*. This gave a 51% increment in protein concentration and a reduction of 55% in the lignocellulosic matrix.

Table 4 Comparative parameters of different fermentation processes.

Fermentation process	Liquid	Semi-solid		Solid-state
Substrate conc. (% w/v)	1	9	12	140
Treatment	Acid pre-treatment + enzymic hydrolysis	Mixed cultures	Acid pre-treatment + enzymic hydrolysis	------------------
Fermentation	*Geotrichum candidum*	*Geotrichum candidum*		Mixed cultures + *Geotrichum candidum*
Final protein content (%)	38.6	32.6	31.6	19.6

Solid-state fermentation

The hydrolytic starter with *Geotrichum candidum* was the inoculum 40% (w/v) with a cellular density of 3.8 x 10^8 colonies/ml of substrate with 140% concentrate (w/v). The SSF was run at room temperature with a moisture content of 60-70% controlled by a stream of moist air (100 l/min). During fermentation aw and RH were controlled while pH ranged from 4.5 to 7.5. Hemicellulose degradation occurred first and was followed by conversion of cellulose. The efficiency of degradation of lignocellulosic material to a constant content of lignin was 54%. An initial attack by the microorganisms on the hemicellulose (64.5% of the initial value) was followed by the utilization of cellulose (53% of the initial value). The different fermentation processes are compared in Table 4.

To assess these results from the point of view of profitability the information in Table 5 must be included.

Table 5. Reactor parameters.

Fermentation process	Liquid	Semi-solid	Solid-state
Agitation (rpm/min)	300	700	-
Fermentation volume /total volume (%)	60	60	80
Total solids/total fermentation vol. (%)	0.6	5.4 7.2	11.2

THE ATTACK ON LIGNOCELLULOSIC MATERIALS BY MICROORGANISMS

The attack on lignocellulosic matrices by microorganisms after 5 days at 30°C was assessed by measuring the reduction in initial cellulose concentration. A reduction of 28.2% was obtained by treatment with *Trichoderma viride,* 14.4% with *Aureobasidium pullulans,* 37.8% with *Sporotrichum* sp. and 28.2% with *Papulospora* sp. Comparative studies on the ultrastructure of tomato skins were carried out and confirmed previous analytical results (Amaral Collaço, 1981, 1984; Amaral Collaço *et al.*, 1988).

REFERENCES

Amaral Collaço, M.T. (1981). Produção de S.C.P. a partir de repiso de tomate a bagaço de uva com *Geotrichum candidum.* Pré-tratamento e hidrólise enzimática. In *1º Simposio Nacional NOPROT,* M.T. Amaral-Collaço *et al.*, eds., NOPRAI, Lisboa, pp. 194-201.

Amaral Collaço, M.T. (1984). Protein enrichment of tomato residues for animal feeding. In *Proceedings of the Third European Congress on Biotechnology,* Vol. II, Dechema, Verlag-Chemie, München, FRG, p. 563.

Amaral Collaço, M.T. (1988). Physiological studies on microbial bioconversion of tomato pomace syrups. In *Proceeedings of the Eight International Biotechnology Symposium,* IUPAC, Chirac, Paris, A-21, p.96.

Cannel, E. and Moo-Young, M. (1980.) Solid-state fermentation systems. Process Biochem. June/July, 2-7. ibid. Aug/Sept. 24-28.

Domsch, K.H., Gams, W. and Anderson, T.H. (1980). *Compendium of Soil Fungi,*

Academic Press, New York.

Eriksson, K.-E. (1975). Enzyme mechanisms involved in the degradation of wood components. In *Symposium on Enzymatic Hydrolysis of Cellulose,* M. Bailey, T.-M. Enari and M. Linko (eds.), SITRA, Helsinki, pp. 263-280.

Georgiou, G. and Shuler, M.L. (1986). A computer model for the growth and differentiation of a fungal colony on solid substrate. Biotechnol. Bioeng. **28**, 405-416.

Gritzali, M.K. and Brown, R.D. (1979). The cellulase system of *Trichoderma.* Relationships between purified extracelluklar enzymes from induced or cellulose-grown cells. Adv. Chem. Ser. **181**, 237-260.

Jeffries, T. W. and Lins, C. W. (1989). Effects of *Aureobasidium pullulans.* xylanase on properties of aspen thermomechanical and Kraft pulps. In *Proceedings of the Fourth International Conference on Biotechnology in the Pulp and Paper Industry,* H.-M. Chang and T.K. Kirk (eds.), in press.

Pitt, J. I. and Hocking, A.D. (1985). Fungi and food spoilage. In *Food Science and Technology.* A Series of Monographs, Academic press, New York.

Smith, R.E. (1977). Rapid tube test for detecting fungal cellulase production. Appl. Environ. Microbiol. **33**, 980-981.

BIOCONVERSION OF LIGNOCELLULOSIC RESIDUES BY MEMBERS OF THE ORDER APHYLLOPHORALES

N. Teixeira Rodeia
Departamento de Biologia Vegetal, Faculdade de Ciências
Universidade de Lisboa, Bloco C2, Campo Grande, 1700 Lisboa, Portugal

Fifty seven strains of thirty eight species of the order Aphyllophorales, viz. of the families Fistulinaceae, Ganodermataceae and Polyporaceae, were tested for their ability to produce enzymes that catalyze the degradation of soluble and insoluble cellulose (e.g. Avicel, Whatman and dyed cotton cellulose, newspaper strips and paper mill sludge) and lignin (Westvacco and sawdust). Some species grew well on culture media containing phenolic compounds and in some cases effected the oxidation of these compounds. The kinetic parameters of partially purified components of the cellulase systems of 4 species, viz. *Dichomitus squalens, Fistulina hepatica, Ganoderma resinaceum* and *Phellinus pini,* grown on different substrates, were determined. The Aphyllophorales species studied produced large quantities of mycelium, secreted a variety of enzyme activities and promoted significant bioconversion of different lignocellulosic substrates depending on the quality and concentration of the carbon and nitrogen sources used in the growth media.

INTRODUCTION

We have collected Aphyllophorales growing on trees, dead trunks, mine supports and timber in Portugal. Some of these were identified for the first time by Melo (1978, 1980, 1981a, b) and by Melo *et al.* (1980, 1983). In preliminary work, we screened 57 strains of 38 species belonging to the Families, Fistulinaceae, Ganodermataceae and Polyporaceae, of the Order, Aphyllophorales, so as to separate those that preferentially use cellulose from those that effect the simultaneous degradation of cellulose and lignin (Rodeia, 1983). Many of the Aphyllophorales promote white- or brown-rot of monocotyledonous or dicotyledenous plants. Some degrade the heartwood and others the sapwood of forest plants and some degrade timber or dead trees. Accordingly, we studied the cellulolytic and phenol oxidase activities produced by these species when growing on a variety of substrates. The latter included: Avicel, Whatman and Walseth celluloses, blue-dyed cotton; sawdust of *Eucalyptus sp., Populus sp., Pinus pinea* and

Pinus pinaster; newspaper strips; indulin Westvacco; paper mill sludge; and a number of phenolic compounds, viz. tannic and gallic acids, cathecol, guaiacol, hydroquinone, naphthol, phlorglucinol, resorcinol and tyrosine (Rodeia, 1983; Rodeia *et al.*, 1984; Rodeia and Gonçalves, 1986; Rodeia and Martins, 1986).

MATERIALS AND METHODS

Fungi and culture conditions

The cultures listed earlier were obtained from mycelia of fruitbodies, collected on trees, dead trunks, mine supports and timber, and were maintained by sub-culture on potato dextrose agar or on malt agar at 25°C. *F. hepatica,* which grew slowly under these conditions, was subcultured on cherry agar at 19°C. Subcultures on agar plates were inoculated with a disc cut (using a 12 mm cork-borer) from cultures growing on agar media. In the case of liquid media (100 ml), 25 such discs were macerated by blending in 25 ml of sterile water, left for 3 min, and the supernatant was used as the inoculum. Liquid media (Norkrans and Hammarstrom, 1963; Mandels and Weber, 1969), with or without supplementation with biotin and thiamine, contained various carbon (newspaper strips, paper mill sludge or CM-cellulose) and nitrogen (hydrolyzed casein, peptone, ammonium phosphate, ammonium nitrate, yeast extract or corn steep liquor) sources.

Assay procedures

Cellulase production was evaluated by the appearance of clearing zones (Rautela and Cowling, 1966; Nilsson, 1973, 1974a, b) on agar containing soluble (CM-cellulose) or insoluble cellulose (Avicel, Whatman), by the release of blue dye from cellulose azure (Petterson *et al.*, 1963; Smith, 1977) or by the appearance of halos around the mycelium (Teather and Wood, 1982). Cellulase activity was quantified from the direct relationship between log enzyme concentration and the diameter of the clearing zones brought about by growth of the mycelium or by diffusion of enzyme in culture filtrates placed in wells (in agar) containing 0.1% CM-cellulose.

CMCase activity was also determined by a modification of the method of Norkrans (1967) by measuring the release of reducing sugars (Nelson, 1944; Somogyi, 1952) following incubation of 1 ml of neat or diluted culture filtrate with CM-cellulose (0.1-1.0%, w/v) in phosphate buffer for 1 h at 50°C. Activities were also measured at pH values between 4 and 8, temperatures between 30 and 75°C, using filtrates from cultures harvested between 7 and 40 days. In addition to the above, CMCase activity was also determined by measuring the relative decrease in viscosity (Ostwald viscosimeter) following incubation of 6 ml of filtrate with 30 ml of 0.1% (w/v) buffered CM-cellulose for 1 h by comparison with that given by a standard CMCase preparation (10 mg/ml) from *Trichoderma reesei* under the optimum conditions of pH, temperature and age of culture (Rodeia *et al.*, 1984; Rodeia and Gonçalves, 1986; Rodeia and Martins, 1986).

Xylanase activity was determined according to the method of Sumner and Somers (1944) by measuring the release of reducing sugars following incubation of culture filtrate with xylan for 1 h at 50°C.

ß-Glucosidase activity was analyzed by monitoring the release of o-nitrophenol from o-nitrophenyl-ß-D-glucoside [ONPG (Wood, 1968; Yeoh et al., 1984)] following incubation of 1 ml of ONPG with 1 ml of filtrate and 1 ml of acetate buffer, pH, 4 for 30 min. Activities were also assayed at pH values from 3.58-7.0, temperatures between 37-75°C and culture filtrates aged from 7-50 days. Enzyme stability was assessed by carrying out incubations for 5 h. The inhibition of ß-glucosidase activity by glucose, using ONPG as substrate, was determined in assays at 50°C for 1 h. The K_i value was determined from Dixon and Webb (1964) plots. Kinetics constants for the ß-glucosidases of *D. squalens* (grown on different celluloses) and of *F. hepatica* (grown on paper mill sludge) with ONPG as substrate were also carried out at 50°C for 1 h.

The enzyme components of the cellulase systems were fractionated as follows: The culture filtrates were concentrated by acetone precipitation or by vacuum filtration through a 10,000 cut-off UF F membrane and then chromatographed on Sephadex G-75, Ultrogel UG-54 or Ultrogel AcA-202. The protein content of chromatographed fractions was determined by the method of Lowry *et al.* (1951).

Phenol oxidase activity was tested on malt agar medium containing one of the nine phenolic compounds listed earlier at concentrations of 0.5% (w/v) or 1 mM. The growth of the mycelium after 5 days and the extent of the surrounding discoloured zone were recorded (Bavendamm, 1928; Davidson *et al.*, 1938; Nobles, 1958; Káarik, 1965; Kirk and Kelman, 1965; Gilbertson *et al.*, 1975). In order to evaluate the ability of *P. pini* to degrade lignin, a liquid medium containing 0.5% (w/v) of the appropriate compound (Cortizo *et al.*, 1982; Rodeia, 1986a). On harvesting, the culture filtrates were analyzed by NMR, GC and TLC (Rodeia, 1986a, 1987a).

The resistance of the cellulose and lignin components of sound wood to hydrolysis by *P. pini* or *D. squalens* were examined as follows: Three different wood specimens, cut tangentially or radially, were added to the fungal cultures growing on malt agar in Kolle flasks. After 4 months incubation, the wood specimens were subjected to tests for tensile and compressive strengths and rupture force tests. For comparative purposes, wood specimens incubated in Kolle flasks under the same controlled conditions of temperature and humidity (Rodeia, 1986a; Leão and Rodeia, 1986), but without fungal mycelia, were tested in the same way. The changes in the deflection of wood beams kept in contact with mycelium for 11 weeks were also investigated. Samples of the wood specimens, after resistance testing, were subjected to scanning electron microscopy (SEM).

RESULTS

Cellulase and ß-glucosidase activities

The activities of *Coriolus versicolor, Dichomitus squalens, Fistulina hepatica, Ganoderma resinaceum, Lenzites betulina* and *Phellinus pini* against soluble and insoluble celluloses were measured (Rodeia *et al.*, 1984; Rodeia and Gonçalves, 1986; Rodeia and Martins, 1986; Resende *et al.*, 1986, 1987; Sàágua *et al.*, 1986; Dias *et al.*, 1988 a, b; Lemos *et al.*, 1988; Melo and Rodeia, 1988; Pereira and Rodeia, 1988; Rodeia, 1989; Sàágua and Rodeia, 1987; Sàágua *et al.*, 1989).

C. versicolor

C. versicolor, when grown on agar containing soluble cellulose, produced a zone of clearing of 40 mm diameter. Culture filtrates harvested after 15 days contained 294 µg/ml reducing sugars and 231 µg/ml protein and 850 µg/ml of CMCase activity. These culture filtrates also effected the greatest relative decrease in viscosity of solutions of CM-cellulose (Rodeia and Gonçalves, 1986; Lemos *et al.*, 1988).

D. squalens

D. squalens produced a clear zone on agar containing insoluble cellulose and showed good growth on and high activity against CM-cellulose. The ß-glucosidase is the best studied of the enzymes produced by this species. It was very stable between 37-70°C and was most active at pH 5.6. With ONPG as substrate, the V_{max} was 4.09 to 7.75 $mmol.min^{-1}$ and the K_m was 0.97 to 1.1 mM. The K_i for glucose was 3.85 mM. The endoglucanases (CMCases) of this species had high activity and stability at pH values from 4.0 to 7.0 (Resende *et al.*, 1986, 1987; Dias *et al.*, 1988a; Sàágua *et al.*, 1989).

F. hepatica

The amount of protein secreted by this species was greatest after 7 days of cultivation on media containing corn steep liquor and Avicel as the nitrogen and carbon sources, respectively. However, extracellular ß-glucosidase activity was greatest after 21 days cultivation on medium containing ammonium nitrate and Avicel as nitrogen and carbon sources, respectively (Melo and Rodeia, 1988; Pereira and Rodeia, 1988). With ONPG as the substrate for ß-glucosidase, the K_m was 0.74 to 0.91 mM, the V_{max} was 5 $mmol.min^{-1}$, and glucose (K_i 2 to 6 mM) was a non-competitive inhibitor (Rodeia, 1989). The freeze-dried filtrate of a 17-day culture on 1%(w/v) Whatman cellulose contained 2506 µg/ml protein, 134 mIU/ml cellobiohydrolase, 127 mIU/ml endoglucanase and 32 mIU/ml ß-glucosidase. Four peaks of cellobiohydrolase activity and 1 of endoglucanase were detected following chromatography of this material on Ultrogel UG-54.

G. resinaceum

This species gave a 36 mm zone of clearing when grown on agar containing insoluble cellulose. Extracellular ß-glucosidase activity was greatest in cultures grown on CM-cellulose. In addition to ß-glucosidase, 2 endoglucanases were detected following chromatography of the concentrated filtrate on Sephadex G-75.

L. betulina

Growth of *L. betulina* on agar containing Avicel or Whatman cellulose was accompanied by the production of clearing zones of 39 and 24 mm diameter, respectively. Dye was quickly released from cellulose azure by the growing fungus. Filtrates obtained from cultures grown on CM-cellulose contained 457 µg/ml protein, 230 µg/ml reducing sugars and, under the conditions of the assay, brought about a relative decrease of 89.9% in the viscosity of a solution of CM-cellulose. The CMCase activity, measured by the method of Teather and Wood (1982), corresponded to 1550 µg/ml (Rodeia and Martins, 1986).

P. pini

This species grew slowly on agar containing Avicel or Whatman cellulose and clearing zones were not detected. Extracellular enzyme activities were maximal after 44 days of cultivation. The relevant filtrates contained 132 µg/ml reducing sugars, exhibited 500 µg/ml of CMCase activity and effected a relative decrease of 92.4% in the viscosity of solutions of CM-cellulose. By contrast with enzyme activity, the protein content of filtrates was greatest (288.4 µg/ml) after 8 days cultivation at 25°C (Rodeia, 1986a). The optimum pH for CMCase production was 7.4. Filtrates from cultures grown for 8 days at this pH contained 116 µg/ml CMCase activity. Freeze-dried filtrates from cultures grown for 40 days (the optimum time) on Avicel (1%, w/v), Avicel (1%, w/v) plus indulin (1%, w/v), or sawdust (1%, w/v) contained 995 µg/ml protein, 39 mU/ml CMCase, 48.4 mU/ml ß-glucosidase and 3 U/ml of filter paper activity (Rodeia et al., 1984; Rodeia

1986a). One ß-glucosidase, 1 cellobiohydrolase and 1 endoglucanase were detected on chromatography of such extracts on Sephadex G-75.

Interspecies synergism

P. pini, a white-rot species, and *F. hepatica,* a brown-rot species, were co-cultured in order to determine whether they could interact synergistically to effect the degradation of cellulose. The following results were obtained: (a), Cellobiohydrolase and CMCase activities were detected following chromatography on Ultrogel AcA-202 of freeze-dried filtrates of co-cultures grown on Whatman cellulose (1%, w/v) or indulin (1%, w/v); (b), Cellobiohydrolase and CMCase activities were also detected following chromatography on Ultrogel UG-54 of freeze-dried filtrates of co-cultures grown on *Pinus pinaster* sawdust for 17 days; (c), the freeze-dried filtrate of a 9-day co-culture on indulin (1%, w/v) plus Whatman cellulose (1%, w/v) contained 32 mg/ml protein, 560 mIU/ml cellobiohydrolase, 749 mU/ml endoglucanase and 7.3 mU/ml ß-glucosidase; (d), the freeze-dried filtrate of a 17-day co-culture on *P. pinaster* sawdust (1%, w/v) contained 8.2 mg/ml protein, 5,909 mIU/ml cellobiohydrolase, 3,055 mIU/ml endoglucanase and 70.5 mIU/ml ß-glucosidase.

Xylanase activity

Extracellular xylanase activity of *F. hepatica* was greater than 200 U/ml when Avicel and ammonium nitrate were the carbon and nitrogen sources, respectively (Melo and Rodeia, 1988; Pereira and Rodeia, 1988).

Lignin degradation

Of the 57 strains of 35 different species tested for their ability to utilize lignin, some effected the oxidation of one or other of the 9 phenolic compounds listed earlier and some grew on agar containing phenols (Rodeia, 1983; Rodeia et al., 1984; Rodeia and Gonçalves, 1986; Rodeia and Martins, 1986).

C. versicolor effectively oxidized all of the phenols in question with the exception of naphthol but did not grow in the presence of cathecol or naphthol (Rodeia, 1983; Rodeia and Gonçalves, 1986). *D. squalens* oxidized neither hydroquinone nor tyrosine nor did it grow on media containing these phenols (Rodeia, unpublished observations). *F. hepatica* did not oxidize any of the phenols in question nor did it grow in the presence of naphthol (Rodeia, unpublished observations). *G. resinaceum* did not oxidize hydroquinone, resorcinol or tyrosine and did not grow in the presence of naphthol (Rodeia, unpublished observations). *L. betulina* oxidized all of the phenolic compounds with the exception of hydroquinone and tyrosine but did not grow in the presence of naphthol (Rodeia, unpublished observations; Rodeia and Martins, 1986).

P. pini did not oxidize tyrosine and growth on the other phenols depended on their concentration in the media used. This species also grew slowly on basal medium supplemented with glucose, lignin or glucose plus lignin. In the presence of these substrates, the highest dry weight of mycelium produced (689 mg) was on medium containing 0.1% (w/v) glucose plus 0.4% (w/v) lignin (Rodeia, 1986a). After 44 days of cultivation on such a medium, 16 phenolic compounds derived from lignin were detected in filtrates by TLC. Analysis of these filtrates by TLC, NMR and GC showed that the liberated phenolic compounds included vanillic acid, 2,6-dimethoxybenzoic acid, m-methoxybenzoic acid, methoxybenzaldehyde and 3,4-dimethoxybenzaldehyde (Rodeia, 1986a, 1987a).

Table. 1. Changes in the properties of wood incubated with fungi.

Property	Incubated with *P. pini*	Incubated with *D. squalens*	Control *P. pini*	Control *D. squalens*
Tensile strength ($kgf.cm^{-2}$)				
Heartwood	19	25	13	22
Sapwood	23	24	20	26
Compression strength ($kgf.cm^{-2}$)				
Heartwood	531	560	544	523
Sapwood	534	498	573	527
Rupture force				
Heartwood	10	13	7	15
Sapwood	10	11	8	13
Deflection change				
Heartwood	0.19	0.20	0.16	0.16
Sapwood	0.18	0.23	0.19	0.19

Resistance of sound wood to fungal attack

Wood specimens, cut tangentially or radially, incubated with *P. pini* or *D. squalens* for 4 months were tested for tensile and compressive strengths and rupture force tests. These properties were compared with those of wood specimens maintained under the same conditions except for the presence of fungi (Table 1).

CONCLUSIONS

Until the inception of this work, little had been carried out on the cellulose hydrolysis by members of the Order, Aphyllophorales. We have found that the extracellular ß-glucosidase activities of *Coriolus versicolor, Dichomitus squalens* and *Ganoderma resinaceum,* grown on various carbon sources are higher than those of *Trichoderma reesei,* one of the most thoroughly studied of cellulolytic fungi (e.g. since Reese, 1950). Culture filtrates of some members of the Aphyllophorales, when grown on CM-cellulose, also exhibited high endoglucanase activities. We have not succeeded in obtaining good production of cellobiohydrolase by members of this Order nor have we succeeded in separating and characterizing this enzyme, although we followed the kinetic procedures described by Parr (1983). Preliminary electrophoretic studies indicate the existence of 2 or 3 endoglucanases in filtrates of *G. resinaceum* and 1 or 2 ß-glucosidases in those of *F. hepatica, D. squalens* and *G. resinaceum.*

The preliminary results of investigations on lignin utilization by *P. pini* and *F. hepatica* indicate that different analytical procedures, e.g. the use of ^{14}C-labelled phenolic compounds (Crawford, 1981; Buswell *et al.*, 1982), will be required if we are to obtain more definitive answers.

REFERENCES

Bavendamm, W. (1928). Uber das Vorkommen und den Nachweis von oxidasen bei holzzerstorenden Pilzen, I. Mitteilung. Z. Pflanzenkrankh. Pflanzenchutz, **38**, 257-276.

Buswell, K.A., Eriksson, K.-E., Gupta, J.K., Hamp, S.G. and North, I. (1982). Vanillic acid metabolism by selected soft-rot, brown-rot and white-rot fungi. Arch. Microbiol. **131**, 366-374

Cortizo, M., Ferraz, J.F.P. and Melo, E.M.P.F. (1982). Utilización de la lignina como fuente de carbono por *Rosellinia necatrix.* Garcia de Orta, Serie de Estudos Agronómicos, Vol. **9**, No. 1 & 2. Revista da Junta de Investigações Científicas do Ultramar, Lisboa, 179-184.

Crawford, D.L. (1981). Microbial conversions of lignin to useful chemicals using a lignin-degrading *Streptomyces.* Biotechnol. Bioeng. Symp. 11, 274-291.

Davidson, R.W., Campbell, W.A. and Blaisdell, D.J. (1938). Differentiation of wood-decaying fungi by their reactions on gallic or tannic acid medium. J. Agric. Res. 57 (9), 683-695.

Dias, A., Carolino, Ma. M., and Rodeia, N.T. (1988a). Produção de celulases por *Dichomitus squalens* (Karst.) Reid. In *Proc. IV Congresso Nacional de Biotechnologia,* Abstract no. FM-14, Coimbra, Portugal.

Dias, A., Resende, Ma. E., Saagua, Ma C., Carolino, Ma M. and Rodeia, N.T. (1988b). Actividade xilanásica de *Dichomitus squalens* (Karst.) Reid induzida por vários substratos. In *Proc. IV Congresso Nacional de Biotechnologia,* Abstract no. FM-16, Coimbra, Portugal.

Dixon, M. and Webb, E. (1964). *Enzymes,* 2nd. edn. Longmans, London.

Gilbertson, R.L., Lombard, F.F. and Canfield, E.R. (1975). Gum guaiac in field tests for extracellular phenoloxidases of wood-rotting fungi. For. Prod. Lab. For. Serv. USDA (23 pp.)

Káarik, A. (1965). The identification of the mycelia of wood-decaying fungi by their oxidation reactions with phenolic compounds. Stud. For. Suec. 31, 1-80.

Kirk, T.K. and Kelman, A. (1965). Lignin degradation as related to the phenoloxidases of selected wood decaying Basidiomycetes. Phytopathol. 55, 739-745

Leão, L.M.F. and Rodeia, N.T. (1986). Alterações mecanicas e biologicas da madeira de pinho por *Dichomitus squalens* (Karst.) Reid. In *Proc. 3rd Encontro Nacional de Biotecnologia,* Outubro de 1986, Lisboa, Portugal, in press.

Lemos, P., Taborda, A.T., Carolino, Ma M. and Rodeia, N.T. (1988). Efeito da temperatura na producão de celulases por *Coriolus versicolor* (L.) Quel., In *Proc IV Congresso Nacional de Biotechnologia,* Abstract no. FM-15, Coimbra, Portugal, (Poster abstract).

Lineweaver, H. and Burk, D. (1934). Determination of enzyme dissociation constants. J. Amer. Chem. Soc. **56**, 658-666.

Lowry, O.H., Rosebrough, N.J., Farr, A.L. and Randall, R.J. (1951). Protein measurements with the Folin phenol reagent. J. Biol. Chem. **193**, 265-275.

Mandels, M. and Weber, J. (1981). The production of cellulases. Adv. Chem. Ser. **95**, 391-414.

Melo, I.M. (1978). *Buglssoporus pulvinus* (Pers. ex Pers.) Donk e *Polyporus mori* Poll. ed Fr., duas espécies de *Polyporaceae* novas para Portugal. Bol. Soc. Broteriana (2nd Ser.), 52, 277-283.

Melo, I.M. (1980). Sete especies de *Polyporaceae* novas para Portugal. Bol. Soc. Broteriana (2nd Ser.), **53**, 647-662.

Melo, I.M. (1981a). Fungi decaying mining timber. A preliminary survey of the Panasqueira mines. Port. Acta Biol. (Ser. B), **13**, 5-12.

Melo, I.M. (1981b). *Incrustoporia percandida* (Malenc. & Bert.) Donk e *Spongipellis spumeus* (Sow. ex Fr.) Pat., novas colheitas de fungos em Portugal. Port. Acta Biol. (Ser. B), **13**, 119-125.

Melo, I.M., Correia, M. and Cardoso, J. (1980). Acerca das *Polyporaceae* de Portugal. II, Bol. Soc. Broteriana (2nd Ser.), **53**, 675-723.

Melo, I.M., Correia, M. and Cardoso, J. (1983). Acerca das *Polyporaceae* de Portugal. III. Rev. Biologia 12, 109-120.

Melo, E.P. and Rodeia, N.T. (1988) Influência da fonte de azoto na indução de ß-glucosidase de *Fistulina hepatica* (Huds.) Fr. In *Proc. IV Congresso Nacional de Biotecnologia,* Abstract no. FM-12, Coimbra, Portugal, (Poster abstract no.?).

Nelson, N. (1944). A photometric adaptation of the Somogyi method for the determination of glucose. J. Biol. Chem. **153**, 375-380.

Nilsson, T. (1973). Studies on wood degradation and cellulolytic activity of microfungi. Stud. F. Suec. **104**, 1-29.

Nilsson, T. (1974a). Comparative study on the cellulolytic activity of white-rot and brown-rot fungi. Material und Organismen, 9(3), 173-198.

Nilsson, T. (1974b). The degradation of cellulose and the production of cellulase, xylanase, mannanase and amylase by wood-attacking microfungi. Stud. F. Suec. 114, 1-61.

Nobles, M.K. (1958). A rapid test for extracellular oxidase in cultures of wood-inhabiting Hymenomycetes. Can. J. Bot. 36, 91-99.

Norkrans, B. (1967). Cellulose and cellulolysis. Adv. Appl. Microbiol. 9, 91-129.

Norkrans, B. and Hammarstrom, A. (1963). Studies on growth of *Rhizina undulata* Fr. and its production of cellulose- and pectin-decomposing enzymes. Physiol. Plant. 16(1), 1-10.

Parr, S.R. (1983). Some kinetic properties of the ß-D-glucosidase (cellobiase) in a commercial cellulase product from *Penicillium funiculosum* and its relevance in the hydrolysis of cellulose. Enzyme Microb. Technol. 5, 457-462.

Pereira, M.C. and Rodeia, N.T. (1988). Crescimento, degradação de celulose cristalina, actividade celulásica e xilanásica de *Fistulina hepatica* (Huds.) Fr. In *Proc. IV Congresso Nacional de Biotecnologia,* Abstract no. FM-13, Coimbra, Portugal.

Petterson, G.E., Cowling, B. and Porath, J. (1963). Studies on cellulolytic enzymes. I. Isolation of a low-molecular-weight cellulase from *Polyporus versicolor.* Biochim. Biophys. Acta, 67, 1-8.

Rautela, G.S. and Cowling, E.B. (1966). Simple cultural test for relative cellulolytic activity of fungi. Appl. Microbiol. 14 (6), 892-898.

Reese, E.T., Siu, R. and Levinson, H.S. (1950). The biological degradation of soluble cellulose derivatives and its relationship to the mechanism of cellulose hydrolysis. J. Bacteriol. 59(4), 485-497.

Resende, Ma E., Lemos Carolino, Ma M. and Teixeira Rodeia, N. (1986). Propriedades cinéticas de uma ß-glucosidase de *Dichomitus squalens* (Karst.) Reid. In *Proc.*

3rd Encontro Nacional de Biotechnologia, Outubro de 1986, Lisboa, Portugal, Poster no. P-41, in press.

Resende, Ma E., Lemos Carolino, Ma M. and Teixeira Rodeia, N.T. (1987). *Dichomitus squalens* and its relevance in the hydrolysis of cellulose: ß-glucosidase, 1,4-ß-glucan-glucanohydrolase and 1,4-ß-glucancellobiohydrolase. In *Proc. FEMS Symp. No. 43, Biochemistry and Genetics of Cellulose Degradation,* eds. J.-P. Aubert, P. Béguin and J. Millet, September, Institut Pasteur, Paris, France, (Poster abstract no. P1-13).

Rodeia, N.T. (1983). Polyporaceae, Actividade celulásica. Rev. Biol. **12**, 435-448.

Rodeia, N.T. (1986a). *Phellinus pini* (Brot. per Fr.) A. Ames. Sua actividade lenhino-celulásica *in vitro* e a sua influência no comportamento físico-mecânico da madeira de pinho. ICT, Informação Cientifica, Estruturas, INCES 1, Laboratório Nacional de Engenharia Civil, 150 pp.

Rodeia, N.T. (1986b). Mechanical and cellulasic alterations induced by *Phellinus pini* in sound wood of *Pinus pinaster* Sol. ed Aiton, In *Proc. Symposium on Biodegradation of Lignocellulosic Materials,* November, Paisley, Scotland, (Poster abstract no. P4).

Rodeia, N.T. (1987a). Chromatographic fractionation of lignocellulosic extracts of *Phellinus pini* (Brot. per Fr.) A. Ames. In *Proc. FEMS Symp. No. 43, Biochemistry and Genetics of Cellulose Degradation,* eds. J.-P. Aubert, P. Béguin and J. Millet, September, Institut Pasteur, Paris, France, (Poster abstract no. P4-02).

Rodeia, N.T. (1987b). Synergism between cellulases of *Phellinus pini* (Brot. per Fr.) A. Ames and *Fistulina hepatica* (Schaeffer) per Fr. involved in the solubilization of native cellulose. In *Proc. 8º Congresso Nacional de Bioquímica,* Póvoa de Varzim, (28 Nov. - 1 Dez.), Ciência Biológica, 12 (5A), Suplemento (Poster abstract no. S6-21).

Rodeia, N.T. (1989). Sludge from one paper mill influencing the production of cellulases by *Fistulina hepatica* (Schaef.) Fr., In *Proc. Tricel '89, International Symposium on Trichoderma cellulases,* September 14-16, Vienna, Austria, (Poster abstract no. P28).

Rodeia, N.T., Carolino, Ma M., Lemos and Gonçalves, A.M.S. (1984). Cellulasic and phenoloxidasic activities of *Phellinus pini* (Brot. per Fr.) A. Ames, In *Bioenergy 84, III,* , eds. H. Egneus and A. Ellegard, Elsevier Applied Science, London, pp. 263-268.

Rodeia, N.T. and Gonçalves, A.M.S. (1986). Polyporaceae: cellulasic and phenoloxidasic activities. II. Enzymatic activities of *Coriolus versicolor* (L. ex Fr.) Quel. (742). Boletim da Sociedade Broteriana, **LIX** (2nd Ser.), 43-57.

Rodeia, N.T. and Martins, Ma T. (1986). Polyporaceae: cellulasic and phenoloxidasic activities. III. Enzymatic activity in *Lenzites betulina* (L. ex Fr.) Fr. (strain 379). Boletim Soc. Broteriana, LIX (2nd Ser.), 67-76.

Sáàgua, M.C., Carolino, Ma M. and Rodeia, N.T. (1986). Purificação parcial de uma ß-glucosidase de *Ganoderma resinaceum* (Pers. ex Gray) Pat. In *Proc 3º Encontro Nacional de Biotechnologia,* 6-9 de Outubro, Lisboa, Portugal, (Poster abstract no. P35), in press)

Sáàgua, M.C. and Rodeia, N.T. (1987). Gel filtration chromatography and electrophoresis of cellulases of *Ganoderma resinaceum* Boud. Apud. Pat., In *Proc. FEMS Symp. No. 43, Biochemistry and Genetics of Cellulose Degradation,* eds. J.-P. Aubert, P. Béguin and J. Millet, September, Institut Pasteur, Paris, France, (Poster abstract no. P1-14).

Sáàgua, M.C., Resende, Ma E., Carolino, Ma M. and Rodeia, N.T. (1989). Cellulases from *Polyporaceae.* In *Proc. Tricel '89, International Symposium on Trichoderma cellulases,* September 14-16, Vienna, Austria, (Poster abstract no. P27).

Smith, R.E. (1977). Rapid tube test for detecting fungal cellulase production. Appl. Environ. Microbiol. **33** (4), 980-981.

Somogyi, M. (1952). Notes on sugar determination. J. Biol. Chem. **195**, 19-23.

Sumner, J.B. and Sommers, G.F. (1944). *Laboratory experiments in biological Chemistry*. Academic Press, New York.

Teather, R.M. and Wood, R.J. (1982). Use of Congo red polysaccharide interactions in enumeration and characterization of cellulolytic bacteria from the bovine rumen. Appl. Environ. Microbiology, **43** (4), 777-780.

Walseth, C.S. (1952). Occurrence of cellulases in enzyme preparations from microorganisms. TAPPI, **35**, 228-233.

Wood, T.M.. (1968). Cellulolytic enzyme system of *Trichoderma koningii*. Separation of components attacking native cotton. Biochem. J. **109**, 217-227.

Yeoh, H.H., Tan, T.K. and Tian, K.E. (1984). Cellulolytic enzymes of fungi isolated from wood materials. Mycopathol. **87**, 51-55.

REACTOR FOR ENZYMIC HYDROLYSIS OF CELLULOSE

P. Thonart, E. Auguste, D. Roblain, R. Rikir and M. Paquot *

Centre Wallon de Biologie Industrielle

Unite d'Enseignement et de Recherche des Bioindustries, FSAG

passage des déportés, 5800 Gembloux

Service de Technologie Microbienne, ULq, Sart-Tilman, B40, 4000 Liege, Belgium

and

*UER de Technologie des Industries Agro-Alimentaires, FSAG

passage des déportés, 5800 Gembloux, Belgium

The objective of this publication is to describe the monitoring rather than the modelling of the enzymic reactor. The design and monitoring of an enzymic cellulose hydrolysis reactor must take into account not only the study of the interaction between the enzyme and the substrate, and the design of different models, but also the problems of measuring the reaction progress, detecting trouble during the reaction and recovering the enzymes.

INTRODUCTION

Cellulose, the most abundant biopolymer on earth, comprises almost one half of the dry weight of plant biomass. The biodegradation of cellulose represents an interesting and challenging research problem both from the basic and applied standpoints.

The enzymic hydrolysis of cellulose to fermentable sugars is not practical at present largely because of the high cost of production of the necessary hydrolytic enzymes. It is, therefore, economically important when designing the reactor to maximize recycling of the cellulases from both the hydrolysate and the unhydrolyzed substrate. Moreover, the relation between the properties of the substrate (crystallinity, pore size, chemical composition) and the biodegradability of these substrates as a function of reaction time is an important parameter in the yield from a cellulose hydrolysis reactor .

This publication also describes the effects of shear and air-liquid interface on cellulase inactivation and the monitoring of the hydrolysis reaction by measurement of physico-chemical properties.

MATERIALS AND METHODS

The following have been described previously: characteristics of the cellulosic substrate (Paquot and Thonart, 1982; Thonart *et al.*, 1983; Parajo *et al.*, 1988); the enzyme complex or system and its adsorption on the substrate (Desmons *et al.*, 1983; Paquot *et al.*, 1984a); effect of agitation on cellulase activity (Parajo *et al.*, 1988).

CHARACTERISTICS OF THE CELLULOSIC SUBSTRATE AND THE HYDROLYSIS RESIDUES

Hydrolysis of cellulose by enzymic methods depends primarily on the accessibility of the enzymes to the substrates. Such accessibility is largely determined by the lignin/carbohydrate interactions and the greater or lesser degree of compactness of the cellulose itself. Crystallinity has long been known to be a strong determinant of its biodegradability. However, the weaker correlations between crystallinity and degradation rate in the *Clostridium* or *Trichoderma* systems suggest that crystallinity is less important or a less limiting factor in cellulose degradation (Weimer and Wayde, 1985; Paquot and Thonart, 1982; Table 1). Enzymic accessibility is correlated with other properties such as porosity, specific surface area and degree of hydration. These properties also provide a basis for characterizing the hydrolysis residues. Results obtained during hydrolysis of a hardwood sulphate pulp are shown in Table 1.

Table 1. Changes in the characteristics of the residue from enzymic hydrolysis of a hardwood sulphate pulp (Paquot and Thonart, 1982; Paquot *et al.*, 1984b).

Properties	Enzymic hydrolysis	
	Before ± 60%	After ± 60%
Reducing power	Increased	None
Microporosity	Increased	Decreased
Macroporosity	Increased	Increased
Solubility in 18% NaOH	Much increased	Increased
Degree of polymerization	Same or decreased	Increased
Index of crystallinity	None	None
Ash content	Increased	Increased

In addition to the problems of the accessibility with regard to cellulose *per se,* enzymic hydrolysis of various agricultural substrates is impeded by the presence of lignin. Studies were carried out with various substrates (straw, maize stalks, bran, brewery draff, beet pulp, jute, sisal, etc.) and various pretreatments, both chemical (cooking in the presence of lime, neutral sulphite, etc.) or mechanical (wet grinding, refining, etc.).

These treatments cause the fibres to swell in water and so increase their accessibility. Complete delignification is not required: all that is necessary is that the structure of lignin and its relationship with the cellulosic compounds be altered.

Cooking with lime-soda proved most effective for maize stalks, straw, sisal and jute. Neutral sulphite treatment was particularly beneficial for maize and beet pulps. Both treatments permitted hydrolysis yields in the region of 90% after 48-72 h saccharification but potentially useful results were obtained after as little as 8 h (Table 2). The results for maize, bran, draff and pulps are not improved by mechanical treatment in the form of wet grinding. This process does, however, enhance straw treatment with lime by about 50% (Paquot *et al.*, 1984a, b; Parajo *et al.*, 1988).

Several bark samples (*Pinus pinaster*) have been submitted to alkalis, organic solvents, oxidizers or steam explosion. Residues contain 38.8-72.3 % polysaccharide; enzymic hydrolysis yields vary from a few percent to 44% (Parajo *et al.*, 1988).

Table 2. Degree of hydrolysis of various substrates after 8 h saccharification (substrate conc.3% (w/v); 0.3 IU cellulase.ml^{-1}; 5 IU cellobiase.ml^{-1}).*

Type of pulp	Maize	Straw	Bran	Draff	Beet pulp	Jute	Sisal	Pine bark
Mechanical	16	15	21	17	9	-	-	-
Semi-chemical (lime soda)	79	61	45	46	5	67	63	-
Semi-chemical (neutral) sulphite)	84	60	67	66	87	-	-	-
Chemical	-	-	-	-	-	-	-	44

* Hydrolysis is given as a % relative to α-cellulose and pentosan contents

The results given in Table 2 are in the form of a percentage hydrolysis relative to α-cellulose and pentosan contents. They must be weighted in order to assess residual lignin content or the pretreatment losses. In this respect, the most efficient pretreatment is not always the most effective.

Generally speaking, if cooking yields are taken into account, the best overall degree of hydrolysis is obtained by cooking with lime-soda. The yield obtained by lime-soda cooking of straw, for example, is 82% (straw/liquor ratio = 1/12; liquor composition, 10% $Ca(OH)_2$ + 1% NaOH; 80°C, 5 h).

When cooking yields are considered, it appears that this type of pretreatment is too exhaustive and, hence, not suitable for brewery draff, beet pulps and bran. Mechanical treatment in the form of refining has proved to be a very useful aid to improving the rate and overall degree of hydrolysis. Fig. 1 shows that refining of straw with powdered lime improves the degree of hydrolysis by at least 10% and that the treatment is effective after the first few min of the refining process.

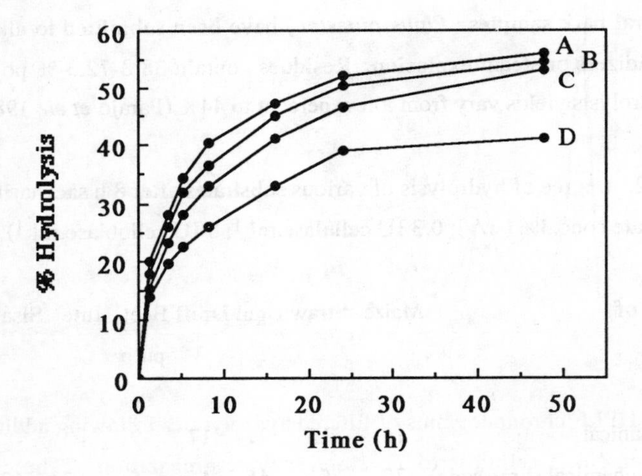

Fig. 1. Hydrolysis of straw/powdered lime pulp (conc. 3%, w/v) as a function of time and of refining treatment in the presence of cellulase (0.3 IU) and cellobiase (5 IU). A, 60 min refining; B, 15 min; C, 5 min; D, unrefined.

ADSORPTION OF THE ENZYME COMPONENTS ON THE SUBSTRATE

The enzymic hydrolysis of cellulose is not practical at present largely because of the high cost of cellulases and because of the relatively large amounts of enzyme needed for reasonable hydrolysis rates. Therefore, it is economically important that recycling of the cellulases be maximized.

The proportion of enzymes remaining adsorbed to the residual cellulosic substrate after hydrolysis varies from 40-90% depending on temperature, enzyme-substrate ratio and hydrolysis time (Desmons *et al.*, 1983; Paquot *et al.*, 1984a; Otter *et al.*, 1989).

The study of the composition of the enzyme complex described in the paper by Desmons *et al.*, (1983) made it possible to isolate, although not totally, various exocellulase, endocellulase, ß-glucosidase, ß-xylosidase and xylanase activities. Adsorption of the enzyme complex on various cellulosic substrates was monitored, in particular by HPLC (Fig. 2).

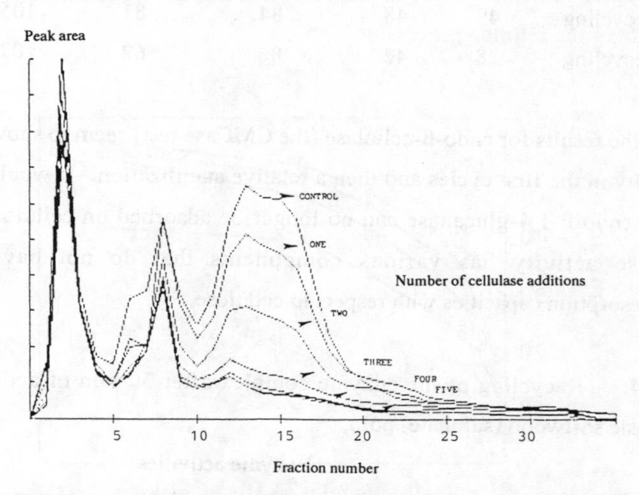

Fig. 2. HPLC chromatograms of filtered hydrolysates following additions of Whatman No. 1 cellulose to *T. reesei* enzyme preparation. The proteins were separated as described previously (Desmons *et al.*, 1983; Paquot *et al.*, 1984a).

The results in Tables 3 and 4 show that exo-ß-1,4-cellobiohydrolase and endo-ß-1,4-glucanase activities are sharply decreased when the enzyme system is recycled. These two enzymes are very rapidly adsorbed on the cellulosic substrate in the first minutes of the hydrolytic process. By contrast, there was little change in xylanase, ß-glucosidase and ß-xylosidase activities. The results for adsorption of the cellulolytic enzymes clearly demonstrate that exo-ß-1,4-cellobiohydrolase activity is the limiting factor in the design of a reactor that allows recycling of the enzyme system.

Table 3. Recycling of the enzyme complex after 30 min of action on a cellulosic hardwood (sulphate) pulp.

	Enzyme activities				
Sample	FPase	CMCase	ß-glucosidase	xylanase	ß-xylosidase
Commencement	100	100	100	100	100
1st recycling	90	70	93	102	107
2nd recycling	80	70	88	104	85
3rd recycling	45	63	96	65	105
4th recycling	49	48	84	81	105
5th recycling	28	42	86	62	107

In addition, the results for endo-ß-cellulase (the CMCase test) seem to show a very rapid loss of activity in the first cycles and then a relative stabilization. It would appear that some of the endo-ß-1,4-glucanase can no longer be adsorbed on cellulose or that the endocellulase activity has various components that do not have the same adsorption/desorption capacities with respect to cellulose.

Table 4. Recycling of the enzyme complex after 30 min of action on a cellulosic softwood (sulphate) pulp.

	Enzyme activities				
Sample	FPase	CMCase	ß-glucosidase	xylanase	ß-xylosidase
Commencement	100	100	100	100	100
1st recycling	81	74	96	97	100
2nd recycling	46	48	94	79	98
3rd recycling	19	41	91	69	96
4th recycling	8	25	95	60	96
5th recycling	1	26	86	52	91

Attempts to desorb the cellulases have included multiple washing with water, washing with dilute solutions of Tween 80, grinding the residual cellulose and washing with a wide range of chemical desorbents. Urea, guanidine-HCl, dimethylsulphoxide and n-propanol were all successful in eluting cellulase activity at pH 5.0 and 30°C, but the high concentrations needed (3M or more) made this approach impractical on a large-scale. The most promising technique is desorption by alkaline treatment. Carefully-controlled conditions of pH (10.0), low temperatures (0°C) and short time (10 min) enabled recovery of up to 60% of the original cellulase activity.

EFFECTS OF AGITATION ON CELLULASE ACTIVITY

Since cellulose is a water-insoluble solid, mixing must be provided in the enzymic hydrolysis process to promote contact between the enzyme and the substrate. When cellulases are exposed to agitation, deactivation is observed. However, shear forces alone are not as effective in causing enzyme inactivation as it is generally believed. Effects occurring in conjunction with high shear forces can evidently be important.

The existence of more complex conditions in industrial reactors means that further studies taking into account these additional factors are needed. An air-liquid interface is created in the 5-l reactor by the incorporation of air bubbles sucked into the liquid through a vortex. Two different methods were used to study the relative contributions of shear and air-liquid interface.

First, shear was applied to a solution of cellulase in a coaxial cylindrical viscosimeter at 48°C and shear rate of up to 2106 sec^{-1}. There was no loss of activity during 4 h of total shearing. Since shear alone does not affect enzyme activity, the air-liquid interface seems to be a major factor in cellulase inactivation.

To confirm this hypothesis, enzyme deactivation was assayed in the reactor while suppressing the incorporation of air. The reactor was then completely filled with the enzymic solution. In this case, no loss of activity was observed. Other reports indicate that shear does not cause inactivation (Thomas and Dunnill, 1979; Thomas *et al.*, 1979). Charm and Wong (1981) suggest that shear sometimes causes inactivation and that this process is enhanced by exposure to air.

Fig. 3 shows the results of the FPase inactivation observed when different impeller diameters and different speeds of agitation were used. It seems that the geometrical and working parameters that control the extent of the inactivation are: impeller diameter (d, m), impeller rotational speed (N, rev/sec), vessel diameter (D, m), time (t, sec). The mean circulation velocity (expressed by ln $(t.Nd^2.D^{-1})$ is correlated with losses of FPase activity as in eqn (i)[see also Fig. 4]:

$$\text{Loss of activity} = -32.34 + 6.53 \ln (Nd^2.t.D^{-1}) \qquad \text{(i)}$$

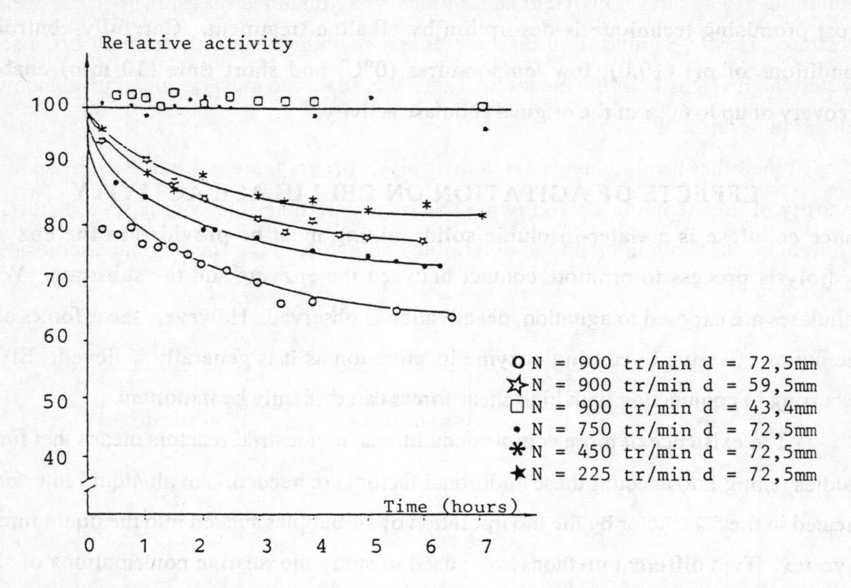

Fig. 3. FPase activity as a function of time with different impellers (d, m) and different impeller rotational speeds (N. rev/sec).

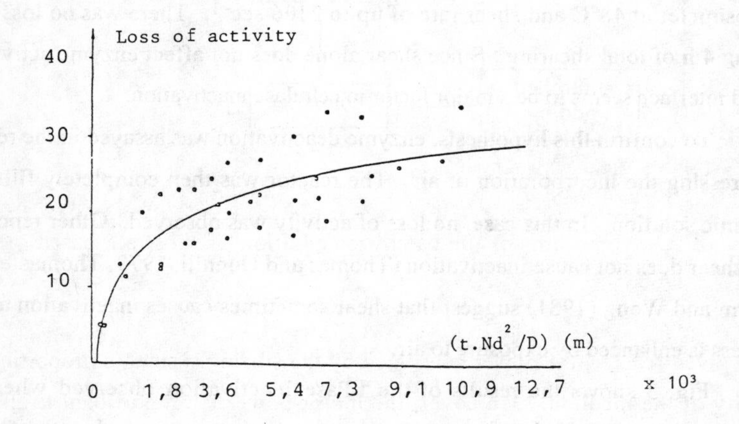

Fig. 4. Loss of activity (FPase) as a function of ln $(t.Nd^2.D^{-1})$.

INDIRECT SENSORS FOR ENZYME REACTOR

Substrate addition

In order that hydrolysis be efficient the contents, viz. enzyme and substrate, of the reactor must be thoroughly mixed. The hydrolysis yield must reach 70%; the resulting juice containing 7% sugars. This yield can, in theory, be obtained by using 10% dry weight substrate. However, in scaling-up the process, stirring becomes a problem because of the very high viscosity of the medium. Therefore, substrate must be added in a batchwise manner rather than all at once.

On the other hand, when enzymic hydrolysis occurs there is a quick decrease in the viscosity of the medium. By following this latter parameter, i.e. viscosity, it is possible to automate substrate input and to determine continuously the rheological conditions of the reactor.

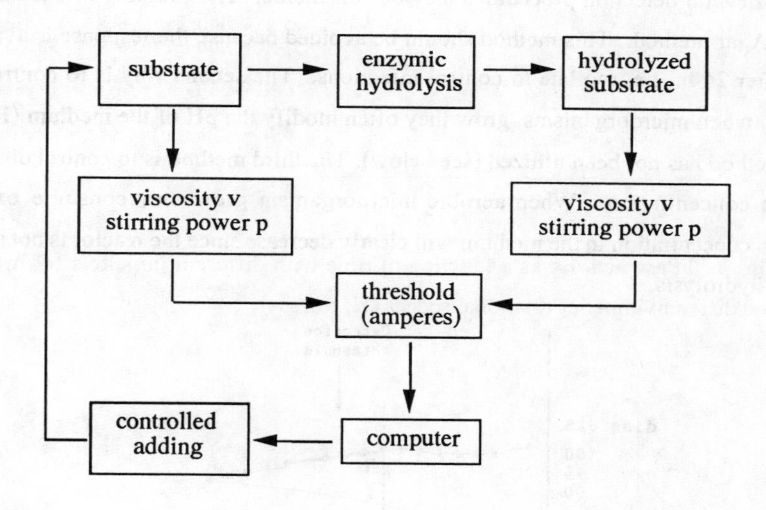

Fig. 5. Algorithm for the batchwise addition of substrate based on the

For a given reactor, the power consumed by the stirring system is proportional to the fluidity of the medium. Moreover, the motor power is proportional to its electrical alimentation. Therefore, we can correlate the amperage and the mixing properties. The addition of substrate is carried out between two viscosity thresholds: the maximum corresponds to the concentration at which the stirring system is choked; the minimum is the lowest tolerated concentration of substrate that allows of an efficient hydrolysis. These two thresholds have been correlated with two amperage levels; with a computer control, we determined when it was necessary to add substrate (Fig. 5).

Microbial contamination

During enzymic hydrolysis of cellulose the concentrations of low molecular weight sugars, viz. glucose and cellobiose, increase. However, if microbial contamination occurs, sugars are consumed and the saccharification yield decreases drastically. In order to avoid these contaminations, there are two possible solutions. First, knowing that contaminations are primarily caused by the substrate flora, it is possible to sterilize substrate before use. However, this is a very expensive solution. Secondly, it is possible to suppress infections by adding anti-microbial substances. Accordingly, it is necessary to act as soon as possible, i.e. to detect contamination early during the process.

Several detection procedures may be considered. The classical way is the Plate Count Agar method. This method should be avoided because the response is available only after 24 h., i.e. too late to control infections. The second way is to control pH. Indeed, when microorganisms grow they often modify the pH of the medium (Fig. 6). This method has not been utilized (see below). The third method is to control dissolved oxygen concentration. When aerobic microorganism grow they consume oxygen. Oxygen concentration in the medium will clearly decrease since the reactor is not aerated during hydrolysis.

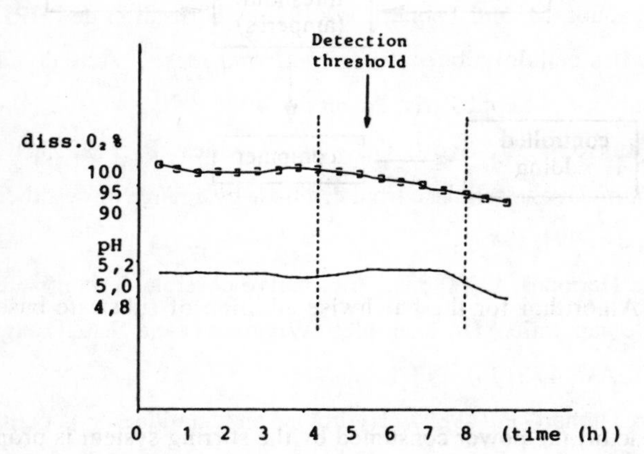

Fig. 6. Alteration in pH and oxygen concentration after microbial contamination of a cellulose hydrolysis reactor.

We chose this third method to detect contamination. Like the second one, it is direct and easy, but its response time is shorter (Fig. 6). Indeed, at the beginning of the contamination, there is a decrease in the amount of dissolved oxygen while pH varies only if contamination is more extensive. Moreover, the buffer strength of the medium has to be low if that method of detection is to be used.

CONCLUSION

As stated earlier, the objective of this publication was to describe a procedure for monitoring reactor performance rather than modelling the process. Hydrolysis of cellulose depends on the accessibility of the substrate; this property is correlated with porosity, specific surface area, degree of hydration, crystallinity and lignin/carbohydrate interactions. To recycle the enzymes, it is necessary to understand the mechanisms of adsorption/desorption and of enzyme denaturation during the process. In order to achieve high and reproducible hydrolysis yields, we propose that the batchwise addition of substrate addition be regulated by the viscosity threshold method described above and that microbial contamination be monitored by measurement of oxygen consumption.

REFERENCES

Charm, S.E. and Wong, B.L. (1981). Shear effects on enzymes. Enzyme Microb. Technol. **3**, 111-118.

Desmons, P., Paquot, M. and Thonart, P. (1983). Utilisation de l'HPLC pour séparer les activités cellulolytiques de *Trichoderma reesei*. Acte du colloque Pasteur Biosciences, 6-9 Sept., Paris, Poster no. 49.

Otter, P.E., Munro, P.A., Scott, G.K. and Geddes, R. (1989). Desorption of *Trichoderma reesei* cellulase from cellulose by a range of desorbents. Biotechnol. Bioeng. **34**, 291-298.

Paquot, M. and Hermans, L. (1983). Alternative possible de la mise en décharge des boues de papeterie. Un exemple: Wiggins Teape S.A. (Belgium). Tribune CEBEDEAU, **473**(36), 147-155.

Paquot, M. and Thonart, P. (1982). Hydrolyse enzymatique de la cellulose régénérée. Holzforschung, **36**, 177-181.

Paquot, M., Foucart, M., Desmons, P. and Thonart, P. (1983). Conversion des dechets agricoles et industriels pour l'hydrolyse de la cellulose. In *Production and*

Feeding of Single-Cell Protein, eds. M.P. Ferranti and A. Fiechter, Elsevier Applied Science, London, pp. 115-117.

Paquot, M., Foucart, M., Desmons, P. and Thonart, P. (1984a). Etude comparative de l'hydrolyse enzymatique et de l'hydrolyse par voie acide de la cellulose. Partie II: Morphologie du substrat au cours de l'hydrolyse acide. Holzforschung, **38**, 185-190.

Paquot, M., Thonart, P., Foucart, M., Desmons, P. and Mottet, A. (1984). Improvement of pretreatments and technologies for enzymatic hydrolysis of cellulose from industrial and agricultural refuse and comparison with acid hydrolysis. In *Anaerobic Digestion and Carbohydrate Hydrolysis of Waste,* eds. G.L. Ferrero, M.P. Ferranti and H. Naveau, Elsevier Applied Science, London, pp. 112-124.

Parajo, J.C., Paquot, M., Foucart, M., van Rolleghem, I., Vasquez, G. and Thonart, P. (1988). Hydrolyse enzymatique de residus d'extraction alcaline d'ecorces de *Pinus pinaster.* Belgian J. Food Chem. Biotechnol. **43**, No. 2, pp. 51-57.

Thomas, C.R. and Dunnill, P. (1979). Action of shear on enzymes: studies with catalase and urease. Biotechnol. Bioeng. **21**, 2279-2302.

Thomas, C.R., Nienow, A.W. and Dunnill, P. (1979) Action of shear on enzymes: studies with alcohol dehydrogenase. Biotechnol. Bioeng. **21**, 2263-2278.

Thonart, P., Marcoen, J.M., Desmons, P., Foucart, M. and Paquot, M. (1983). Etude comparative de l'hydrolyse enzymatique et de l'hydrolyse par voie acide de la cellulose. Partie I: Morphologie du substrat en cours d'hydrolyse enzymatique. Holzforschung, **37**, 173-178.

Weimer, P.J. and Wayde, M.W. (1985). Relationship between the fine structure of native cellulose and its degradability by the cellulase complexes of *Trichoderma reesei* and *Clostridium thermocellum.* Biotechnol. Bioeng. **27**, 1540-1547.

CHAIRMAN'S REPORT ON SESSION III

Bioconversion of lignocellulosic materials in submerged and solid-state cultivation and in reactors.

Tuohy and colleagues presented the results of their studies on the production of polysaccharide-degrading enzyme systems - including those active against cellulose, hemicellulose and pectin - by the thermophilic aerobic fungus, *Talaromyces emersonii.* In particular, they compared enzyme productivity by the organism when grown on various straws and pulps by liquid- and solid-state procedures. The effects of pH control and time of harvesting on enzyme yields, types of enzyme found and their thermostability were also examined.

Fernandez and coworkers discussed their investigations on upgrading - by enrichment with protein - sunflower seed shell, considerable quantities of which are generated annually in Spain. *Trichosporon penicillatum* proved to be most suitable of the several species of fungus examined for upgrading by solid-state fermentation. It provided the greatest yield of protein in a relatively short cultivation period, and used all of the glucose and about half of the xylose and fat present in the shell during the fermentation process. The data obtained will be of use in the design of an industrial-scale process for the purpose in question.

Collaço and her colleagues at LNETI studied the extent of degradation of a variety of lignocellulosic residues, and the yield of protein obtained, when cultivated with several different fungi and mixtures thereof. Liquid, semi-solid and solid-state cultivation procedures were compared and the use of enzymes for the degradation of lignocellulosic materials was also examined. The ultimate aim of their studies is to transfer this small-scale technology to farm-scale processes and to develop the necessary skills on site.

Teixeira Rodeia described the ability of a large number of species and strains from the order Aphyllophorales to act on substrates relevant to the hydrolysis of lignin, cellulose and hemicellulose. Since these species have, in the main, been collected from their natural habitats, trees, timbers, dead trunks and mine supports, it is hoped that, among the species examined, some may be of use in the applied bioconversion of plant materials.

The final paper by Thonart and colleagues described some of their experiences with the use of enzyme reactors for cellulose hydrolysis. Account was taken in these studies of the various parameters that affect the operational efficiency and stability of the enzymes involved and procedures for their recovery or re-use were investigated. However, much of their presentation was concerned with the measurements needed to monitor the progress of the saccharification reactions and on procedures for detecting trouble during such reactions.

M.P. Coughlan

for M.T. Amaral Collaço

SESSION IV

Bioconversion of lignocellulosic materials *in vivo*

MODELLING *IN VITRO* AND *IN VIVO* RUMEN PROCESSES

J. France, R. C. Siddons, M. K. Theodorou, D. E. Beever and P. J. van Soest *

AFRC Institute for Grassland and Animal Production
Hurley, Maidenhead, Berks SL6 5LR, UK
and
*Department of Animal Science
Cornell University, Ithaca, New York, USA

The elementary principles of mechanistic modelling are described and illustrated in relation to estimation of fungal biomass, volatile fatty acid production, the rate and extent of digestion, and whole-rumen function. We suggest that modelling could play a key role in understanding the mechanisms of rumen processes at various levels of organization by interpreting kinetic data, subjecting current concepts to critical scrutiny and suggesting further avenues of experimentation.

INTRODUCTION

The reticulo-rumen and omasum form a gastrointestinal unit that has evolved uniquely to process and extract nutrients from fibrous feeds. The evolutionary adaptation of ruminants has been to retain slowly-digesting fractions for maximal extraction of available nutrients. This has allowed ruminant herbivores to compete with larger non-ruminant herbivores since Miocene times (van Soest, 1982). Grazed and conserved forage and other fibrous feeds are heavily relied upon in most ruminant livestock production systems. Both the level of forage voluntarily consumed by the animal and the availability of absorbed nutrients for production are significantly influenced by processes occurring in the reticulo-rumen. Thus, the importance of the rumen ecosystem has been recognized for some time (Hungate, 1966). Research on *in vivo* and *in vitro* digestion of ingested nutrients within the rumen and passage of undigested residues from the rumen, on the metabolism of rumen microbes, and on production of microbial biomass and volatile fatty acids have all contributed significantly to our knowledge of rumen processes.

As our understanding of quantitative aspects of the rumen ecosystem has improved, the use of mechanistic modelling as a research tool has increased. Such modelling is now used routinely in analyzing and interpreting experimental data that describe aspects of a particular process or system (e.g. Grovum & Williams, 1973; Ørskov & McDonald, 1979). It has also been used, albeit at a more complex level, to simulate whole systems so as to evaluate the adequacy of current understanding, and where this is found to be inadequate, modelling may help to identify critical experiments (e.g. Mazanov & Nolan, 1976; Black *et al.*, 1981). This paper describes and illustrates the principles of dynamic, deterministic and mechanistic modelling in relation to rumen processes. Four specific areas are considered, viz. estimation of fungal biomass, volatile fatty acid production, rate and extent of digestion, and whole-rumen function.

PRINCIPLES OF DYNAMIC, DETERMINISTIC AND MECHANISTIC MODELLING

Dynamic models have time as the independent variable in their mathematical formulation. They are needed for studying non-steady-state behaviour. This generally gives more information about a system and provides a more stringent test of underlying concepts than steady-state analysis alone. Deterministic models give exact predictions for quantities, without any associated probability distributions. Mechanistic models seek to understand causation. A mechanistic model is constucted by looking at the structure of the system under investigation, dividing it into its key components, and analyzing the behaviour of the whole system in terms of its indivdual components and their interactions with one another. Thus, it is the connections that interrelate the components that make a model mechanistic. Mechanistic modelling follows the traditional reductionist method of the physical and chemical sciences.

The mathematically-standard way of representing dynamic, deterministic and mechanistic models is called the rate:state formalism. The system under investigation is defined at time t by q state variables: $X_1, X_2, ..., X_q$. These variables represent properties or attributes of the system, such as the weight of microbial mass, the quantity of substrate, etc.. The model then comprises q first order differential equations that describe how the state variables change with time:

$$dX_i/dt = f_i(X_1, X_2, ..., X_q; P); \; i = 1, 2, ..., q \qquad (i)$$

where P denotes a set of parameters, and the function f_i gives the rate of change of the state variable X_i. The function f_i comprises terms that represent the rates of processes (with dimensions of state variables per unit time), and these can all be calculated from the values of the state variable alone, with of course the values of any parameters and constants. The rate:state equations (equations i), are not as restrictive as might first be thought, as any higher order differential equation can be written as, and many partial differential equations well approximated by, a series of first-order differential equations. In this type of mathematical modelling, the rate:state equations are formed through direct application of the laws of science (e.g. the law of mass conservation, the first law of thermodynamics, Newton's second law of motion), or by application of a continuity equation derived from more fundamental scientific laws (e.g. Bernoulli's equation in hydrodynamics). Rate:state equations are central to the physical sciences. For example, Schrodinger's equations are the rate:state equations for quantum mechanics. The rate:state equations can sometimes be solved analytically. However, most models are too complex, and only numerical solutions can be obtained. This can be conveniently achieved by using one of the many computer software packages available for tackling such problems. Further discussion of the principles of mechanistic modelling can be found in France & Thornley (1984).

ESTIMATING FUNGAL BIOMASS

Until recently, it was assumed that the degradation and fermentation of plant biomass in the rumen was mediated only by populations of anaerobic bacteria and protozoa. However, evidence accumulated since the mid-1970s leaves little doubt as to the involvement of obligate anaerobic fungi in rumen cellulolysis (Orpin, 1975; Lowe *et al.*, 1987b). These unique microorganisms, collectively known as the anaerobic or rumen fungi, are highly cellulolytic, zoospore-producing fungi belonging to the class Chytridiomycetes (chytrids). Along with their aerobic counterparts, the chytrids are generally regarded as being among the most primitive of eucaryotic organisms (Barr, 1983).

The existence of cellulolytic fungi in the rumen has led to a general reappraisal of the accepted function of particle-associated microorganisms in the digestive tract ecosystem. The precise role and overall contribution of anaerobic fungi to the degradation of plant biomass in the rumen is unknown. This is so largely because of the difficulty of measuring the amount and activity of fungal biomass associated with particulate substrates in a mixed-population ecosystem. However, with fibrous diets, a substantial proportion of the plant fragments that enter the rumen are rapidly and extensively colonized by large populations of rumen fungi (Bauchop, 1979). Thus, the rumen fungi may participate in the initial colonization of plant cell-walls and assist in ruminal cellulolysis by increasing the accessibility of particulate substrates to invasion by other rumen microorganisms (Bauchop, 1979; Theodorou *et al.*, 1988).

The morphologies and life cycles of the rumen fungi are very similar to those described for aerobic Chytridiomycetes. Apart from one or two polycentric forms (Borneman *et al.*, 1989), the majority of rumen fungi have a classical monocentric type of life cycle, in which motile zoospores alternate with particle-associated fungal thalli. The thallus dies when the zoosporangium undergoes endogenous development and zoosporogenesis (Lowe *et al.*, 1987a). For example, the life cycle of *Neocallimastix* sp., strain R1, lasts about 30 h at 39°C and culminates in the release of an average of 88 zoospores per zoosporangium (Fig. 1; Lowe *et al.*, 1987a).

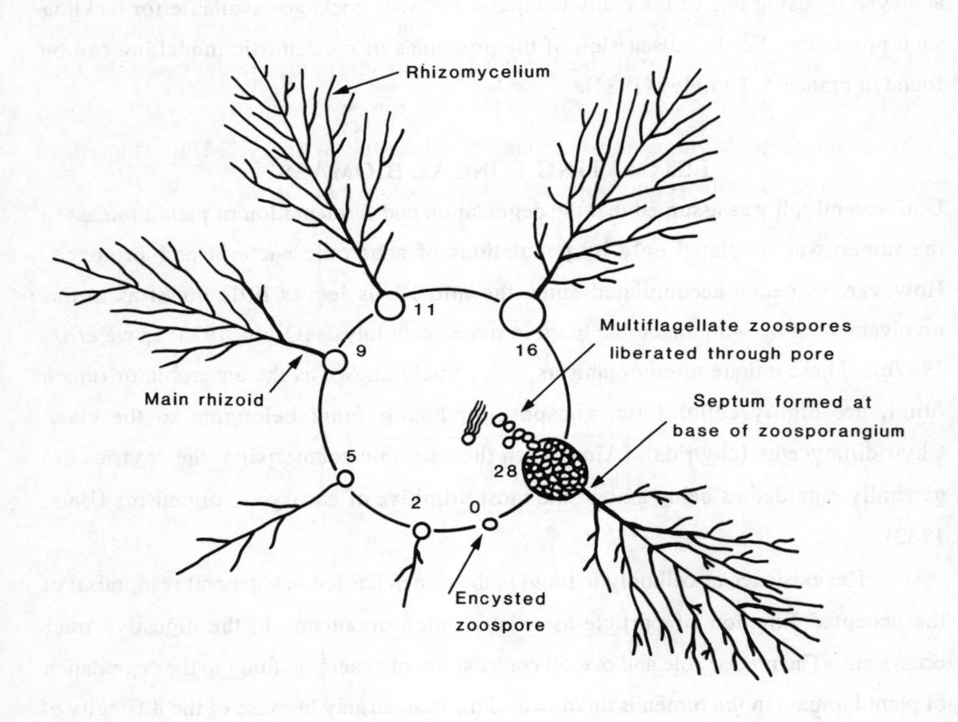

Fig. 1. Diagrammatic representation of the life cycle of a monocentric rumen fungus (*Neocallimastix* sp., strain R1). The numbers represent hours after encystment of the zoospore. Constructed from Lowe *et al.* (1987a).

As a consequence of their finite life span, these monocentric fungi require the presence of both zoospores and fungal thalli in the rumen for the continued production of biomass. Thus, the combined attributes of ingestion of nutrients, absorption of fermentation end-products and passage of digesta contents in the rumen are essential for the survival of monocentric fungi. The open (continuous culture) nature of the rumen serves to prevent the development of adverse physiological conditions and ensures a relatively constant (steady-state) environment. In the absence of substantial environmental perturbation, populations of zoospores and thalli are able to co-exist indefinitely in a state of dynamic equilibrium. Although the population density of zoospores in rumen fluid can be determined by light microscopy or roll-tube techniques (Orpin, 1977; Joblin, 1981), that of the particle-associated fungal thalli remains unknown. However, the size of each population is likely to be dependent on a variety of factors, including substrate type and availability, the respective swimming and encystment times of zoospore populations, the maturation time of fungal thalli, the yield of zoospores from mature thalli and the flow properties of rumen fluid and solid digesta from the rumen. From a knowledge of some of these factors, the diagrammatic representation of the fungal life cycle (Fig. 1) has been converted to an equivalent compartmental scheme as shown in Fig. 2.

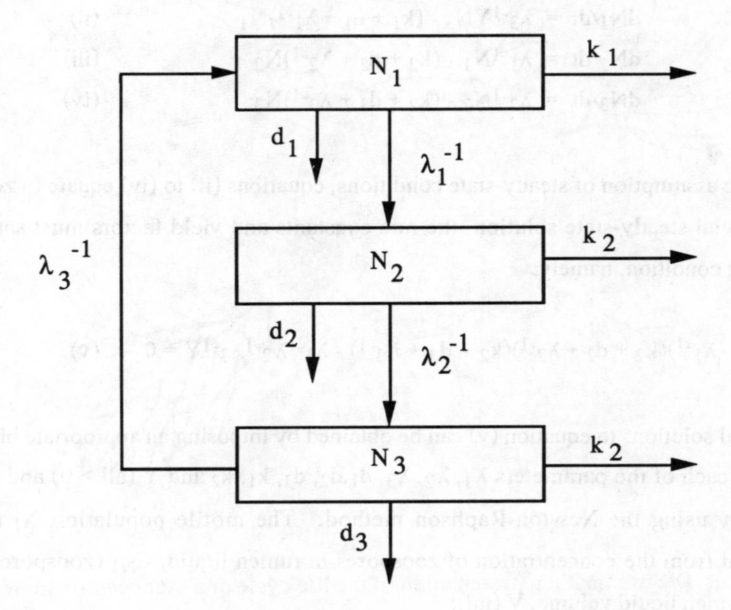

Fig. 2. Compartmental scheme for estimating the biomass of anaerobic fungi in the rumen.

For the purpose of this model, the fungal population, N, is considered in terms of 3 mutually-exclusive sub-populations, N_1, N_2, and N_3, where N_1 denotes the number of free-swimming zoospores, N_2 the number of particle-associated (attached) thalli undergoing vegetative growth, and N_3 the number of particle-associated mature thalli between septation and zoosporogenesis. It is assumed that the time spent in each phase of the life cycle (by those individuals to enter the next phase) follows an exponential distribution with respective known means λ_1, λ_2 and λ_3 (h). On completion of the life cycle at zoosporogenesis, Y motile zoospores are released back into the liquid phase. In Fig. 2, the state variables N_1, N_2, and N_3 are depicted as compartments (boxes) in the rumen, and the fluxes between compartments and out of the system are shown as arrowed lines. To effect a solution, all passage is taken to be a first-order process with a rate constant k_1 for liquid and k_2 for particulate matter (both per h). Similarly, death in each phase (N_1, N_2, and N_3) is assumed to be first-order with respective rate constants d_1, d_2, and d_3 (all per h). The rate constant pertaining to each flux is shown against the corresponding arrowed line. Applying the principle of mass conservation, the differential equations describing the dynamics of the system are:

$$dN_1/dt = \lambda_3^{-1}YN_3 - (k_1 + d_1 + \lambda_1^{-1})N_1 \quad \text{(ii)}$$
$$dN_2/dt = \lambda_1^{-1}N_1 - (k_2 + d_2 + \lambda_2^{-1})N_2 \quad \text{(iii)}$$
$$dN_3/dt = \lambda_2^{-1}N_2 - (k_2 + d_3 + \lambda_3^{-1})N_3 \quad \text{(iv)}$$

Under the assumption of steady-state conditions, equations (ii) to (iv) equate to zero. For a non-trivial steady-state solution, the rate constants and yield factors must satisfy the following condition, namely:

$$(k_1 + d_1 + \lambda_1^{-1})(k_2 + d_2 + \lambda_2^{-1})(k_2 + d_3 + \lambda_3^{-1}) - \lambda_1^{-1}\lambda_2^{-1}\lambda_3^{-1}Y = 0 \quad \text{(v)}$$

Numerical solutions to equation (v) can be obtained by imposing an appropriate biological range on each of the parameters λ_1, λ_2, λ_3, d_1, d_2, d_3, k_1, k_2 and Y (all > 0) and solving iteratively using the Newton-Raphson method. The motile population N_1 may be calculated from the concentration of zoospores in rumen liquid, C_{N1} (zoospores.ml^{-1}), and the rumen liquid volume, V (ml):

$$N_1 = C_{N1}V \quad \text{(vi)}$$

Having determined appropriate values for the parameters λ_1, λ_2, λ_3, d_1, d_2, d_3, k_1, k_2 and Y by solving equation (v), the particle-associated thallus populations N_2 and N_3 can then be evaluated using equations (ii) and (iii) as follows:

$$N_2 = \lambda_1^{-1} N_1 / (k_2 + d_2 + \lambda_2^{-1}) \quad \text{(vii)}$$

$$N_3 = (k_1 + d_1 + \lambda_1^{-1}) N_1 / (\lambda_3^{-1} Y) \quad \text{(viii)}$$

The total count of thallus-forming units (TFUs, viz. zoospores and particle-associated thalli) in the rumen is given by the sum:

$$N = N_1 + N_2 + N_3 \quad \text{(ix)}$$

Therefore, the total population of TFUs in the rumen, including the number of particle-associated thalli, can be estimated by solving non-linear equations (v) to (ix) numerically, subject to appropriate bounds. A TFU may be defined as a zoospore, or a collection of zoospores (inside a zoosporangium), that has the ability to produce a fungal thallus in culture. From a knowledge of the biomass associated with individual zoospores and fungal thalli, the total amount of fungal biomass in the rumen can be determined.

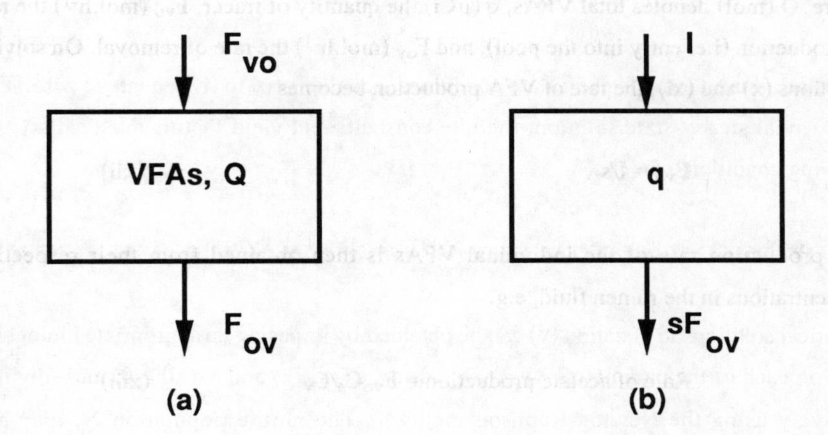

Fig. 3. Single-compartment model for estimating VFA production: (a) tracee, (b) tracer.

VOLATILE FATTY ACID PRODUCTION

Volatile fatty acids (VFAs), predominantly acetate, propionate and butyrate, produced by microbial fermentation of cellulose and other organic matter in the rumen, are the major sources of absorbed energy in ruminant animals. Quantitative estimates of their rates of production in and removal from the rumen can be obtained using compartmental models to interpret the results of isotope dilution studies. A relatively simple approach, that assumes steady-state conditions as imposed by continuous feeding, was proposed by Weller *et al.* (1967). In this approach, total VFAs are considered to behave as a homogenous pool and can, therefore, be represented as a single-compartment model (Fig. 3).

The isotopic form of any one of the individual VFAs, or a mixture of the VFAs, is infused into the rumen at a constant rate, I (μCi.h^{-1}), and the plateau specific activity of the total VFAs, s (μCi.mol^{-1}), is subsequently determined from the isotope concentration (μCi.ml^{-1}) and total VFA concentration (mol.ml^{-1}) in rumen fluid. The rate:state equations, based on mass conservation principles, for this steady-state scheme are

$$dQ/dt = F_{vo} - F_{ov} = 0 \qquad \text{(x)}$$

$$dq/dt = I - sF_{ov} = 0 \qquad \text{(xi)}$$

where Q (mol) denotes total VFAs, q (μCi) the quantity of tracer, F_{vo} (mol.h^{-1}) the rate of production (i.e. entry into the pool), and F_{ov} (mol.h^{-1}) the rate of removal. On solving equations (x) and (xi), the rate of VFA production becomes

$$F_{vo} = I/s \qquad \text{(xii)}$$

The production rate of the individual VFAs is then obtained from their respective concentrations in the rumen fluid, e.g.

$$\text{Rate of acetate production} = F_{vo}C_a/C_v \qquad \text{(xiii)}$$

where C_a and C_v (both mol.ml^{-1}) are the concentrations of acetate and total VFAs, respectively.

Weller's method can be adapted for single-dose injection of tracer, rather than constant infusion. Equation (xi) reduces to

$$dq/dt = -sF_{ov} \quad \text{(xiv)}$$

were s is now the instantaneous specific activity. Assuming the rate of VFA removal is first order with rate constant k (.h^{-1}) (i.e. $F_{ov} = kQ$), equation (xiv) simplifies to

$$dq/dt = -kq \quad \text{(xv)}$$

which on integration and division by (constant) Q yields

$$s = (D/Q)e^{-kt} \quad \text{(xvi)}$$

where D (μCi) is the dose injected at time zero. Integration of equation (xvi) gives

$$A = {}_0\int^{\infty} s dt = D/(kQ) \quad \text{(xvii)}$$

where A denotes the area under the VFA specific activity-time curve. As the rate of removal equals that of production in steady state, then

$$F_{vo} = D/A \quad \text{(xviii)}$$

i.e. the rate of VFA production equals dose over area under the specific activity-time curve.

The use of Weller's method is advantageous in that only one infusion (or single injection) experiment need be undertaken and the specific activities of the individual VFAs do not have to be determined. However, it is dependent on the production rate of the acids being proportionally the same as their concentrations in rumen fluid and this may not always be so.

An alternative method for estimating VFA production rates in steady state, that is not dependent on the proportionality between VFA production and concentration and that also provides a more detailed description of VFA metabolism, is to use interchanging compartmental models to interpret isotopic tracer data. The models may be complete [i.e. exchange between all compartments (plus the external environment) included] or incomplete (i.e. exchange between some compartments excluded). Tracer is administered

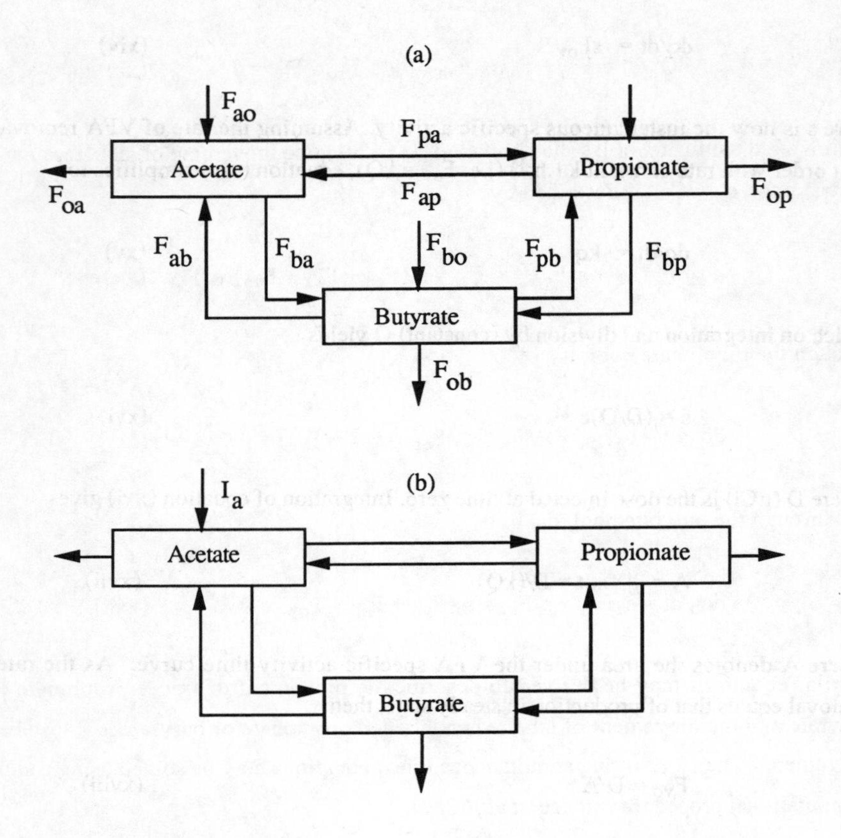

Fig. 4. Fully interchanging 3-compartment model for acetate, propionate and butyrate production: (a) tracee, (b) tracer. The scheme assumes no re-entry of label into the rumen.

into each compartment in turn and on each occasion the specific activities of all compartments are determined. A unique solution to the model is obtained by deriving a series of n simultaneous equations (where n is the number of fluxes included in the model) to describe the movement of tracer and tracee between compartments. Consider the fully-interchanging 3-compartment model for acetate, propionate and butyrate (Fig. 4). Under steady-state conditions, the isotopic form of each VFA is infused into the rumen at a constant rate and, for each infusion, the plateau specific activity (μCi.mol^{-1}) of acetate (s_a), propionate (s_p) and butyrate (s_b) is determined. Since the system is in steady state, the rate:state equations are as follows: The movement of tracee acetate, Q_a (mol), is described by

$$dQ_a/dt = F_{ao} + F_{ap} + F_{ab} - F_{oa} - F_{pa} - F_{ba} = 0 \quad \text{(xix)}$$

Following the infusion of labelled acetate, I_a (μCi.h^{-1}), the movement of label through the acetate pool, q_a (μCi), is described by

$$dq_a/dt = I_a + s_pF_{ap} + s_bF_{ab} - s_a(F_{oa} + F_{pa} + F_{ba}) = 0 \quad \text{(xx)}$$

through the propionate pool, q_p, by

$$dq_p/dt = s_aF_{pa} + s_bF_{pb} - s_p(F_{op} + F_{ap} + F_{bp}) = 0 \quad \text{(xxi)}$$

and through the butyrate pool, q_b, by

$$dq_b/dt = s_aF_{ba} + s_pF_{bp} - s_b(F_{ob} + F_{ab} + F_{pb}) = 0 \quad \text{(xxii)}$$

Similar equations may be derived to describe the movement of tracee propionate and butyrate and the movement of label when labelled propionate or butyrate is infused into the rumen. The 12 resulting simultaneous linear equations may be solved using a simple computational procedure (France *et al.*, 1987).

The method can also be adapted for single-dose injection of tracer. The system is now in non-isotopic steady-state so the rate:state equations for labelled material are non-zero. Movement of label through the acetate pool following injection at time zero of a single dose of labelled acetate, D_a (μCi), is given by

$$dq_a/dt = s_pF_{ap} + s_bF_{ab} - s_a(F_{oa} + F_{pa} + F_{ba}) \quad \text{(xxiii)}$$

through the propionate pool by

$$dq_p/dt = s_aF_{pa} + s_bF_{pb} - s_p(F_{op} + F_{ap} + F_{bp}) \quad \text{(xxiv)}$$

and through the butyrate pool by

$$dq_b/dt = s_aF_{ba} + s_pF_{bp} - s_b(F_{ob} + F_{ab} + F_{pb}) \quad \text{(xxv)}$$

The s's now refer to instantaneous specific activities. Integrating these 3 equations with respect to time between the limits zero and infinity yields

$$-D_a = A_pF_{ap} + A_bF_{ab} - A_a(F_{oa} + F_{pa} + F_{ba}) \quad \text{(xxvi)}$$
$$0 = A_aF_{pa} + A_bF_{pb} - A_p(F_{op} + F_{ap} + F_{bp}) \quad \text{(xxvii)}$$
$$0 = A_aF_{ba} + A_pF_{bp} - A_b(F_{ob} + F_{ab} + F_{pb}) \quad \text{(xxviii)}$$

where A_a, A_p and A_b are the areas under the acetate, propionate and butyrate specific activity-time curves respectively (i.e. $A_a = {}_0\int^{\infty} s_a dt$, etc.). Equations such as (xxvi) to (xxviii) can be derived for the movement of label when labelled propionate and butyrate are injected into the rumen. The system of equations for single dose is, therefore, the same as for constant infusion, but with dose and area replacing infusion rate and plateau specific activity, respectively.

When the system is not in steady-state (i.e. with animals that are not continuously fed), the VFA pool size, Q (g carbon), and the production rate will vary with time. Under these conditions, the instantaneous production rate of the total VFA, F_{vo} (g carbon.h^{-1}), if it behaves as a single homogenous pool, is given by

$$F_{vo} = I/s + sQd(1/s)/dt \quad \text{(xxix)}$$

Equation (xxix) is derived using the rate:state equations for Weller's method in non steady state [i.e. from equations (x) and (xi) not equated to zero] and eliminating the flux F_{ov}, as shown by Morant *et al.* (1978).

The instantaneous production rate may be determined by varying the rate of isotope infusion in synchrony with the rate of VFA production so that the specific activity remains constant. Consequently the differential term in equation (xxix) is equal to zero. Gray *et al.* (1966) used this method to measure VFA production in sheep fed twice-daily but, since it is dependent on prior knowledge of the rate of VFA production, it is unlikely to be of general applicability.

An alternative procedure is to infuse the isotope at a constant rate and to monitor the pool size simultaneously by the administration of 2 liquid-phase markers [either one by continuous infusion, the other as a single dose (Morant *et al.*, 1978) or both as a single dose (France *et al.*, 1990)]. The differential term in equation (xxix) is given by the slope of the curve of inverse specific activity against time. This procedure could be used to determined the instantaneous fluxes in multi-compartment schemes (such as shown in Fig. 4) by including the measured rates of change in both the compartment size and the amount of tracer in the compartments in the simultaneous equations (xix) to (xxii).

RATE AND EXTENT OF DIGESTION

Studies on the rate and extent of ruminal digestion of basal forages and supplementary feeds have utilized enzymes or rumen microbes *in vitro* (e.g. Broderick, 1978; Siddons *et al.*, 1985), or polyester bags containing feed samples placed *in rumeno* (e.g. Ørskov and McDonald, 1979; McDonald, 1981). These systems are normally used to analyze the disappearance of a feed component such as cellulose, organic matter or nitrogen. They assume that substrate is limiting the rate of digestion. If substrate is not limiting, the system ceases to be an assay of the feed, but rather becomes a reflection of the inadequacies of the enzymes or microbes. The latter condition probably prevails in the early stages of digestion, termed the lag. When the substrate becomes saturated with attached microbes or secreted enzymes, the rate is considered to be influenced soley by the amount of substrate. This is assumed to be uniform.

These assumptions can be expressed mathematically as follows: Consider a quantity of a specific feed component (g) entering the rumen at time zero, and let the feed component be divided into 3 fractions: (i) an instantly-digestible fraction W; (ii) a potentially (i.e. more slowly) digestible fraction S; and (iii) an indigestible fraction U.

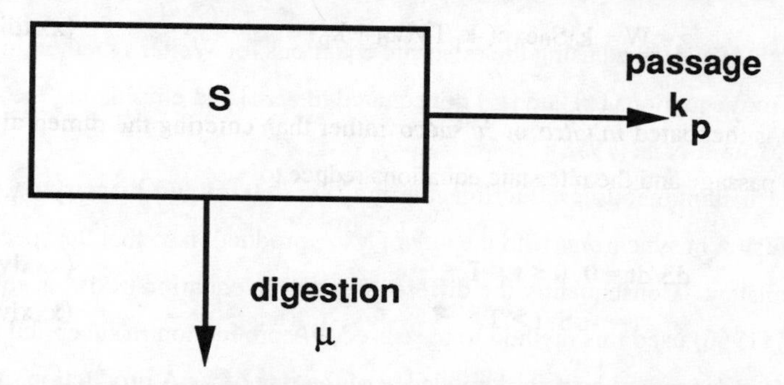

Fig. 5. Scheme for the disappearance from the rumen of the potentially-digestible fraction of a feed component.

A scheme for the kinetics of the S pool is shown in Fig. 5, where k_p ($.h^{-1}$) is the (constant) specific rate of passage from the rumen and μ ($.h^{-1}$) the specific rate of digestion. Applying the law of mass conservation gives the following rate:state equation for S

$$dS/dt = -k_pS, \; 0 \le t \le T \quad \text{(xxxa)}$$
$$= -\mu S - k_pS, \; t > T \quad \text{(xxxb)}$$

where T (h) denotes the lag before digestion commences. As digestion is assumed to be influenced solely by the amount of substrate, then the specific digestion rate is invariant and

$$\mu = k_d \quad \text{(xxxi)}$$

where kd ($.h^{-1}$) is a constant. Equations (xxxa and b) now integrate to give

$$S = S_0\exp(-k_pt), \; 0 \le t \le T \quad \text{(xxxiia)}$$
$$= S_0\exp(-k_pT)\exp[-(k_d + k_p)(t - T)], \; t > T \quad \text{(xxxiib)}$$

permitting determination of the extent of ruminal digestion, E, as

$$E = W + \int_T^{\infty} k_dSdt \quad \text{(xxxiiia)}$$
$$= W + k_dS_0\exp(-k_pT)/(k_d + k_p) \quad \text{(xxxiiib)}$$

If the feed is incubated *in vitro* or *in sacco* rather than entering the rumen directly, then there is no passage and the rate:state equations reduce to

$$dS/dt = 0, \; 0 \le t \le T \quad \text{(xxxiva)}$$
$$= -\mu S, \; t > T \quad \text{(xxxivb)}$$

Putting μ equal to k_d and integrating now gives

$$S = S_0, \; 0 \le t \le T \quad \text{(xxxva)}$$
$$= S_0\exp[-k_d(t-T)], \; t > T \quad \text{(xxxvb)}$$

permitting determination of cumulative disappearance, D, as

$$D = W + {}_T\int^t k_d S dt \qquad \text{(xxxvia)}$$
$$= W + S_0\{1 - \exp[-k_d(t - T)]\} \qquad \text{(xxxvib)}$$

Thus, fitting the non-linear model

$$y = a + b\{1 - \exp[-c(t - T)]\} \qquad \text{(xxxvii)}$$

to serial disappearance data allows direct estimation of W, S_0, k_d and T. Using these estimates, in conjunction with an estimate of k_p obtained from rumen sampling or faecal marker concentration data, permits evaluation of the extent of ruminal digestion from equation (xxxiiib).

The later stages of digestion of lignified material, which contains an indigestible core, may violate the above assumptions. Such material tends to ferment more slowly than would be expected from the amount of digestible substrate. The lignified core increasingly dominates the remaining substrate with time. Thus, digestion of lignified material illustrates a higher-order process in which both the quantity of digestible substrate and the concentration of lignin affect the kinetic rate. Unfortunately, mathematical difficulties are encountered in explicit representation (i.e. as a function of U) of even the simplest of inhibition concepts. However, it is possible to represent inhibition implicitly (i.e. as a function of time). Let the specific digestion rate be

$$\mu = k_d e^{-bt} \qquad \text{(xxxviii)}$$

where k_d and b (both $.h^{-1}$) are constants. On integration, the rate:state equation (xxx) for S now yields

$$S = S_0\exp(-k_p t),\ 0 \le t \le T \qquad \text{(xxxixa)}$$
$$= S_0\exp[k_d(e^{-bt} - e^{-bT})/b - k_p t],\ t > T \qquad \text{(xxxixb)}$$

and the extent of ruminal digestion (equation xxxiiia) becomes

$$E = W + k_d S_0\ {}_T\int^{\infty} \exp[k_d(e^{-bt} - e^{-bT})/b - (k_p + b)t]dt \qquad \text{(xxxx)}$$

If the feed is incubated, rate:state equations (xxxiva and b) apply and these now give the following solution for S, namely

$$S = S_0,\ 0 \le t \le T \quad \text{(xxxxia)}$$
$$= S_0 \exp[k_d (e^{-bt} - e^{-bT})/b],\ t > T \quad \text{(xxxxib)}$$

Cumulative disappearance is then

$$D = W + S_0 \{1 - \exp[k_d (e^{-bt} - e^{-bT})/b]\} \quad \text{(xxxxii)}$$

Thus, the parameters W, S_0, k_d, b and T may be estimated by fitting the Gompertz equation (xxxxii) to serial data describing cumulative disappearance. Having estimated these parameters and the passage rate k_p, the extent of ruminal digestion can then be found by solving equation (xxxx) numerically.

WHOLE-RUMEN FUNCTION

There have been several attempts to construct mechanistic models of whole-rumen function, largely initiated by Baldwin *et al.* (1970). All the models have endeavoured to represent the digestion of ingested nutrients, microbial metabolism and the yield of end-products of digestion, principally the volatile fatty acids and microbial biomass, but they have not necessarily shared common objectives. Some were designed to predict nutrient supply whilst others have attempted to reconcile current knowledge and suggest further experimentation. Generally, the predictive models have failed to simulate reality satisfactorily, but the research models have contributed significantly to the direction of subsequent research programmes.

Whilst not the first to be proposed, the model of Black *et al.* (1981) provides a comprehensive description of rumen function (Fig. 6). Although attempts to improve its mathematical representation have been made (France *et al.*, 1982), the problems involved in whole-rumen modelling can be highlighted through reference to the original proposal. Feed components were divided into eight major categories, with α hexose, β hexose and protein being further classified with respect to their potential digestibilities and maximal rates of digestion as influenced by the chemical and physical characteristics of the diet. Inorganic sulphur was included as an input because of its importance for wool-producing sheep in Australia, and due recognition was given to nitrogenous inputs arising from

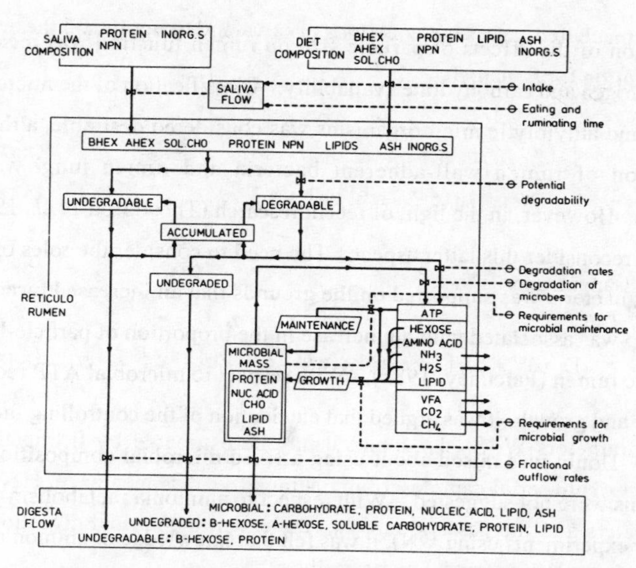

Fig. 6. Representation of whole-rumen function (Black *et al.*, 1981).

saliva production. No distinction between bacterial or protozoal metabolism was attempted and the composition of microbial biomass was assumed to be constant, although the use of ammonia and amino acids as substrates for microbial protein synthesis was included. Distinction between microbial demands for maintenance and growth was considered necessary on the basis of work by Isaacson *et al.* (1975), and an attempt was made to represent microbial recycling, with an equation for catabolism of part of the microbial population to provide fermentable energy for maintenance of the remaining population during periods of dietary-substrate deprivation.

Whilst the outputs were generally realistic, model representation was not totally adequate in certain areas. The model consistently under-predicted duodenal protein supply on low-nitrogen forages, which was attributed to an inadequate representation of microbial ATP maintenance and the factors controlling nitrogen recycling. Furthermore, sensitivity analysis by Beever *et al.* (1981) showed that contemporary knowledge of the factors affecting degradation rates within the rumen and fractional outflow rates from the rumen was insufficient to permit accurate representation of these processes.

The major issues relating to representation of nitrogen metabolism in models of whole-rumen function was subsequently addressed by Beever (1984). Improved characterization of the dietary inputs was stressed, and consideration was given to the spatial distribution of the nitrogen constituents in forage plants (e.g. leaves versus stems) and its bearing on their subsequent attack by microbes. The need for greater comprehension of the effects of particle size on rumen function was recognised, with respect to nitrogen and carbohydrate availability. Classification of the microbial pool into cellulolytic and amylolytic microorganisms was considered desirable, although explicit representation of rumen wall-adherent bacteria and rumen fungi was felt to be unnecessary. However, in the light of recent research (Theodorou *et al.*, 1988) it may be necessary to reconsider this latter aspect. The need to consider the roles of attached and free-floating bacteria was supported on the grounds that an increased turnover of rumen fluid in sheep was associated with an increase in the proportion of particle-bound bacteria present in the rumen (Faichney, 1985). With respect to microbial ATP requirements for maintenance and growth, it was argued that elucidation of the controlling factors was still inadequate. Doubt was raised about using a fixed microbial composition but specific improvements were not suggested. With respect to ammonia metabolism (in the light of recent tracer experiments using ^{15}N), it was felt that knowledge of ammonia kinetics was considerably enhanced and it was now appropriate for these concepts, to describe ammonia production, assimilation, absorption and outflow, to be incorporated into any major revision of rumen models.

Many of these inadequacies were addressed in the model (Baldwin *et al.*, 1987) of rumen function in dairy cows. The model has 16 inputs and 12 major state variables, plus a representation of the distribution of large and small particles and water. Depending upon the nature of the feedstuff, nutrients could enter either the large particle pool, a small particle pool, or water-soluble pool, with loss from the large particle pool occurring by rumination only. The model consists of a series of rate:state equations, derived from conservation principles, in which the fluxes are given either mass action or Michaelis-Menten forms. Both microbial growth requirements and the stoichiometric yields of rumen VFAs were represented more explicitly, but apart from a spatial association of microbes with large or small particles or water, no further refinement of the microbial population was included. When the model was validated against *in vivo* data not used in its construction (Sutton, 1985), the overall agreement between observation and simulatation was satisfactory for diets containing 60% cereal. This was particularly so with respect to amino acid, long chain fatty acid and glucose absorption, but whilst total VFA energy yields were comparable, the model over-predicted propionate and under-predicted butyrate supply. By contrast, for diets containing 90% cereal, acetate yields and total VFA yields were over-predicted and the results for amino acid supply were inconsistent. Also for the 90% maize cereal diet, the extent of starch digestion in the rumen, which is reduced on such diets, was over-predicted. Baldwin *et al.* (1987) concluded that such discrepancies may be related to lack of provision in the model for a depression in microbial activity at reduced pH.

Recently, Dijkstra *et al.* (1990) have constructed another model of rumen fermentation, at a level of aggregation similar to that used in early attempts. Four specific areas of improved representation are included, viz. microbial classification, microbial recycling, microbial passage, and chemical composition of microbes. As regards microbial classification, 2 bacterial groups were represented. Distinction was made on the basis of those (cellulolytic) utilizing hexose derived from structural carbohydrates and those (amylolytic) using hexose from non-structural carbohydrates such as soluble sugars, starch and pectin. To accomodate microbial recycling, equations representing the engulfment of bacteria by protozoa and the lysis of protozoa were included. Recognizing that the different microbial species occupy different niches within the rumen ecosystem, the cellulolytic bacteria were assumed to adhere to particulate material in the rumen and flow out at the same rate as the undigested dietary material. Amylotytic bacteria were assumed to be unattached (i.e. free-floating) and, hence, to flow out of the rumen with the liquid phase. However, on the basis of several reports suggesting selective retention of protozoa in the rumen (e.g. Leng, 1984), such organisms were only removed from the protozoal pool by death, with subsequent lysis and recycling of the released nutrients within the rumen. As regards variation in the chemical composition of rumen microbes, a large part of this can be accounted for by changes in their storage polysaccharide contents (Czerkawski, 1976). In order to accomodate this, suitable relationships were included in the model to allow the polysaccharide contents of amylolytic bacteria to vary in response to the availability of non-structural carbohydrate. Dijkstra *et al.* (1990) have yet to complete the evaluation of their model. However, sensitivity analysis has revealed a realistic relationship between protozoal and cellulolytic bacterial numbers, coupled with an overall decline in total microbial concentration at high protozoal numbers, suggesting increased bacterial engulfment resulting in a reduction in duodenal non-ammonia nitrogen supply.

REFERENCES

Baldwin, R. L., Lucas, H. L. and Cabrera, R. (1970). Energetic relationships in the formation and utilization of fermentation endproducts. In *Physiology of*

Digestion and Metabolism in the Ruminant, ed. A.T. Phillipson, Oriel Press, Newcastle-upon-Tyne, pp. 319-334.

Baldwin, R.L., Thornley, J.H.M. and Beever, D.E. (1987). Metabolism of the lactating cow. II. Digestive elements of a mechanistic model. J. Dairy Res. **54**, 107-131.

Barr, D.J.S. (1983). The zoosporic grouping of plant pathogens. In *Zoosporic Plant Pathogens,* ed. S.T. Buczacki, Academic Press, New York, pp. 161-192.

Bauchop, T. (1979). Rumen anaerobic fungi of cattle and sheep. Appl. Environ. Microbiol. **38**, 148-158.

Beever, D.E. (1984). Some problems of representing nitrogen metabolism in mathematical models of rumen function. In *Proceedings of the 2nd International Workshop on Modelling Ruminant Digestion and Metabolism,* ed. R.L. Baldwin and A.C. Bywater, University of California, Davis, pp. 54-58.

Beever, D.E., Black, J.L. and Faichney, G.J. (1981). Simulation of the effects of rumen function on the flow of nutrients from the stomach of sheep. Part 2. Assessment of computer predictions. Agric. Systems, **6**, 221-241.

Black, J.L., Beever, D.E., Faichney, G.J., Howarth, B.R. and Graham, N. McC. (1981). Simulation of the effects of rumen function on the flow of nutrients from the stomach of sheep. Part 1. Description of a computer program. Agric. Systems, **6**, 195-219.

Borneman, W.S., Akin, D.E. and Ljungdahl, L.G. (1989). Fermentation products and plant cell wall-degrading enzymes produced by monocentric and polycentric anaerobic ruminal fungi. Appl. Environ. Microbiol. **55**, 1066-1073.

Broderick, G.A. (1978). *In vitro* procedures for estimating rates of ruminal protein degradation and proportions of protein escaping the rumen undegraded. J. Nutr. **108**, 181-190.

Czerkawski, J.W. (1976). Chemical composition of microbial matter in the rumen. J. Sci. Food Agric. **27**, 621- 632.

Dijkstra, J., Neal, H.D.St. C., Gill, M., Beever, D.E. and France, J. (1990). Representation of microbial metabolism in a mathematical model of rumen fermentation. In *Procedings of the 3rd International Workshop on Modelling Digestion and Metabolism in Farm Animals,* eds. B. Robson and D.P. Poppi, Lincoln University Press, Canterbury, New Zealand, in press.

Faichney, G.J. (1985). The kinetics of particulate matter in the rumen. In *Control of Digestion and Metabolism in Ruminants,* eds. L.P. Milligan, W.L. Grovum and A. Dobson, Prentice-Hall, Englewood Cliffs, NJ, pp 173-195.

France, J., Gill, M., Dhanoa, M.S. and Siddons, R.C. (1987). On solving the fully-interchanging N-compartment model in steady-state tracer kinetic studies with reference to VFA absorption from the rumen. J. Theor. Biol. **125**, 193-211.

France, J. and Thornley, J.H.M. (1984). *Mathematical Models in Agriculture.* Butterworths, London.

France, J., Thornley, J.H.M. and Beever, D.E. (1982). A mathematical model of the rumen. J. Agric. Sci. (Cambridge), **99**, 343-353.

France, J., Thornley, J.H.M., Siddons, R.C. and Dhanoa, M.S. (1990). Determination of rumen volume in non-steady-state using digesta markers: a dynamic, deterministic, mechanistic approach. In *Procedings of the 3rd International Workshop on Modelling Digestion and Metabolism in Farm Animals,* eds. B. Robson and D.P. Poppi, Lincoln University Press, Canterbury, New Zealand, in press.

Gray, F.V., Weller, R.A., Pilgrim, A.F. and Jones, G.B. (1966). The rate of production of volatile fatty acids in the rumen. III. Measurement of production *in vivo* by two isotope dilution procedures. Australian J. Agric. Res. **17**, 69-80.

Grovum, W.L. and Williams, V.J. (1973). Rate of passage of digesta in sheep. 4. Passage of marker through the alimentary tract and the biological relevance of rate constants derived from changes in the concentration of marker in faeces. Brit. J. Nutr. **30**, 313-329.

Hungate, R.E. (1966). *The Rumen and its Microbes.* Academic Press, New York.

Isaacson, H.R., Hinds, F.C., Bryant, M.P. and Owens, F.N. (1975). Efficiency of energy utilization by mixed rumen bacteria in continuous culture. J. Dairy Sci. **58**, 1645-1659.

Joblin, K.N. (1981). Isolation, enumeration and maintenance of rumen anaerobic fungi in roll tubes. Appl. Environ. Microbiol. **42**, 1119-1122.

Leng, R.A. (1984). Microbial interactions in the rumen. In *Ruminant Physiology, Concepts and Consequences,* eds. S.K. Baker, J.M. Gauthorne, J.B. Macintosh and D.B. Purser, University of Western Australia, Nedlands, pp. 161-173.

Lowe, S.E., Griffith, G.W., Milne, A., Theodorou, M.K. and Trinci, A.P.J. (1987a). The life cycle and growth kinetics of an anaerobic rumen fungus. J. Gen. Microbiol. **133**, 1815-1827.

Lowe, S.E., Theodorou, M.K. and Trinci, A.P.J. (1987b). Cellulases and xylanase of an anaerobic rumen fungus grown on wheat straw, wheat straw holocellulose, cellulose, and xylan. Appl. Environ. Microbiol. **53**, 1216-1223.

Mazanov, A. and Nolan, J.V. (1976). Simulation of the dynamics of nitrogen metabolism in sheep. Brit. J. Nutr. **35**, 149-174.

McDonald, I. (1981). A revised model for the estimation of protein degradability in the rumen. J. Agric. Sci. (Cambridge), **96**, 251-252.

Morant, S.V., Ridley, J.L. and Sutton, J.D. (1978). A model for the estimation of volatile fatty acid production in the rumen in non-steady-state conditions. Brit. J. Nutr. **39**, 451-462.

Orpin, C.G. (1975). Studies on the rumen flagellate *Neocallimastix frontalis.* J. Gen. Microbiol. **91**, 249-262.

Orpin, C.G. (1977). Invasion of plant tissue in the rumen by the flagellate *Neocallimastix frontalis.* J. Gen. Microbiol. **98**, 423-430.

Ørskov, E.R. and McDonald, I. (1979). The estimation of protein degradability in the rumen from incubation measurements weighted according to rate of passage. J. Agric. Sci. (Cambridge), **92**, 499-503.

Siddons, R.C., Paradine, J., Gale, D.L. and Evans, R.T. (1985). Estimation of the degradability of dietary protein in the sheep rumen by *in vivo* and *in vitro* procedures. Brit. J. Nutr. **54**, 545-561.

Sutton, J.D. (1985). Digestion and absorption of energy substrates in the lactating cow. J. Dairy Sci. **68**, 3376-3393.

Theordorou, M.K., Lowe, S.E. and Trinci, A.P.J. (1988). Fermentative characteristics of anaerobic rumen fungi. BioSystems, **21**, 371-376.

van Soest, P.J. (1982). *Nutritional Ecology of the Ruminant.* O & B Books, Corvallis OR.

Weller, R.A., Gray, F.V., Pilgrim, A.F. and Jones, G.B. (1967). The rates of production of volatile fatty acids in the rumen. IV. Individual and total volatile fatty acids. Australian J. Agric. Res. **18**, 107-118.

DEGRADATION OF LIGNOCELLULOSIC FORAGES BY ANAEROBIC RUMEN FUNGI

G. Fonty[1,2], A. Bernalier[1] and Ph. Gouet[1]
[1]Laboratoire de Microbiologie
Centre de Recherche de Clermont-Ferrand Theix
63122 Saint Genès-Champanelle, France
and
[2]Laboratoire de Biologie comparée des Protistes
URA CNRS 138, Université Blaise Pascal
63170 Aubière, France

In vitro, anaerobic fungi are capable of solubilizing a high proportion of the dry weight of even the most highly-lignified tissues of plant fragments. They produce all of the enzymes necessary both for the depolymerization of cellulose and hemicelluloses in plant cell wall and for the hydrolysis of free oligosaccharides. Carbohydrates are fermented by anaerobic fungi via a mixed acid-type fermentation with the production of acetate, formate, lactate, ethanol, CO_2 and H_2. In the rumen, fungal populations are particularly abundant in animals fed high-fibre diets. *In vivo,* these fungi have been shown to colonize preferentially the lignocellulosic tissues of plant particles.

INTRODUCTION

Microbial digestion of cellulose and other plant fibre components is central to the utilization of natural diets by the ruminant animal. Until recently, the dense complex microbial population of the rumen was believed to comprise only bacteria and protozoa. More recently, strictly anaerobic fungi have been shown to form part of this microbiota (Orpin, 1975). These fungi have been found in the rumen of a wide range of ruminants (sheep, cattle, deer, goat, reindeer, etc.) and also in the gut of various hindgut-fermenters (horse, elephant, rhinoceros). The distribution of these anaerobic fungi appears to be limited to the gut of herbivores. Despite extensive sampling of terrestrial and aquatic environments no evidence has been obtained for the existence therein of anaerobic fungi of the type found in herbivores (Orpin and Joblin, 1988).

In the rumen, the anaerobic fungi colonize the lignocellulosic tissues of the plant particles. This close association of fungi with plant fibres suggests that these microorganisms may have the ability to degrade plant cell wall components. Confirmation of the degradation of cellulose and hemicellulose has been observed with pure cultures (see reviews by Citron *et al.*, 1987; Fonty *et al.*, 1988a; Orpin and Joblin, 1988).

The present paper will focus on the metabolism of these fungi and on their role in plant cell wall degradation *in vitro* and *in vivo*.

TAXONOMY AND CHARACTERISTICS OF RUMEN FUNGI

The life cycle of the rumen anaerobic fungi consists of an alternation between a zoospore stage and a vegetative stage carrying the fruiting body (sporangium) attached by a rhizoid to plant particles. The zoospores are actively motile. After fixation on plant fragments and germination they develop into the vegetative stage. Rhizoids invade plant cell walls to obtain fermentable carbohydrates and develop sporangia that, on maturation, release zoospores to establish another cycle. In the case of monocentric species, the rhizoidal system bears one sporangium whereas in the polycentric species several sporangia are associated with a rhizoid.

A variety of morphological types have been isolated but the taxonomy status of rumen fungi is not yet clear. Monocentric fungi can be grouped into the 3 morphological types originally described and named by Orpin: *Neocallimastix* sp. with a polyflagellated spore and a highly branched rhizoid (Orpin, 1975), *Piromonas* sp. with a monoflagellated spore and a branched rhizoid (Orpin, 1977) and *Sphaeromonas* sp. with a monoflagellated zoospore and a bulbous rhizoid (Orpin, 1976). Recently, Gold *et al.* (1988), Barr *et al.* (1989) proposed renaming the genera *Sphaeromonas* and *Piromonas* to *Caecomyces* and *Piromyces*, respectively. In the interests of continuity we have retained the existing nomenclature in this paper.

The recent discoveries of polycentric fungi in buffalo, cattle and in sheep have indicated that the range and diversity of fungi found in ruminants is greater than was apparent initially . On the basis of flagellum number, Breton *et al.* (1989) included a polycentric strain in the genus *Neocallimastix* (*N. joyonii*) while Barr *et al.* (1989), propose to include the polycentric fungi in a new genus, *Orpinomyces*.

PLANT CELL WALL DEGRADATION AND FERMENTATION OF CARBOHYDRATES BY RUMEN FUNGI *IN VITRO*

Plant cell wall degradation

In vitro, fungi are capable of solubilizing a high proportion of the dry weight of even the most highly-lignified tissues of plant fragments. A mixed fungal population can degrade up to 60% of plant material placed in incubation (Akin *et al.*, 1989). According to the origin of the plant material, the efficiency of degradation by the fungal population may be equal to or higher than that of the total microbial population of the rumen (Windham and Akin, 1984; Akin and Rigsby, 1987; Akin *et al.*, 1989). Theodorou *et al.* (1989) showed that a strain of *Neocallimastix* removed c. 53% (w/w) of the cell walls or c. 75% (w/w) of the structural polysaccharides in Italian rye-grass hay, after six days of incubation. The four major cell wall monosaccharides (glucose, xylose, arabinose, galactose) were removed simultaneously and at similar rates.

Table 1. Dry matter disappearance (%) from three lignocellulosic substrates in pure cultures of rumen fungi.*

	N. frontalis (MCH3)	*N. joyonii* (NJ1)	*P. communis* (FL)	*S. communis* (FG10)
Wheat straw in fragments	30.0	25.5	16.0	6.0
Milled NH_3-treated w. straw	33.0	28.5	25.0	5.5
Milled rye-grass hay	32.0	43.0	39.0	11.0

*After 10 days of incubation

Probably because their rhizoids are capable of penetrating plant tissues, *Neocallimastix* and *Piromonas* are better at degrading plant cell walls *in vitro* than is *Sphaeromonas communis*, a non-filamentous rhizoid species (Table 1). Transmission electron microscopy (TEM) has revealed that the tissues of maize stem and of straw were similarly degraded by pure fungal cultures (*N. frontalis, P. communis, N. joyonii*) and by the major rumen cellulolytic bacteria, *Ruminococcus flavefaciens* and *Fibrobacter succinogenes* (Grenet *et al.*, 1989d, Grenet and Fonty, unpublished observations).

In a semi-continuous *in vitro* system, simulating the rumen (Rusitec), the dry matter disappearance of wheat straw increased by approximately 15% when a rumen fungus *Neocallimastix* sp. was included. Inclusion of the fungus led to an increase in losses of ADF and of NDF (Hillaire *et al.*, 1990).

Fermentation of carbohydrates

Rumen fungi are able to use a wide range of soluble sugars as energy sources. The pattern of sugar utilization varies according to the species and the strains (Table 2).

Table 2. Utilization of soluble sugars by four fungal species.*

Sugar	*N. frontalis* (MCH3)	*N. joyonii* (NJ1)	*P. communis* (FL)	*S. communis* (FG10)
D-glucose	+	+	+	+
D-mannose	+/-	-	+	-
L-fructose	+	+	+	+
L+arabinose	-	-	-	-
Galactose	-	-	-	-
D+xylose	+	+	+/-	+/-
Fucose	-	-	-	-
D+cellobiose	+	+	+	+
D+maltose	+	+	+	+
D+raffinose	+/-	+	+	-
Lactose	+	+	-	+/-
Sucrose	-	+	+	-
Gentobiose	+/-	+	+	+/-

*From Bernalier *et al.* (1990b)

Rumen fungi are able to use various plant polysaccharides, with the exception of pectin and polygalacturonate. All of the *Neocallimastix* strains use, with a few exceptions, the same polymers: cellulose (microcrystalline and amorphous), xylan, pullulan, pustulan, inulin and starch (Table 3). Utilization of polysaccharides by *Piromonas* strains is more variable. *Sphaeromonas* appear to utilize a more restricted range of polysaccharides than do *Neocallimastix* or *Piromonas* isolates (Table 3). All of the *Sphaeromonas* isolates examined so far, used xylan, but only some degraded pure cellulose (Fonty *et al.*, 1988a; Gordon and Phillips, 1989a,b; Bernalier *et al.*, 1990a, b).

Table 3. Polysaccharide utilization by rumen anaerobic fungi.*

Polysaccharide	*Neocallimastix*	*Piromonas*	*Sphaeromonas*
Cellulose	+	+	±
CM-cellulose	-	-	-
Inulin	+	-	-
Pectin	-	-	-
Polygalacturonate	-	-	-
Pullulan	+	±	-
Pustulan	+	±	-
Starch	+	±	±
Xylan	+	+	+

* From Bauchop and Mountfort (1981), Lowe *et al.* (1987c), Phillips and Gordon (1988), Williams and Orpin (1987a, b); Breton *et al.* (1989) ± variable character.

Rumen fungi produce all of the enzymes necessary both for the depolymerization of cellulose and hemicelluloses and for the hydrolysis of free oligosaccharides. All of the fungal species produce an extremely wide range of polysaccharidases, viz. endo-ß-1, 4-glucanase, exoglucanases, xylanase, cellodextrinase, and of glycosidases, including, ß-glucosidase, ß-fructosidase, ß-xylosidase, α-L arabinofuranosidase, etc. (Mountfort and Asher, 1985; Pearce and Bauchop, 1985; Lowe *et al.*, 1987b; Hebraud and Fèvre, 1988a, b; Williams and Orpin, 1987a, b; Borneman *et al.*, 1989; Gordon and Phillips, 1989b). These enzymes are mainly extracellular and are produced by the vegetative stage and by the zoospores of the fungi (Williams and Orpin, 1987a, b). They exhibit maximal activity over wide ranges of pH and temperature. The extracellular cellulase of *N. frontalis* cultured in the presence of methanogenic bacteria had a greater activity against crystalline cellulose than that of a mutant of *Trichoderma reesei* that produces one of the most active cellulase (Wood *et al.*, 1986). Cellulase and xylanase are apparently constitutive in some rumen fungi (Lowe *et al.*, 1987b; Williams and Orpin, 1987b) but may be repressed in the presence of glucose or xylose (Mountfort and Asher, 1985; Hebraud and Fevre, 1988a and b). Rumen fungi also produce an extracellular metalloprotease (Wallace and Joblin, 1985).

Table 4. The extent of degradation of filter paper and the amount of fungal biomass produced during growth of rumen fungi in monoculture and in coculture with the methanogen, *Methanobrevibacter ruminantium*. *

	N. frontalis	*P. communis*	*S. communis*
F.P degradation (% D.M.)			
fungi alone	68.6	87.6	9.7
fungi + methanogens	79.9	90.3	19.0
Fungal biomass (μg protein/ml)			
fungi alone	78.8	74.6	30.7
fungi + methanogens	98.5	118.2	64.9

*From Bernalier *et al.*, unpublished data.

Carbohydrates are fermented by anaerobic fungi via a mixed acid-type fermentation (Bauchop and Mountfort, 1981; Lowe *et al.*, 1987c; Fonty *et al.*, 1987b; Borneman *et al.*, 1989). In the case of *N. patriciarum,* glycolysis seems to be the sole mechanism for hexose fermentation (Yarlett *et al.*, 1986). The end-product profiles, viz. formate, acetate, lactate, ethanol, CO_2 and H_2, of all strains examined to-date appear to be similar.

However, ethanol and formate were not produced by *N. patriciarum* (Orpin and Munn, 1986) and succinate has been found to be produced by some fungal strains (Borneman *et al.*, 1989; Fonty *et al.*, unpublished observations). In *N. patriciarum*, H_2 production occurs in specialized organelles termed "hydrogenosomes" (Yarlett *et al.*, 1986, 1987), similar to those described in the H_2-producing rumen protozoa. Hydrogenase, pyruvate: ferredoxin oxidoreductase, NADPH: ferredoxin oxidoreductase and a malic enzyme are involved in this H_2 production (Yarlett *et al.*, 1986).

Because of their high rate of production of H_2 the rumen fungi interact with methanogenic bacteria (Bauchop and Mountfort, 1981; Mountfort *et al.*, 1982; Fonty *et al.*, 1988b; Stewart and Richardson, 1989). *In vitro,* the association between fungi and methanogens led to an increase in the fungal biomass (Bernalier *et al.*, 1990a) and to an increase in the rate and extent of cellulose (filter paper) hydrolysis [(Bauchop and Mountfort, 1981; Fonty *et al.*, 1988b); Table 4].

Table 5. Fermentation of cellulose by *Neocallimastix frontalis* in the presence and absence of methanogenic bacteria.

Product	*N. frontalis*	*N. frontalis* + methanogens
	(mol produced/100 mol hexose fermented)	
Acetate	72.7	134.7
Lactate	67.0	2.9
Ethanol	37.4	19.0
Methane	0.0	58.7
Carbon dioxide	37.6	88.7
Hydrogen	35.3	0.05
Formate	83.1	1.0

*From Bauchop and Mountfort (1981).

Cellulase activity was also higher in cocultures than in monocultures (Mountfort and Asher, 1985; Wood *et al.*, 1986). This increased productivity was also apparent with lignocellulosic substrates but was less pronounced than with filter paper (Joblin, personal communication). In coculture, with *Methanobrevibacter ruminantium,* the metabolism of the fungus was shifted towards a greater production of acetate at the expense of the production of more reduced compounds, such as lactate and ethanol (Table 5).

In vitro, rumen fungi also interact with various rumen bacterial species, particularly *Ruminococcus flavefaciens*, a cellulolytic species, *Selenomonas ruminantium* and *Eubacterium limosum* (Bernalier *et al.*, 1988, 1990a).

ROLE OF ANAEROBIC FUNGI IN FORAGE DEGRADATION *IN VIVO*

Effects of the diet on rumen fungal populations

Anaerobic fungi appear in the rumen of flock-reared lambs 8-10 days after birth, i.e. before the ingestion of solid feed. However, the subsequent development and survival of these fungi depend on the composition of the diet. They disappeared from 80% of lambs fed a rich concentrated diet but remained stable in those fed a dehydrated alfalfa diet (Fonty *et al.*, 1987a, Fonty and Gouet, 1989).

In adult ruminants, in general, the higher the fibre content in the diet the more numerous the rumen fungal population (Orpin, 1984; Bauchop, 1979, 1981, 1989; Grenet *et al.*, 1988, 1989a, c). Lucerne hay and meadow hay are particularly favourable to growth of the fungal population (Bauchop, 1981; Grenet *et al.*, 1989a, c). By contrast, such fungi are found only in small numbers in animals fed sugar beet (Grenet *et al.*, 1989a), flush young pasture or soft leafy plants (Bauchop, 1979, 1981, 1989) or cereals (Bauchop, 1979, 1981; Grenet *et al.*, 1988, 1989a). They are absent from animals continuously grazed on pasture (Bauchop, 1989) and in animals fed a diet of seaweeds (Greenwood *et al.*, 1983) and from those fed with of straw treated with sodium chlorite (Elliot *et al.*, 1987).

The nutritional status of the forage ingested by the animals may also be important for the development of rumen fungi. Akin *et al.* (1983) found that forage grown in soils fertilized with sulphur stimulated fungal development in the rumen of sheep. By contrast, Millard *et al.* (1987) noted that the rumen fungal population was greatest when sheep were fed forage grown under S-limitation.

High-fibre diets that favour the development of fungi in the rumen are also those that favour their occurrence in the hindgut (Grenet *et al.*, 1989c). *Sphaeromonas* was the dominant genus in the intestine.

Colonization and degradation of plant tissues

Following the ingestion of plant material by the ruminant, fungal zoospores are released from mature sporangia attached to plant particles already present in the rumen fluid. The zoospores are then attracted by chemotaxis to the newly ingested particles. They attach, preferentially to damaged tissues, including even minor lesions, and to stomata (Bauchop, 1981; Akin *et al.*, 1983; Fonty *et al.*, 1988a). After germination, the zoospores produce rhizoids that penetrate the plant tissues, thereby allowing access by the fungi to fermentable carbohydrates not immediately available to bacteria. The mechanisms that allow penetration of the fungi into the plant walls are unknown. Unlike rumen cellulolytic bacteria, rumen fungi are proteolytic (Wallace and Joblin, 1985). This probably makes it easier for the rhizoid to penetrate the proteinaceous layer of the plant materials. The sporangium develops gradually from the zoospore and reaches maturity about 24-32 h after encystment (Joblin, 1981; Lowe *et al.*, 1987a). Rumen fungi predominantly colonize lignified tissues, such as sclerenchyma, xylem, vascular bundles, that are retained longest in the rumen (Bauchop, 1979, 1981, 1989; Akin and Benner, 1988; Grenet and Barry, 1988; Akin *et al.*, 1989) and thick-walled tissues, such as the palisade layer of soya bean hulls (Grenet and Barry, 1988).

In vitro, fungi have been shown to reduce particle size (Orpin, 1984), to decrease the tensile strength of plant tissues (Akin *et al.*, 1983, 1989) and to disrupt plant physical structure (Joblin, 1989; Akin *et al.*, 1989). These *in vitro* observations suggest that fungi assist in the physical breakdown of plant particles in the rumen. This conclusion has recently been reinforced by *in vivo* studies in which it was found that the particle size distribution of digesta leaving the rumen of sheep differed depending on whether the sheep rumens in question were populated with fungi or not (Calderon-Cortes *et al.*, 1989).

Although fungi preferentially colonize the lignocellulosic tissues, their role in the degradation of lignin is not clear. *In vitro,* they are able to solubilize part of the lignin of the plant cell walls (Orpin, 1984; Akin and Benner, 1988; Gordon and Phillips, 1989b) but there is no evidence that they are able to utilize lignin as a carbon source.

Quantitatively, little is known of the contribution of rumen fungi to the degradation of plant cell walls and to the overall fermentative processes. Elimination of fungi from the rumen markedly reduced the DM digestibility of straw *in sacco* and resulted in an elevated proportion of propionic acid in rumen fluid (Ford *et al.*, 1987).

Subsequent inoculation of these sheep with a pure culture of fungus decreased propionate concentrations within 3 days (Elliott *et al.*, 1987). The introduction of fungi in Rusitec was also associated with a greater production of acetate at the expense of propionate and with a lower CO_2/CH_4 ratio (Hillaire *et al.*, 1990).

Table 6. Cellulose (ADF-lignin, van Soest) degraded from rye grass hay and wheat straw in the rumen of isolated lambs harbouring anaerobic fungi as the sole cellulolytic microorganisms.*

		Cellulose degraded (%)#			
		Isolated lambs inoculated with		Control lambs (conventional)	
Substrate	Incubation time (h)	*N. frontalis*	*P. communis*	Mixed fungal population	
Ryegrass hay	48	36.3	35.0	ND	ND
Wheat straw	48	21.2	19.0	23.2	36.3
	72	24.4	23.3	25.1	39.3

*From Fonty and Gouet (1989) and Fonty *et al.*, unpublished data.
#Following incubation for 48 or 72 h in nylon bags in the rumen.

In gnotobiotic lambs, harbouring only fungi as cellulolytic microorganisms, approximately 25% and 36% of cellulose (ADF-lignin) from wheat straw and rye grass hay, respectively, was degraded (Table 6). The rate of degradation of stem fragments of maize and straw, observed by scanning electron microscope (SEM), was slightly slower in these gnotobiotic animals than in those reared conventionally (Grenet *et al.*, 1989b). In both cases, fungi preferentially colonized the lignified tissues of the ingesta.

CONCLUSION

Anaerobic rumen fungi, first observed at the beginning of the century, have only recently been studied in detail. Further studies on their potential need to be done. *In vitro*, they are strongly cellulolytic, but relatively is known about their role in plant cell wall degradation *in vivo*. Rumen fungi are not essential to rumen function *per se*, as they are present only in very low numbers, or even absent, in ruminants fed low-fibre diets. However, it is likely that they play an important role in the digestion of poor quality roughages. An assessment of their contribution to fibre digestion requires a method for accurate determination of fungal biomass and comparative studies on animals having and lacking fungi in their rumen microflora. A greater understanding of their extracellular enzyme systems is also essential to a fuller understanding of the functioning of these microorganisms.

REFERENCES

Akin, D.E. and Benner, R. (1988). Degradation of polysaccharides and lignin by ruminal bacteria and fungi. Appl. Environ. Microbiol. **54**, 1117-1125.

Akin, D.E. and Rigsby, L.L. (1987). Mixed fungal populations and lignocellulosic tissues degradation in the bovine rumen. Appl. Environ. Microbiol. **53**, 1987-1995.

Akin, D.E., Gordon, G.L.R. and Hogan, J.P. (1983). Rumen bacterial and fungal degradation of *Digitaria pentzii* grown with and without sulphur. Appl. Environ. Microbiol. **46**, 738-748.

Akin, D.E., Lyon, C.E., Windham, W.R. and Rigsby, L.L. (1989). Physical degradation of lignified stem tissues by ruminal fungi. Appl. Environ. Microbiol. **55**, 611-616.

Barr, D.J.S., Kudo, H., Jakober, K.D. and Cheng, K.J. (1989). Morphology and development of rumen fungi: *Neocallimastix* sp., *Piromyces communis* and *Orpinomyces* gen. nov, sp. nov. Can. J. Botany, **67**, 2815-2824.

Bauchop, T. (1979), Rumen anaerobic fungi of cattle and sheep. Appl. Environ. Microbiol. **38**, 148-158

Bauchop, T. (1981). The anaerobic fungi in rumen fibre digestion. Agric. Environ. **6**, 339-348.

Bauchop, T. (1989). Colonization of plant fragments by protozoa and fungi. In *The Roles of Protozoa and Fungi in Ruminant Digestion.* eds. J.V. Nolan, R.A. Leng and D.I. Demeyer, Penambul Books, Armidale, Australia, pp. 83-96.

Bauchop, T. and Mountfort, D.O. (1981). Cellulose fermentation by a rumen anaerobic fungus in both the absence and the presence of rumen methanogens. Appl. Environ. Microbiol. **42**, 1103-1110.

Bernalier, A., Fonty, G. and Gouet, Ph. (1988). Dégradation et fermentation de la cellulose par *Neocallimastix* sp. MCH_3 seul ou associé à quelques espèces bactériennes du rumen. Reprod. Nutr. Develop. **28**, 75-76.

Bernalier, A., Fonty, G. and Gouet, Ph. (1990a). Cellulose degradation by two rumen anaerobic fungi in monoculture or in coculture with rumen bacteria. Anim. Feed. Sci. Technol. in press.

Bernalier, A., Fonty, G. and Gouet, Ph. (1990b). Fermentation properties of four strictly anaerobic rumen fungal species: H_2-producing microorganisms. FEMS Symposium. September 1989. in press.

Borneman, W.S., Akin, D.E. and Ljungdahl, L.G. (1989). Fermentation products and plant cell wall degrading enzymes produced by monocentric and polycentric anaerobic ruminal fungi. Appl. Environ. Microbiol., **55**, 1066-1073.

Breton, A., Bernalier, A., Bonnemoy, F., Fonty, G., Gaillard, B. and Gouet, Ph. (1989). Morphological and metabolic characterization of a new species of strictly anaerobic rumen fungus, *Neocallimastix joyonii.* FEMS Microbiol. Lett. **58**, 309-314.

Calderon-Cortes, F.J., Elliott, R. and Ford, C.W. (1989). Influence of rumen fungi on the nutrition of sheep fed forage diets. In *The Roles of Protozoa and Fungi in Ruminant Digestion.* eds. J.V. Nolan, R.A. Leng and D.I. Demeyer, Penambul Books, Armidale, Australia, pp. 181-187.

Citron, A., Breton, A. and Fonty, G. (1987). The rumen anaerobic fungi. Bull. Inst. Pasteur. **85**, 329-343.

Elliott, R., Ash, A.J., Calderon-Cortes, F., Norton, B.W. and Bauchop, T. (1987). The influence of anaerobic fungi on rumen volatile fatty acid concentrations *in vivo.* J. Agr. Sci. Camb. **109**, 13-17.

Fonty, G. and Gouet, Ph. (1989). Establishment of microbial populations in the rumen. Utilization of an animal model to study the role of the different cellulolytic micro-organisms *in vivo.* In *The Roles of Protozoa and Fungi in Ruminant Digestion.* eds. J.V. Nolan, R.A. Leng and D.I. Demeyer, Penambul Books, Armidale, Australia, pp. 39-49.

Fonty, G., Gouet, Ph., Jouany, J.P. and Senaud, J. (1987a) Establishment of the microflora and anaerobic fungi in the rumen of lambs. J. Gen. Microbiol., **123**, 1835-1843.

Fonty, G., Breton, A., Fevre, H., Citron, A., Hebraud, M. and Gouet, Ph. (1987b) Isolement et caractérisation des champignons anaérobies stricts du rumen de moutons. Premiers résultats. Reprod. Nutr. Develop. **27**, 107-108.

Fonty, G., Grenet, E., Fevre, M., Breton, A. and Gouet, Ph. (1988a) Ecologie et fonctions des champignons anaerobies du rumen. Reprod. Nutr. Develop., **28**, Suppl. No. 1, 1-18.

Fonty, G., Gouet, Ph. and Sante, V. (1988b) Influence d'une bactérie méthanogène sur l'activité cellulolytique de deux espèces de champignons du rumen, *in vitro.* Resultats preliminaires. Reprod. Nutr. Develop. **28**, 133-134.

Ford, C.W., Elliott, R. and Maynard, P.J. (1987). The effect of chlorite delignification on digestibility of some grass forages and on intake and rumen microbial activity in sheep fed barley straw. J. Agr. Sci., Camb. **108**, 129-136.

Gold, J.J., Heath, I.B. and Bauchop, T. (1988). Ultrastructural description of a new chydrid genus of caecum anaerobe, *Caecomyces equi* gen. nov, sp. nov, assigned to the Neocallimasticaceae. Biosystems, **21**, 403-415.

Gordon, G.L.R. and Phillips, M.W. (1989a). Comparative fermentation properties of anaerobic fungi from the rumen. In *The Roles of Protozoa and Fungi in Ruminant Digestion.* eds. J.V. Nolan, R.A. Leng and D.I. Demeyer, Penambul Books, Armidale, Australia, pp. 127-138.

Gordon, G.L.R. and Phillips, M.W. (1989b). Degradation and utilization of cellulose and straw by three different anaerobic fungi from the ovine rumen. Appl. Environ. Microbiol., **55**, 1703-1710.

Greenwood, Y., Hall, F.J., Orpin, C.G. and Paterson, I.W. (1983). Microbiology of seaweed digestion in Orkney sheep. J. Physiol. **343**, 121p.

Grenet, E. and Barry, P. (1988). Colonization of thick-walled plant tissues by anaerobic fungi. Anim. Feed Sci. Technol. **19**, 25-31.

Grenet, E., Breton, A., Fonty, G., Barry, P. and Remond, B. (1988). Influence du régime alimentaire sur la population fongique du rumen. Reprod. Nutr. Develop. **28**, 127-128.

Grenet, E., Breton, A., Barry, P. and Fonty, G. (1989a). Rumen anaerobic fungi and plant substrates colonization as affected by diet composition. Anim. Feed Sci. Technol. **26**, 55-70.

Grenet, E., Fonty, G. and Barry, P. (1989b). SEM study of the degradation of maize and lucerne stems in the rumen of gnotobiotic lambs harbouring only fungi as cellulolytic micro-organisms. In *The Roles of Protozoa and Fungi in Ruminant Digestion.* eds. J.V. Nolan, R.A. Leng and D.I. Demeyer, Penambul Books, Armidale, Australia, pp. 265-267.

Grenet, E., Fonty, G., Jamot, J. and Bonnemoy, F. (1989c). Influence of diet and monensin on development of anaerobic fungi in the rumen, duodenum, caecum and faeces of cows. Appl. Environ. Microbiol. 55, 2360-2364.

Grenet, E., Jamot, J., Fonty, G. and Bernalier, A. (1989d). Kinetics study of the degradation of wheat straw and maize stem by pure cultures of anaerobic rumen fungi observed by scanning electron microscopy. AJAS. 2, 456-457.

Hebraud, M. and Fevre, M. (1988a) Caractérisation des hydrolases secrétées par les champignons anaérobies du rumen. Reprod. Nutr. Develop. 28, 131-132.

Hebraud, M. and Fevre, M. (1988b) Characterization of glycosides and polysaccharide hydrolases secreted by the rumen anaerobic fungi *Neocallimastix frontalis, Sphaeromonas communis* and *Piromonas communis.* J. Gen. Microbiol. **134**, 1123-1129.

Hillaire, M.C., Jouany, J.P. and Fonty, G. (1990). Wheat straw degradation in Rusitec, in the presence or absence of rumen anaerobic fungi. Proc. Nutr. Soc. in press.

Joblin, K.N. (1981). Isolation, enumeration and maintenance of rumen anaerobic fungi in roll tubes. Appl. Environ. Microbiol. **42**, 1119-1122.

Joblin, K.N. (1989). Physical description of plant fibre by rumen of the *Sphaeromonas* group. In *The Roles of Protozoa and Fungi in Ruminant Digestion.* eds. J.V. Nolan, R.A. Leng and D.I. Demeyer, Penambul Books, Armidale, Australia, pp. 259-260.

Lowe, S.E., Griffith, C.G., Milne, A., Theodorou, M.K. and Trinci, A.P.J. (1987a). The life cycle and growth kinetics of an anaerobic rumen fungus. J. Gen. Microbiol. **133**, 1815-1827.

Lowe, S.E., Theodorou, M.K. and Trinci, A.P.J. (1987b). Cellulases and xylanase of an anaerobic rumen fungus grown on wheat-straw lignocellulose, wheat-straw holocellulose, cellulose and xylan. Appl. Environ. Microbiol. 53, 1216-1223.

Lowe, S.E., Theodorou, M.K. and Trinci, A.P.J. (1987c). Growth and fermentation of an anaerobic rumen fungus on various carbon sources and the effects of temperature on development. Appl. Environ. Microbiol. 53, 1210-1215.

Millard, P., Gordon, A.H., Richardson, A.J. and Chesson, A. (1987). Reduced ruminal degradation of rye-grass caused by sulphur limitation. J. Sci. Food Agric., 40, 305-314.

Mountfort, D.O. and Asher, R.A. (1985). Production and regulation of cellulase by two strains of the rumen anaerobic fungus *Neocallimastix frontalis.* Appl. Environ. Microbiol. 49, 1314-1322.

Mountfort, D.O., Asher, R.A. and Bauchop, T. (1982). Fermentation of cellulose to methane and dioxide by a rumen anaerobic fungus in a triculture with *Methanobrevibacter* sp. strains RA_1 and *Methanosarcina barkeri.* Appl. Environ. Microbiol. **44**, 128-134.

Orpin, C.G. (1975). Studies on the rumen flagellate *Neocallimastix frontalis.* J. Gen. Microbiol. **91**, 249-262.

Orpin, C.G. (1976). Studies on the rumen flagellate *Sphaeromonas communis.* J. Gen. Microbiol. **94**, 270-280.

Orpin, C.G. (1977). The rumen flagellate *Piromonas communis:* its life-history and invasion of plant material in rumen. J. Gen. Microbiol. **99**, 107-117.

Orpin, C.G. (1984). The role of ciliate protozoa and fungi in the rumen digestion of plant cell walls. Anim. Feed Sci. Technol. **10**, 121-143.

Orpin, C.G. and Munn, E.A. (1986). *Neocallimastix patriciarum* sp. nov, a new member of the Neocallimasticaceae inhabiting the rumen of sheep. Trans. of the Brit. Mycol. Soc. **86**, 178-181.

Orpin, C.G. and Joblin, K.N. (1988). The rumen anaerobic fungi. In *The Rumen Microbial Ecosystem,* ed. P.N. Hobson, Elsevier Applied Science, London, pp. 129-150.

Pearce, P.D. and Bauchop, T. (1985). Glycosidases of the rumen anaerobic fungus *Neocallimastix frontalis* grown on cellulosic substrates. Appl. Environ. Microbiol. **49**, 1265-1269.

Phillips, M.W. and Gordon, G.L.R. (1988). Sugar and polysaccharide fermentation by rumen anaerobic fungi from Australia, Britain and New Zealand. Biosystems, **21**, 377-383.

Stewart, C.S. and Richardson, A.J. (1989). Enhanced resistance of anaerobic rumen fungi to the ionophores monensin and lasalocid in the presence of methanogenic bacteria. J. Appl. Bacteriol. **66**, 85-93.

Theodorou, M.K., Longland, A., Dhanoa, M.S., Lowe, S.E. and Trinci, A.P.J. (1989). Growth of *Neocallimastix* sp. strain R_1 on Italian ryegrass hay: removal of neutral sugars from plant cell walls. Appl. Environ. Microbiol. **55**, 1363-1367.

Wallace, R.J. and Joblin, K.N. (1985). Proteolytic activity of a rumen anaerobic fungus. FEMS Microbiol. Lett. **29**, 19-25.

Williams, A.G. and Orpin, C.G. (1987a). Polysaccharide-degrading enzymes formed by three species of anaerobic rumen fungi on a range of carbohydrate substrates. Can. J. Microbiol. **33**, 418-426.

Williams, A.G. and Orpin, C.G. (1987b). Glycoside hydrolase enzymes present in the zoospore and vegetative growth stages of the rumen fungi, *Neocallimastix partriciarum, Piromonas communis* and an identified isolate, grown on a range of carbohydrates. Can. J. Microbiol. **33**, 427--434.

Windham, W.R. and Akin, D.E. (1984). Rumen fungi and forage fibre degradation. Appl. Environ. Microbiol. **48**, 473-476.

Wood, T.M., Wilson, C.A., McCrae, S.I. and Joblin, K.N. (1986). A highly active extracellular cellulase from the anaerobic rumen fungus *Neocallimastix frontalis.* Microbiol. Lett. **34**, 37-40.

Yarlett, N., Orpin, C.G., Munn, E.A., Yarlett, N.C. and Greenwood, C.A. (1986). Evidence for hydrogenosomes in the rumen fungus *Neocallimastix patriciarum.* Biochem. J. **236**, 729-739.

Yarlett, N., Rowlands, C., Yarlett, N.C., Evans, J.C. and Lloyd, D. (1987). Respiration of the hydrogenosome-containing fungus *Neocallimastix patriciarum.* Arch. Microbiol. **148**, 25-28.

ENHANCED DEGRADABILITY OF WHEAT STRAW FOLLOWING FERMENTATION WITH *PLEUROTUS OSTREATUS* AND THE CONTRIBUTION OF RUMEN FUNGI AND BACTERIA TO STRAW DEGRADATION *IN VITRO*

C.S. Stewart, J.A. Akoyo and F. Zadrazil *

Rowett Research Institute, Bucksburn, Aberdeen AB2 9SB, Scotland

and

*Institute für Bodenbiologie, Bundesforschungsanstalt für Landwirtschaft

Bundesallee 50, D-3300 Braunschweig, FRG

Treatment of wheat straw by fermentation with *Pleurotus ostreatus* increased degradability *in sacco* and upon incubation in rumen liquor *in vitro.* Inhibition of the growth of rumen fungi by cycloheximide caused a relatively minor depression of straw degradability *in vitro.* Mixed laboratory cultures of cellulolytic rumen bacteria (*Ruminococcus albus, R. flavefaciens* and *Bacteriodes succinogenes*) degraded both treated and untreated straw more rapidly than did mixed laboratory cultures of rumen fungi (*Neocallimastix frontalis* and *Piromonas communis*). Although the highly interactive nature of the rumen microflora renders interpretation difficult, it seems that, by comparison with the rumen bacteria, anaerobic fungi play a relatively minor role in straw degradation by rumen liquor *in vitro.*

INTRODUCTION

The lignin-degrading activities of white-rot fungi have been used for the development of processes designed to increase the degradability of straw and other lignified plant materials in the rumen (Zadrazil, 1977; Zadrazil and Brunnert, 1981, 1982; Reade and Macqueen, 1983; Vuillet *at al.*, 1988; Kamra and Zadrazil, 1988). In particular, a large scale process for the solid-state fermentation of wheat straw by *Pleurotus* species and other fungi has been developed in Germany. It is not clear whether the fermentation of straw by *Pleurotus* is likely to change its relative susceptibility to attack by different kinds of rumen microorganisms. The major digestible components of straw are degraded in the rumen by anaerobic bacteria, fungi and protozoa (reviewed by Orpin, 1984; Cheng *et al.*,

1984). It is known that these microbial populations interact extensively through predation (of bacteria by protozoa), competition and the cross-feeding of nutrients and substrates such as hydrogen (reviewed by Stewart *et al.*, 1986). However, the relative contributions of these different types of cellulolytic microorganisms to the degradation of plant cell wall materials are not well understood. Windham and Akin (1984) used antibiotics to suppress the growth of either rumen fungi or bacteria in incubations of bermuda grass or alfalfa hay with rumen contents. Inhibiting the growth of rumen fungi (with cycloheximide) had relatively little effect on fibre digestion, and it was concluded that bacteria probably play a much more significant role in fibre digestion than do the fungi. However, when bacterial growth was inhibited by antibiotics, the fungi increased in number and significant degradation of forage still occurred. The objectives of the present work were: (i) to measure the effect of fermentation with *Pleurotus ostreatus* on the degradability of wheat straw *in sacco;* (ii) to compare the effects of cycloheximide and antibacterial antibiotics on the production, *in vitro*, of methane from untreated and *Pleurotus*-treated straw; (iii) to compare the degradation of treated and untreated straw by mixed populations of cellulolytic rumen bacteria and mixed populations of rumen fungi.

MATERIALS & METHODS

Cultures and microbial counts

Anaerobic cellulolytic bacteria, *Methanobrevibacter smithii* strain PS, and the anaerobic fungi were from the culture collection maintained at the Rowett Institute. The anaerobic methods used were similar to those described by Bryant (1972), in which the media were prepared and maintained either under O_2-free CO_2 (Ruminococci, *Bacteroides succinogenes* and anaerobic fungi) or under H_2/CO_2 (80:20) (*M. smithii*) in screw-capped 'Hungate' tubes fitted with septum stoppers (Bellco Glass Inc., N.J.). Cultures of *M. smithii* were maintained on medium M2 (Hobson, 1969) minus the sugars and lactate. The other microorganisms were maintained on medium M2 modified to contain 0.25% (w/v) cellobiose as sole energy source. Fungal and bacterial counts were performed by standard anaerobic methods using roll tubes of RGCA medium (Holdeman *et al.*, 1977).

Straw

Wheat straw was incubated with *Pleurotus ostreatus* on a commercial (100 tons/day) scale for 21 days at 25-30°C by Pleurotus GmbH, Weiden, FRG (Zadrazil and Heltay,

1989). After harvesting the fruiting bodies, the straw residues were dried in air, ground in a hammer mill without a screen and then sieved (dry) to remove particles smaller than 150 μm in diameter.

Incubations *in sacco*

The nylon bag technique of Mehrez and Ørskov (1977) was used. Nylon mesh bags, 12 x 8 cm, pore size 50 to 80 μm, contained 1 g of the plant material under study. The bags were suspended in the rumen of the test animals and were withdrawn at the intervals stated in the text. The bags were then washed under running tap water until the water was clear. The contents of the bags were transferred to pre-weighed aluminium cups and dried to constant weight at 80°C.

Experimental animals

Three sheep and a cow, each fitted with rumen cannulas were used. Cow (Institute no. 591) received 1 kg grass cubes, 1 kg hay and 1 kg concentrate supplement (85% ground maize, wheat & oats, 7% each fishmeal and linseed meal, plus vitamins and minerals) twice daily. Sheep (Institute nos. 125, 5173 and 6190) were fed twice daily with 800 g hay plus a supplement containing 10 g urea plus vitamins and minerals.

Incubations with rumen liquor *in vitro*

The two stage technique of Tilley and Terry (1963), scaled down to a total volume of 25 ml, was used. The initial weight of straw was 0.5 g, and the rumen liquor was obtained from cow 591. After incubation for the period stated in the text, the residues were further digested with pepsin for 48 h as described by Tilley and Terry (1963).

Methane production *in vitro*

Medium M2 (Hobson, 1969), 4.5ml, without sugars or lactate was prepared under O_2-free CO_2 in Hungate tubes containing 30 mg milled straw. After autoclaving (121°C, 15 min) the medium, 1 ml of rumen fluid (from cow 591) plus sterile solutions of the desired antibiotics were added to the tubes. The tubes were then incubated at 38°C for 11 days, prior to measurement (by gas chromatography) of the methane content of the gas phase (Stewart and Richardson, 1988).

Incubations of straw with bacteria and fungi

The procedures described by Kolankaya *et al.* (1985) were followed, except that the initial weight of straw (hammer-milled to particle sizes of between 1.18 and 2.36 mm) was 50 mg.

Antibiotics

The antibiotics used were from Sigma Chemical Co., St. Louis, MO, USA. Stock solutions (50 x conc.) were prepared in distilled water and sterilised by filtration through 0.2 μm membrane filters prior to use.

RESULTS

Fermentation of straw with *Pleurotus ostreatus* increased the degradability of wheat straw upon incubation in the rumen. However, when nylon mesh bags containing straw were incubated in distilled water with mechanical agitation in the laboratory (i.e. in the absence of rumen microorganisms), much of the increased susceptibility of the straw appeared to be due to solubilization of its dry matter by *Pleurotus* and to the escape of small particles from the bags. For example, during 48 h incubation in water, around 29% of the dry matter of treated straw and 12% of the untreated straw was lost from the bags. In both cases, around 2/3 of the material lost from the bags could be recovered from the wash water in the form of small particles.

Table 1. Losses in weight of untreated and *Pleurotus*-treated wheat straw upon incubation in the rumen *in sacco,* corrected for losses occurring during incubation in water.

	% loss in weight *							
Straw	*Pleurotus*-treated				Untreated			
Animal	cow	------ sheep -------			cow	------- sheep --------		
Time (h)	591	125	5173	6190	591	125	5173	6190
8	-	12	11	9	-	10	9	9
24	34	36	36	29	22	28	29	25
48	45	47	44	42	31	41	41	29
72	49	48	50	47	47	47	39	42
96	57	49	50	48	56	47	46	47

*Loss in weight of straw incubated in the rumen minus loss in weight of straw incubated in distilled water (see text for details). Results are the average of two determinations. -, not studied.

Table 1 shows the losses in weight of straw incubated in nylon mesh bags in the rumen, corrected for the losses that occurred when the straw was incubated in the laboratory in the absence of rumen microorganisms. It is clear that when examined in this way, the main effect of fermentation with *Pleurotus* was to increase the rate of straw degradation.

In an attempt to reduce the losses of particles from the bags, a further experiment was carried out *in sacco* using bags of 5 μm pore size. However, the losses in weight of straw samples incubated in water were again high and it seemed that even with bags of very small pore size, extension of the nylon mesh during incubation allows the escape of particles that are larger than the measured pore size of the bags.

In order to asses the effects on straw degradation of inhibiting growth of rumen fungi, the losses in weight of straw occurring during the two stage *in vitro* procedure of Tilley and Terry (1963) were measured, with and without the addition of cycloheximide (final concentration 100 μg/ml) to the first (rumen liquor digestion) stage of the procedure. The results showed that cycloheximide depressed straw degradation only slightly (Fig. 1). In order to confirm the effect of cycloheximide on the fungi, the same concentration of cycloheximide was added to cultures of *Neocallimastix frontalis* strains PK2 and RE1 and *Piromonas communis* strain P which had been pre-grown (in the absence of drugs) for 48 h. When samples of these treated cultures were then transferred to roll tubes (see Methods) containing cycloheximide, no viable colonies developed.

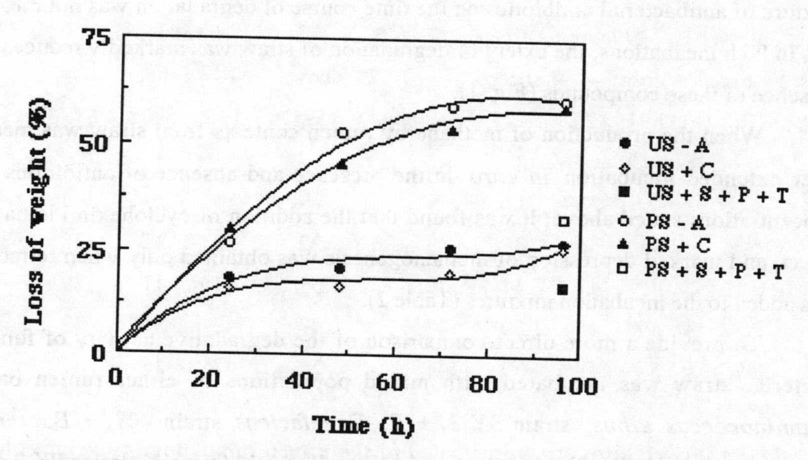

Fig. 1. Effect of antibiotics on straw digestion *in vitro.* Lettering as follows: US, untreated straw; PS, *Pleurotus*-treated straw; - A, without antibiotic; C, cycloheximide; S, streptomycin; P, penicillin; T, tetracycline.

Cycloheximide inhibits protein synthesis in a wide range of eukaryotic cells, including at least some protozoa (Gale *et al.*, 1981). Observations on the fate of protozoa were not made during the experiments described above. However, an additional experiment was performed to monitor the effect of cycloheximide on the rumen protozoa. Examination of samples by light microscopy showed that the predominant ciliate protozoa present were mainly entodiniomorphs. Holotrichs, *Dasytricha* and *Isotricha,* were also present as were small numbers of *Polyplastron* (i.e. a type A ciliate population according to the definition of Eadie, 1967). After incubation for 24 h, many active protozoa remained, although some of the larger types, including the *Polyplastron,* appeared to be devoid of contents. By 48 h, most of the protozoa were no longer motile. The presence of cycloheximide did not appear to affect these changes.

The straws were next incubated with rumen contents in the presence of a mixture of streptomycin and penicillin (final concentrations 40 and 250 μg/ml, respectively) to test the effect of the inhibition of bacterial growth on the degradation of the substrate. However, these drugs clearly failed to inhibit bacterial growth completely, and large numbers of viable bacteria (10^7/ml) could be detected in samples of the incubation mixture. The experiment was repeated with the further addition of tetracycline (20 μg/ml). The mixture of three antibacterial antibiotics was found to reduce the numbers of viable rumen bacteria in the incubation mixtures to around 10^3/ml. The effect of this mixture of antibacterial antibiotics on the time course of degradation was not measured, but, in 96 h incubations, the extent of degradation of straw was markedly reduced in the presence of these compounds (Fig. 1).

When the production of methane by rumen contents from straw was measured after extended incubation *in vitro* in the presence and absence of antibiotics at the concentrations stated above, it was found that the addition of cycloheximide had little effect, and marked depression of methanogenesis was obtained only when tetracycline was added to the incubation mixtures (Table 2).

To provide a more direct comparison of the degradative activity of fungi and bacteria, straw was incubated with mixed populations of either rumen bacteria (*Ruminococcus albus,* strain SY 3, + *R. flavefaciens,* strain 007, + *Bacteroides succinogenes,* strain BL 2) or rumen fungi (*Neocallimastix frontalis,* strains PK 2 and RE 1, + *Piromonas communis,* strain P).

Table 2. Effect of cycloheximide and antibacterial antibiotics on methane production (by rumen contents) from *Pleurotus*-treated and untreated wheat straw incubated *in vitro.* *

Antibiotic added	Methane produced (ml)	
	Pleurotus- treated straw	untreated straw
None	2.8	2.7
Cycloheximide	2.4	2.3
Streptomycin + penicillin	2.3	2.3
Streptomycin + penicillin + tetracycline	0.9	0.7

*Incubation was for 11 days at 38°C. Results are the average of two determinations. Concentrations of antibiotics used are cited in the text.

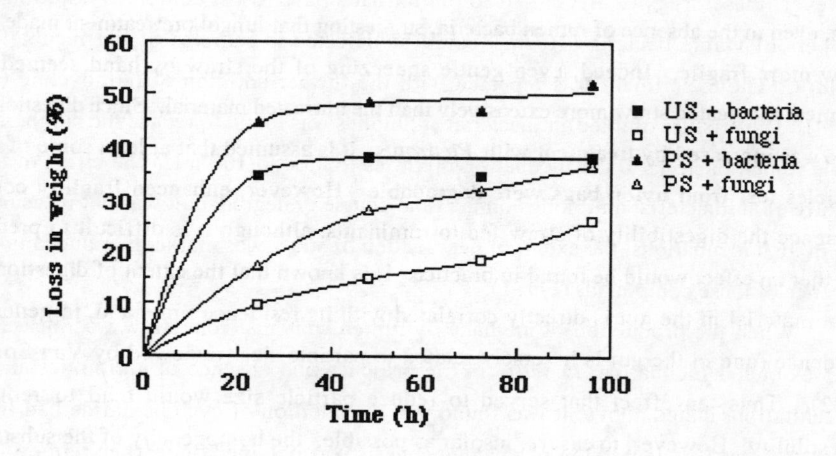

Fig. 2. Degradation of straw by rumen bacteria and fungi. Lettering is as follows: US, untreated straw; PS, *Pleurotus*-treated straw. The mixed bacterial and mixed fungal populations are as defined in the text. *Methanobrevibacter smithii* was present in all incubation mixtures.

Because the presence of methanogens is thought to improve the growth of hydrogen-producing cellulolytic anaerobes (see e.g. Bauchop and Mountfort, 1981), all of the incubations were carried out in the presence of *Methanobrevibacter smithii* strain PS, a sewage sludge methanogenic isolate, physiologically similar to rumen methanogens. Numbers of viable colony forming units (cfu) of bacteria and fungi in the inocula were not measured, but they would typically contain 10^3 to 10^4 fungal cfu/ml and around 10^8 bacterial cfu/ml (A.J. Richardson, unpublished data). This experiment showed that the bacterial co-culture degraded both treated and untreated straw more rapidly than did the co-culture containing the fungi (Fig. 2).

DISCUSSION

The treatment of wheat straw with *Pleurotus* clearly enhanced microbial degradation both *in sacco* and *in vitro*. The enhancement of degradation *in sacco* following treatment with *Pleurotus* appeared to be due in part to an increase in soluble matter. In addition, a significant quantity of small particles of *Pleurotus*-treated straw were lost from nylon bags, even in the absence of rumen bacteria, suggesting that fungal pretreatment made the straw more fragile. Indeed, even gentle squeezing of the straw by hand seemed to fragment the treated straw more extensively than the untreated material. Since digestion *in vitro* was increased by treatment with *Pleurotus*, it is assumed that at least some of the particles lost from nylon bags were degradable. However, enhanced fragility could influence the digestibility of straw fed to ruminants, although it is difficult to predict whether an effect would be found in practice. It is known that the extent of digestion of plant material in the gut is directly correlated with its residence time, and, in general, residence time in the gut is inversely related to particle size (reviewed by Van Soest, 1982). Thus, an effect that served to reduce particle size would tend to reduce digestibility. However, to ensure, insofar as possible, the homogeneity of the substrate the present studies were performed with milled straw. The observations made here might not apply if the straw fed to animals was either in the long form or pelleted.

The effects of antibiotics on straw fermentation in this study are assumed to result mainly from the suppression of bacterial or fungal growth by these compounds. However, rumen protozoa are also thought to contribute to plant cell wall digestion in the rumen, although the extent of their contribution is not known (reviewed by Orpin, 1983). While it was clear, both in the present study and in that of Windham and Akin (1983), that cycloheximide, at the doses used, did not kill the protozoa, it must be recognised that the drug may have reduced the contribution of these organisms to the fermentation. To ascribe the effect of cycloheximide entirely to the inhibition of growth of fungi may therefore be to overestimate the fungal contribution. For future studies, an antibiotic such as nikkomycin, than inhibits synthesis of chitin [a component of fungi but not of protozoa (Gooday *et al.*, 1985)], might prove valuable. Even so, this study suggests that the bacteria are more active in straw digestion than are the fungi. A similar conclusion was reached by Windham and Akin (1983) with respect to the digestion of bermuda grass and alfalfa hay.

The production of methane following prolonged incubation of straw with rumen contents was only slightly reduced in the presence of cycloheximide or streptomycin and penicillin. Tetracycline, to which fungi and methanogens are resistant, strongly inhibited methanogenesis. This suggested that methanogenesis was primarily from substrates (probably hydrogen and formate) provided by tetracycline-sensitive bacteria. There were little differences between the amounts of methane produced from the treated and untreated straw. It is possible that the stoichiometries of the fermentation of the two straws differ, but no direct evidence was sought for this hypothesis.

Direct comparison of straw degradation by cultures of bacteria and fungi is complicated by the marked differences in growth form and life cycles of these organisms. This makes the provision of comparable inocula difficult. In this study, the relative numbers of colony forming units of fungi and bacteria in the inocula would have been similar to the numbers normally detected in the rumen. If this is accepted as a valid basis for comparison, the experiments with mixed laboratory cultures suggest that the bacteria more rapidly degraded *Pleurotus*-treated and untreated straw than did the fungi.

Fungi, bacteria and protozoa in the rumen do not exist in isolation, but function in interacting consortia (Stewart *et al*, 1986). Thus, it will probably never be possible to define precisely the contribution of each group to biomass degradation. For example, we do not know if the fungi produce growth factors for some bacteria, or whether, in the rumen (as opposed to in test tubes), some of the hydrogen produced by fungi could be used in the synthesis of succinate and/or propionate by *Selenomonas* and other rumen bacteria (Henderson, 1980; Richardson & Stewart, 1989). Furthermore, the removal of one population using antibiotics may release constraints on others. For example, the use of antibacterial antibiotics did not eliminate plant cell wall digestion in these studies (Fig.

2). Here it is assumed that the suppression of bacterial growth permitted proliferation of the fungi, a phenomenon reported by Windham and Akin (1983). Thus, although it seems safe to conclude that, under the conditions used here, the direct contribution of the bacteria to straw digestion was greater than that of the fungi, it would be wrong to assume that the fungi are of no importance, or that our attempts to learn about them should cease.

REFERENCES

Bryant, M.P. (1972). Commentary on the Hungate technique for culture of anaerobic bacteria. Am. J. Clin. Nutr. **25**, 1324-1328.

Cheng, K.-J., Stewart, C.S., Dinsdale, D. and Costerton, J.W. (1984). Electron microscopy of bacteria involved in the digestion of plant cell walls. Anim. Fd. Sci. Technol. **10**, 93-120.

Eadie, J.M. (1967). Studies on the ecology of certain rumen ciliate protozoa. J. Gen. Microbiol. **49**, 175-194.

Gale, E.F., Cundliffe, E., Reynolds, P.E., Richmond, M.H. and Waring, M.J. (1981). *The Molecular Basis of Antibiotic Action,* Wiley, London.

Gooday, G., Woodman, J., Casson, E.A. and Browne, C.A. (1985). Effect of nikkomycin on chitin spine formation in the diatom, *Thalassiosire fluviatilis,* and observations on its peptide uptake. FEMS Microbiol. Lett. **28**, 335-340.

Henderson, C. (1980). The influence of extracellular hydrogen on the metabolism of *Bacteroides ruminicola, Anaerovibrio lipolytica* and *Selenomonas ruminantium.* J. Gen. Microbiol. **119**, 485-491.

Hobson, P.N. (1969). The rumen bacteria. In *Methods in Microbiology,* Vol. **3B**, eds. J.R. Norris and D.W. Ribbons, Academic Press, London, pp. 133-149.

Holdeman, L.V., Cato, E.P. and Moore, W.E.C. (1977). *Anaerobic Laboratory Manual,* 4th Edn. Virginia Polytechnic Institute, Blacksburg, VA, USA.

Kolankaya, N., Stewart, C.S., Duncan, S.H., Cheng, K.-J. and Costerton, J.W. (1985). The effect of ammonia treatment on the solubilisation of straw and the growth of cellulolytic rumen bacteria. J. Appl. Bacteriol. **58**, 371-379.

Mehrez, A.Z. and Ørskov, E.R. (1977). A study of the artificial fibre bag technique for determining digestibility of feeds in the rumen. J. Agric. Sci. **8**, 645-650.

Orpin, C.G. (1984). The role of ciliate protozoa and fungi in the rumen digestion of plant cell walls. Anim. Fd. Sci. Technol. **10**, 121-143.

Reade, A.E. and MacQueen, R.E. (1983). Investigations of white-rot fungi for the conversion of poplar into a potential feedstuff for ruminants. Can. J. Microbiol. **29**, 457-463.

Richardson, A.J. and Stewart, C.S. (1989). Hydrogen transfer between *Neocallimastix frontalis* and *Selenomonas ruminantium* grown in mixed culture. In *Microbiology and Biochemistry of Strict Anaerobes Involved in Interspecies Hydrogen Transfer,* ed. J.P. Belaich , Plenum Press, New York, in press.

Stewart, C.S. and Richardson, A.J. (1989). Enhanced resistance of anaerobic rumen fungi to the ionophores monensin and lasolocid in the presence of methanogenic bacteria. J. Appl. Bacteriol. **66**, 85-93.

Stewart, C.S., Gilmour, J. and McConville, M.L. (1986). Microbial interactions, manipulation and genetic engineering. In *New Developments and Future Perspectives in Research on Rumen Function,* ed. A. Niemann-Sørensen, Agriculture - CEC, Luxembourg, pp. 243-257

Sundstøl, F. and Owen, E. (1984). Straw and other fibrous byproducts as feed. Elsevier Applied Science, London.

Tilley, J.M.A. and Terry, R.A. (1963). A two-stage technique for the *in vitro* digestion of forage crops. J. Br. Grassland Soc. **18**, 104-111.

Van Soest, P.J. (1982). *Nutritional Ecology of the Ruminant.* Comstock Publishing Associates, Ithaca, NY, USA.

Vuillet, S., Demarquilly, C., Durand, A., Blachere, H. and Odier, E. (1988). Solid-state fermentation of wheat straw by *Cyathus stercoreus* for improvement of feed value for ruminants. In *Treatment of Lignocellulosics with White-rot Fungi,* eds. F. Zadrazil and P. Reiniger, Elsevier Applied Science, London, pp. 105-114.

Windham, W.R. and Akin, D.E. (1984). Rumen fungi and forage fibre degradation. Appl. Environ. Microbiol. **48**, 473-476.

Zadrazil, F. (1977). The conversion of straw into feed by Basidiomycetes. Eur. J. Appl. Microbiol. **4**, 291-294.

Zadrazil, F. and Brunnert, H. (1981). Investigations of physical parameters important for the solid -state fermentation of straw by white-rot fungi. Eur. J. Appl. Microbiol. Biotechnol. **11**, 183-188.

Zadrazil, F. and Brunnert, H. (1982). Solid-state fermentation of lignocellulose containing plant residues with *Sporotrichum pulverulentum* Nov. and *Dichomitus squalens* (Karst). Eur. J. Appl. Microbiol. Biotechnol. **16**, 45-51.

Zadrazil, F. and Heltay, I. (1989). *In vitro* digestibility of spent *Pleurotus* spp. straw-based substrates proposed as feed for ruminants. in press.

CHAIRMAN'S REPORT ON SESSION IV

Bioconversion of lignocellulosic materials *in vivo*

Ruminants may be thought of as beautifully-adapted anaerobic fermentors that produce biomass and fermentation byproducts useful to the host animal. Although the rumen microorganisms have been studied for more than a century, a recognition of the quantitative importance of the rumen fungi has been relatively recent.

The review paper of Fonty, Bernalier and Gouet describes these fungi and some of their characteristics. As these authors point out, the possession of fungi is not essential to the ruminant. Indeed, fungi are absent from the microbial population of some types of rumen or from the rumen under certain conditions. However, the fact that the rumen fungi are successful, are widely distributed among herbivorous species, and can contribute substantially to the rumen microbial biomass, implies that they may be of importance in the rumen degradation of some materials. The observation that these microorganisms are present only in small numbers, or even absent when the host animal is fed on a diet of rapidly-degradable material, suggests that their importance is likely to be directed to the more slowly-degraded poor-quality roughages.

This makes sense, in that the more slowly a dietary substrate is degraded, the longer is its residence time in the rumen. Thus, the slowly-degraded poor-quality and relatively highly-lignified roughages, will tend to enable a better colonisation by the fungi - these having longer life cycles than have the rumen bacteria or protozoa. Fonty and colleagues refer to work that has shown the fungi to be capable of degrading proteins. However, bacteria and protozoa also degrade protein. More interesting is the observation that some lignin may be degraded by rumen fungi *in vitro.* If low availability of nitrogen encourages the degradation of lignin, then, notwithstanding the ability of the ruminant to recycle nitrogen, any degradation of lignin *in vivo* would again be most likely to occur with poor-quality roughages since these are usually low in nitrogen.

Whether lignin is, or is not, degraded *in vivo,* the observations referred to by Fonty *et al.* i.e. that the rumen fungi are associated with an increase in the fragility of fibre, and of a decrease in the particle size of dry matter outflow from the rumen, could be important. The reasons for this are as follows. It is generally agreed that the rumen contents contain an abundance of particles small enough to leave the rumen without further degradation. It is also well established that reducing the average size of the particles in the rumen increases the rate of outflow of dry matter. Therefore, the effect of rumen fungi one should look for *in vivo* might be an increase in intake, rather than improvement in digestiblity.

The paper of Stewart, Akoyo & Zadrazil describes work that showed that the degradability of straw incubated *in vitro* in rumen liquor was only slightly affected (the rate of degradation was reduced) by the inclusion of a fungicide to remove the rumen fungi. By contrast, removal of the bacteria with the use of antibiotics had a substantial effect (both the rate and the extent of degradation were much reduced). Thus, the main contributors to degradation were the bacteria, with the rumen fungi having only a small effect.

The authors also used as substrate (*in vitro* or *in sacco*) wheat straw that had been treated by fermentation with a white-rot fungus (*Pleurotus* sp.). When incubated *in vitro, Pleurotus*-treatred straw was more extensively degraded than was untreated straw. However, when incubated in nylon bags located in the rumen of an animal (*in sacco*), the differences between the *Pleurotus*- treated and untreated straws were small, and were confined to early stages of the incubation. The *in sacco* degradation losses were corrected for material lost from the nylon bags without degradation (small particles and soluble material). This is important, for such losses, which accounted for 12% of the original material in the untreated straw, were more than doubled in the straw treated with *Pleurotus*. Had the authors not been careful to make this correction, they could have been misled. The production of soluble material was also increased (from about 4% to 9%) by treatment of the straw with *Pleurotus*.

It is not clear why there should have been such large differences between extents of degradation of straws *in vitro* (mostly due to the very poor degradability of the untreated straw *in vitro*) and the relatively small differences *in sacco.* It is probably better to view the *in sacco* results as being more representative of what would happen *in vivo.* If the slightly improved degradation of the *Pleurotus*-treated straw during the initial stages of incubation, measured *in sacco,* was a real effect, then this observation would suggest that feeding of *Pleurotus*-treated straw to animals should result in slightly increased intakes and/or digestibility.

The observation that (as with rumen fungi), treatment with *Pleurotus* increased the fragility of the straw is of interest. This is because, as was suggested above, an increased fragility could have the effect of causing a reduced retention time in the reticulo-rumen and, hence, an increase in voluntary intake by the animal. Therefore, it may be that any useful effect of *Pleurotus*-treatment on the feed value of roughages would be through an effect on voluntary intake. In this case, effects on digestibility due to the increases in soluble material, and on initial degradability, would tend to be offset by the reduction in the retention time in the rumen.

Thus, if the main effect of rumen fungi, and the use of white-rot or other fungi to treat highly lignified forages, is to increase the fragility of the material, then any advantage to the ruminant animal may be through an effect on voluntary intake rather than on improvements in forage digestibility. In the case of fungal treatment of materials before feeding, a further factor is of importance. One must know the losses of material occurring during treatment and whether these losses are of the more degradable dry matter (i.e. that of most value to the animal). We must also be certain that no health hazard to the animal, or its keeper, is introduced by fungal spores or toxins.

Attempts to understand the growth and metabolism of the fungi, and of their function in their ecosystems, have depended upon traditional method of examining a few parameters at a time. The constraints imposed by limited replication mean that this approach will always have to be used in large animal experimentation and with large solid-state fermentors. However, the complexity of biological systems is such that an understanding of the function and interaction of the many factors must ultimately depend on successful mathematical modelling. The paper by France and colleagues demonstrate how the simple integration of basic factors can provide useful information. Many biologists are critical of mathematical modelling, arguing that biological systems are to complex and too variable for useful practical application of models. Part of this criticisr may arise from being faced with three pages of dimly-understood differential calculus However, the power of modern computers means that the effect of changing variables ca be calculated easily and rapidly. The use of such hardware (and software) also make feasible the representation of models in ways that are more easily understood, and i which one can better visualize the dynamics of the system being modelled.

An important benefit to biologists from participation in studies on the modellin process is the necessity to have a very clear concept of the components of the system being modelled. This necessity imposes a valuable discipline on our more generalise 'concept' models. Furthermore, it can force us, as biologists, to think beyond our ow particular field, and create a useful framework for group participation and discussion The need to identify and quantify the components of a system, with subsequen verification of the biology of even relatively simple models, requires tight and goo experimentation. Again, a useful discipline.

Simple models can themselves be useful to biologists. This was illustrated in th paper of France *et al.*, in which a simple model was used to estimate the contribution c the rumen fungi to total rumen microbial biomass. The size of the fungal biomas estimated in this way emphasises that the fungal community must have importance und some conditions. The general comments of France about modelling are also to the poin for the success of any modelling will depend on mathematicians with an empathy for th biology, on biologists with the imagination to exploit what good modelling has to offe and on our ability to communicate one with the other.

F. Dickon Hove

SESSION V

Use of white-rot fungi for food, feed and industrial purposes

THE USE OF WHITE-ROT FUNGI AND THEIR ENZYMES FOR BIOPULPING AND BIOBLEACHING

P. Ander

STFI, Pulp Department
Box 5604, S-114 86 Stockholm, Sweden

Recent progress in biopulping and biobleaching is discussed partly by mentioning new results appearing at the Fourth International Conference on Biotechnology in the Pulp and Paper Industry (Kirk and Chang, 1990). In addition the results with haemoglobin bleaching of pine and birch kraft pulps performed at STFI is discussed. Finally, the possible use of quinone-reducing enzymes such as cellobiose:quinone oxidoreductase (CBQase) to decrease lignin polymerization reactions *in vitro* caused by laccase and different peroxidases is suggested. It was found that kraft lignin polymerization, vanillic acid decarboxylation and veratryl alcohol oxidation by laccase and peroxidases were inhibited, or at least decreased, by CBQase plus cellobiose.

INTRODUCTION

Cellulose, hemicellulose and lignin in wood and straw are important renewable resources. However, the lignin constitutes a hindrance to effective utilization of the polysaccharides and is still poorly used commercially. A great deal of current research is geared towards the better utilization of the different wood components. Examples of such applications are the production of protein and edible fungi (Zadrazil and Reiniger, 1986) and the use of white-rot fungi and their enzymes in the pulp and paper industry (Eriksson, 1988; Eriksson *et al.*, 1990; Kirk and Chang, 1990).

In the above research white-rot fungi are mainly used because they degrade lignin in addition to the other wood components. To produce xylanases and mannanases, however, other fungi or bacteria may also be used. The most important enzymes isolated from white-rot fungi are laccase, manganese-dependent peroxidase and lignin peroxidase. Since the peroxidases contain iron-porphyrin as the active component, such structures are studied in order to achieve lignin degradation and bleaching of pulp fibres (see below). In order to produce the hydrogen peroxide necessary for the peroxidases the white-rot fungi also produce glucose-1-oxidase, glucose-2-oxidase, methanol oxidase and glyoxal oxidase. During lignin degradation by white-rot fungi the lignin is completely degraded to smaller fragments and finally to CO_2. *In vitro*, with laccase or the different peroxidases, however, polymerization is also obtained. It is probable that the fungus uses quinone-reducing enzymes (and possibly phenoxy radical-reducing enzymes) to decrease this polymerization. In this paper some new results, obtained with cellobiose:quinone oxidoreductase (CBQase), about the interaction with laccase and peroxidases will be discussed (Ander *et al.*, 1989).

BIOPULPING

Treatment of wood to produce biological pulps is preferably done by whole organisms and not by enzymes since it is hard for enzymes to penetrate deeply enough into the wood structure. At STFI some effort has been put in to produce active cellulase-less mutants of *Phanerochaete chrysosporium* (Johnsrud, 1986). The new mutants produced by cross-breeding degraded lignin in birch and spruce wood more effectively than did earlier mutants. Although some encouraging results have been obtained regarding energy consumption in mechanical pulp and in kraft pulping after biological treatments, the cost of the fungal pretreatment is still considered to be too high (Eriksson, 1988). Unwanted infections and allergic reactions due to different fungal spores is one serious problem, especially when handling fungus-treated straw and sugar cane bagasse. Otherwise, good results have been obtained after treatment of bagasse with cellulase-less mutants and then treatment in the cold soda process developed in Cuba (Johnsrud *et al.*, 1987). A good quality pulp with a lower requirement for energy in the refining was obtained. The results indicate that Cel$^-$ mutants are better suited for the rotting of wheat straw and bagasse than for wood. However, problems with spore allergy and lung emphysema may occur if the system is not properly handled.

Production of biomechanical pulps is also studied in the USA (Forest Products Lab in Madison) and in Canada (PAPRICAN). Aspen and loblolly pine were used and 35% energy reduction and better strength were obtained after 3-4 weeks treatment with naturally occurring white-rot fungi (Kirk and Chang, 1990).

A Biopulping Consortium, founded in 1987, consists of 17 pulp and paper companies that work together with Forest Products Lab and University of Wisconsin to identify lignin-degrading fungi and to improve them genetically, to characterize important enzymes and to evaluate the pulp produced.

BIOBLEACHING

Bleaching and lignin degradation of both hardwood and softwood kraft pulp by *Coriolus versicolor* are studied by PAPRICAN and NRC in Montreal, Canada. During a 5-10 day treatment lower kappa number and better brightness but lower viscosity were obtained (Kirk and Chang, 1990). One aim of this type of research is to decrease the amount of chlorine/chlorine dioxide in bleaching.

Biomimetic bleaching of pulp using new types of iron-porphyrins is also studied by groups in Vancouver (University of British Columbia), in Lexington, Massachusetts (Repligen Sandoz Research Corporation) and at University of Idaho, Moscow, USA. Since lignin peroxidase contains iron-porphyrin it is said (and has been shown) that iron-porphyrins mimic lignin peroxidase activity. The group in Vancouver obtained new water-soluble porphyrins by coupling chlorine atoms and sulphonic acid groups to the porphyrin. The resulting porphyrin was more stable and had a higher activity. Problems here included the decrease in viscosity indicating cellulose degradation and lower strength pulps.

Treatment of pulp fibres with xylanases and mannanases is studied in Helsinki (VTT) and in New Zealand (Forest Research Institute) in collaboration with Forintek in Ottawa, Canada (Kirk and Chang, 1990). This treatment breaks bonds between hemicellulose and lignin and renders the pulp more easily extractable. In turn less chlorine needs to be used in the bleaching. A similar positive effect has not been obtained with lignin peroxidase.

BLEACHING WITH HAEMOGLOBIN

At STFI several iron-porphyrins like horseradish peroxidase, cytochrome c, haemoglobin, vitamin B_{12}, haematoporphyrin and haeminchloride have recently been investigated for bleaching effect on kraft pulp (Pettersson *et al.*, 1988; Yang, 1989). The best result was obtained with haemoglobin (Table 1) which at room temperature gave a decrease in kappa number of unbleached kraft pulp from 30 to 22 in 26 h, while the viscosity dropped from 1180 to 1051. The brightness increased from 23.5 to 35.6. A positive effect was also obtained when 0_2-bleached pine kraft pulp or unbleached birch kraft pulp was used (Table 1).

The conditions usually used for haemoglobin activation and bleaching is seen in the scheme below:

Activation of Hb

200 mg Hb/10 ml H_2O

| drop by drop

85 ml dioxane
5 ml palmityol chloride (PC)
15 min stirring gives PC-haemoglobin

Bleaching of pulp

15 ml PC-haemoglobin (0.75% Hb/g pulp)
0.1 ml of 30% H_2O_2 (0.75%/g pulp)
200 ml 90% dioxane, or water or phosphate buffer
4 g kraft pulp, kappa 30 = 3.25% lignin

Table 1. Bleaching of unbleached and O_2-bleached pine kraft pulp and unbleached birch kraft pulp by palmityol chloride-treated haemoglobin in dioxane-water (90:10). From Pettersson *et al.* (1988).

Pulp	Property	Bleaching time (h)			
		Ref	2	6	26
Pine kraft pulp	Kappa No.	30	27.3	25.4	22.0
	Viscosity	1180	1151	1117	1051
	Brightness	23.5	34.3	34.7	35.6
O_2-bleached pine kraft pulp	Kappa No.	22.5	19.5	16.4	13.5
	Viscosity	1050	946	886	876
	Brightness	30.5	43.0	43.2	40.7
Birch kraft pulp	Kappa No.	18	14.4	13.2	12.3
	Viscosity	1250	1204	1147	1083
	Brightness	40.0	51.0	51.6	52.3

Before the bleaching chemicals were mixed with the pulp, the haemoglobin was treated in 90% dioxane/water with palmityol chloride [$CH_3(CH_2)_{14}COCl$] which covalently binds to amino groups in the haemoglobin. This renders the resulting PC-haemoglobin more hydrophobic and it can react more easily with the lignin in the fibres. Kappa number 30 corresponds to a lignin content of 3.25% in the fibres as determined with $KMnO_4$-titration.

Later, Yang (1989) found that the kappa number decreased from 30 to 22.6 in 90 min if the temperature was 40°C. A drop in viscosity from 1100 to about 970 was obtained but this decrease was considered to be relatively small.

CBQase INTERACTION WITH LACCASE AND DIFFERENT PEROXIDASES

Laccase, Mn-peroxidase and lignin peroxidase all have some depolymerizing activity but lignin and its degradation products are also polymerized (Haemmerli *et al.*, 1986; Kirk, 1986; Kawai *et al.*, 1988). This is due to formation of quinones, phenoxy radicals and cation radicals which may couple and condense. Cellobiose:quinone oxidoreductase (CBQase), discovered by Westermark and Eriksson (1974a, b), reduces quinones (and possibly phenoxy radicals) during oxidation of cellobiose to cellobionolactone. It is produced by most white-rot fungi (Ander and Eriksson, 1977) and is thought to be one of the enzymes that may regulate the equilibrium between polymerization and depolymerization in cultures growing on wood and cellulose. In cultures grown on glucose several other quinone-reducing enzymes are produced to facilitate further degradation (Buswell *et al.*, 1979; Schoemaker *et al.*, 1989). However, Odier *et al.* (1988) found that CBQase did not reduce the phenoxy radical generated from acetosyringone by lignin peroxidase and that polymerization of guaiacol and synthetic lignin was not prevented by CBQase.

We have now shown that polymerization of kraft lignin by lignin peroxidase decreased in the presence of CBQase plus cellobiose (Ander *et al.*, 1989). We also showed vanillic acid decarboxylation by laccase, Mn-peroxidase, horse radish peroxidase and lignin peroxidase to be strongly inhibited by CBQase plus cellobiose (Figure 1). Thus, laccase-catalyzed decarboxylation decreased from 37.8 to 2.44%, whereas the corresponding figure for Mn-peroxidase was a decrease from 34.3 to 5.54%. Decarboxylation by lignin peroxidase decreased from 24.7% to 0.46% at pH 4, and from 14.8% to 0.58% at pH 3.

In these tests, laccase and peroxidase activities were measured by monitoring the

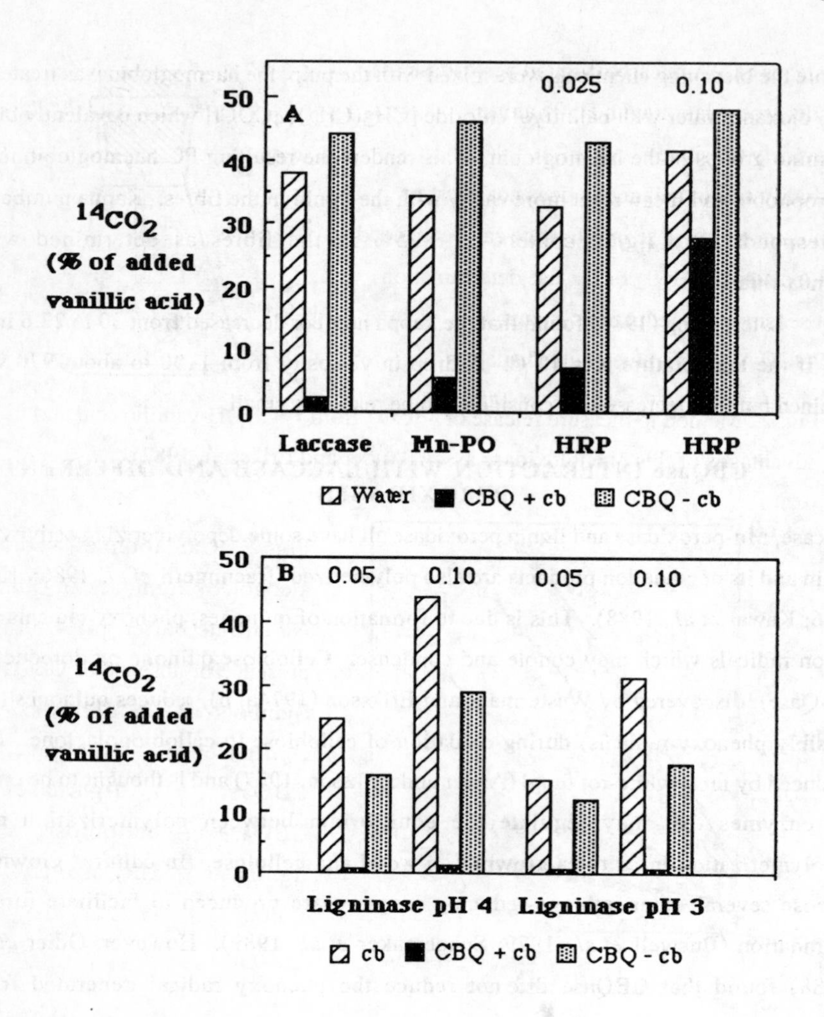

Fig. 1. Decarboxylation of vanillic acid by laccase (A), Mn-peroxidase (A), horse radish peroxidase, HRP (A), and lignin peroxidase (B) in the presence or absence of CBQase ± cellobiose (Ander *et al.*, 1989). In the case of HRP, 0.025 and 0.10 ml were used, while 0.05 and 0.10 ml of lignin peroxidase were used. (The HRP activity was 'strong', giving a non-linear response and a small inhibition by CBQ + cellobiose)

release of $^{14}CO_2$ from ^{14} COOH-vanillic acid (Fig. 2). The method has the advantage of not being dependent on a stable quinone colour. Instead the released $^{14}CO_2$ is immediately trapped in NaOH and counted for radioactivity (Ander and Eriksson, 1987).

Advantages of VA-decarboxylation peroxidase assay.

1. No decarboxylation with H_2O_2 alone
2. Extremely sensitive
3. Turbidity and/or cell debris does not interfere
4. Vanillate hydroxylase interference can be avoided
5. Is good also for H_2O_2 or laccase determination
6. Not sensitive for reducing systems
7. ^{14}COOH-VA is commercially available

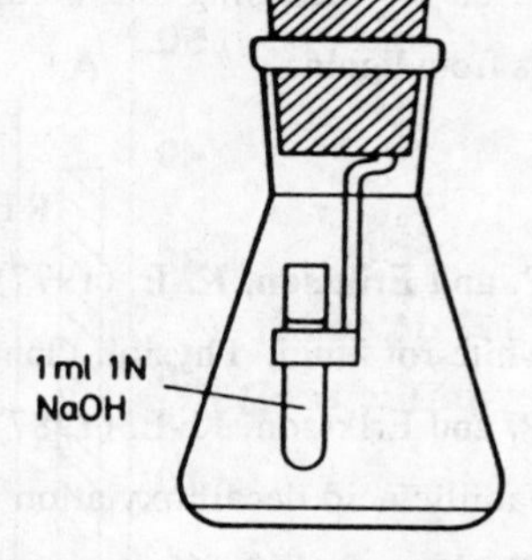

Fig. 2. Method to measure release of $^{14}CO_2$ from ^{14}COOH-vanillic acid and advantages of this phenoloxidase assay (Ander and Eriksson, 1987).

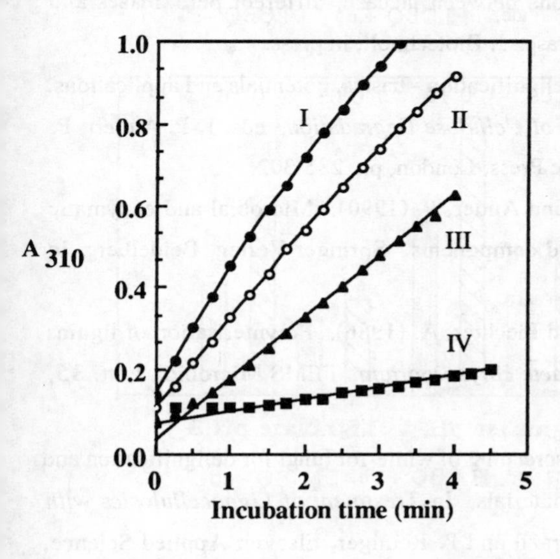

Fig. 3. Oxidation of veratryl alcohol by ligninase in 50mM Na-tartrate buffer, pH 4, in the presence or absence of CBQase. Symbols are as follows: I, ligninase; II, ligninase + CBQase; III, ligninase + CBQase + cellobiose; IV, ligninase + CBQase (2x) + cellobiose (2x) (Ander *et al.*, 1989).

Veratryl alcohol oxidation to veratraldehyde by lignin peroxidase was also measured in the presence of CBQase plus cellobiose. Active CBQase strongly inhibited lignin peroxidase activity (Fig. 3). Only cellobiose had no effect on lignin peroxidase activity.

In addition to these results it was also found that formation of a yellow cation radical from 1, 2, 4, 5-tetramethoxybenzene (TMB) was inhibited by CBQase. Taken together the results indicate that CBQase in some way interacts with laccase and peroxidases. However, it is not yet clear whether CBQase really reduces the phenoxy radicals of vanillic acid or if it interacts with the subsequent attack of activated oxygen species which ultimately cause the release of CO_2. CBQase should probably be one ingredient in an enzyme mixture degrading lignin. In the future such enzyme mixtures may be used for bleaching and modification of pulp fibres or for the production of chemicals from lignin.

REFERENCES

Ander, P. and Eriksson, K.-E. (1977). Selective degradation of wood components by white-rot fungi. Physiol. Plant. **41**, 239-242.

Ander, P. and Eriksson, K.-E. (1987). Determination of phenoloxidase activity using vanillic acid decarboxylation and syringaldazine oxidation. Biotechnol. Appl. Biochem. **9**, 160-169.

Ander, P., Mishra, C., Farrell, R.L. and Eriksson, K.-E.L. (1989). Redox reactions in lignin degradation: Interactions between laccase, different peroxidases and cellobiose:quinone oxidoreductase. J. Biotechnol., in press.

Eriksson, K.-E.L. (1988) Microbial delignification - basics, potentials and applications. In *Biochemistry and Genetics of Cellulose Degradation,* eds. J.-P. Aubert, P. Béguin and J. Miller, Academic Press, London, pp. 285-302.

Eriksson, K.-E.L., Blanchette, R.A. and Ander, P. (1990). Microbial and enzymatic degradation of wood and wood components. Springer Verlag, Beidelberg, in press.

Haemmerli, S.D., Leisola, M.S.A. and Fiechter, A. (1986). Polymerisation of lignins by ligninases from *Phanerochaete chrysosporium.* FEMS Microbiol. Lett. **35**, 33-36.

Johnsrud, S.C. (1986). Selection and screening of white-rot fungi for delignification and upgrading of lignocellulosic materials. In *Treatment of Lignocellulosics with White-rot Fungi,* eds. F. Zadrazil and P. Reiniger, Elsevier Applied Science, London, pp. 50-55.

Johnsrud, S.C. *et al.*(1987). Properties of fungal pretreated high yield bagasse pulps, Nordic Pulp and Paper Res. J., Special Issue 2, 47-52.

Kawai, S., Umezawa, T. and Higuchi, T. (1988). Degradation mechanisms of phenolic ß-1 substructure model compounds by laccase of *Coriolus versicolor.* Arch. Biochem. Biophys. **262**, 99-110.

Kirk, T.K. (1986). The action of ligninase on lignin model compounds and lignin. In *Proc. TAPPI Research and Development Conference,* TAPPI, U.S.A., pp. 73-78.

Kirk, T.K. and Chang, H.-M. (1990). *Proc. Fourth International Symposium on*

Biotechnology in the Pulp and Paper Industry, Raleigh, NC, USA, in press.

Odier, E., Mozuch, M.D., Kalyanaraman, B. and Kirk, T.K. (1988). Ligninase -mediated phenoxy radical formation and polymerization unaffected by cellobiose:quinone oxidoreductase, Biochimie, **70**, 847-852.

Pettersson, B., Yang, J.-L. and Eriksson, K.-E. (1988). Biotechnical approaches to pulp bleaching. Nordic Pulp and Paper Res. J. 3(4), 198-202.

Schoemaker, H.E., Meijer, E.M., Leisola, M.S.A., Haemmerli, S.D., Waldner, R., Sanglard, D. and Schmidt, W.H. (1989). Oxidation and reduction in lignin biodegradation. In *Plant Cell Wall Polymers: Biogenesis and Biodegradation,* eds. N.G. Lewis and M.G. Paice, ACS Symp. Ser. **399**, Washington, DC, pp. 454-471.

Westermark, U. and Eriksson, K.-E. (1974a). Carbohydrate dependent quinone reduction during lignin degradation. Acta Chem. Scand. **B28**, 204-208.

Westermark, U. and Eriksson, K.-E. (1974b). Cellobiose:quinone oxidoreductase, a new wood-degrading enzyme from white-rot fungi. Acta Chem. Scand. **B28**, 209-214.

Yang, J.-L. (1989). *Biotechnical approaches to analysis of mechanical pulp fiber surfaces and pulp bleaching.* Dissertation, STFI and The Royal Institute of Technology, Stockholm.

Zadrazil, F. and Reiniger, P. eds. (1986). *Treatment of Lignocellulosics with White-rot Fungi,* Elsevier Applied Science, London.

CULTIVATION OF EDIBLE FUNGI ON PLANT RESIDUES

J.F. Smith and D.A. Wood

AFRC Institute of Horticultural Research,
Worthing Road, Littlehampton, West Sussex BN17 6LP, UK

In this brief review we emphasize the worldwide increase in total production of edible fungi on lignocellulosic wastes in recent years. We also discuss one of the research projects being carried out at IHR, Littlehampton, and outline a laboratory-scale composting procedure used in trials to assess a wide range of lignocellulosic wastes for use in the culture of *Agaricus bisporus* and closely-related species.

INTRODUCTION

Lignocellulose, the structural complex of terrestrial plants is the most abundant of biological materials on earth (Goldstein, 1981). It consists of 3 major components, cellulose, hemicellulose and lignin. At present, considerable efforts are being made by researchers worldwide to discover economic methods for upgrading low-cost bulk plant wastes such as cereal straws, leaves, wood, bark, etc. into higher value fuel, chemicals and food products (reviewed by Wood, 1985). While several microorganisms can utilize such complexes within intact plant tissue (Kirk, 1983; Tsao and Chiang, 1983), white-rot fungi have been shown to be the most efficient in this regard (Kirk, 1983). The major viable biological processes utilizing significant quantities of lignocelluloses include ruminant digestion, waste digesters and the cultivation of edible mushrooms. Of these processes, mushroom cultivation is the only one capable of utilizing the whole range of complex polymers. Mushrooms can, therefore, be regarded as one of the most economic products that can be produced by exploiting lignocellulosic wastes (Chang and Hayes, 1978; Wood, 1984). Mushroom science is relatively new and the fundamentals of substrate chemistry, microbiology and mushroom nutrition for a few select species have been unravelled only in the last 30 years (Smith *et al.*, 1988a). This has resulted in a tremendous surge in total production as is shown for *Agaricus bisporus* (Table 1). This white, button mushroom accounts for nearly 60% of the world production of fungi (Chang, 1987) and the growth substrate normally used is composted cereal straw.

Table 1. World production (thousands of tonnes) of *A. bisporus* (Delcaire, 1978; Chang, 1987).

	1950	1970	1986	% of EC total
USA	30	80	285	
China	-	2	185	
France	19	68	165	28.7
Holland	0.3	29.5	115	20.0
UK	12	40.4	95	16.5
Italy	0.2	20	75	13.0
Canada	1.5	11.7	51.4	
Poland	0.1	5	45	
Spain	-	4	45	7.8
FRG	0.4	20	38	6.6
Taiwan		39	35	
Yugoslavia	-	0.2	18.5	
Irish Republic	0.2	1.9	16	2.8
Belgium	0.8	5	16	2.8
Australia		5.3	14	
South Africa Rep.	-	1.5	11	
Denmark	0.6	6.5	9.3	1.6
Brazil	-	0.2	6	
Switzerland	0.5	2.8	5.6	
New Zealand	-	1	4.8	
Hungary	0.3	2	4.5	
Austria	0.1	3.2	3	
Greece	-	0.2	1.3	0.2
Sweden	0.4	2.6	1.3	
Czechoslovakia	-	0..7	1.2	
Portugal	-	0.4	?	?
Others	0.1	28.1	20.3	
World total	66.5	381.2	1267.2	

Lentinus edodes (Shiitake mushroom), for many centuries the most favoured mushroom in Japan, China and Taiwan, has traditionally been grown on cut logs of oak or chestnut (Tokimoto and Komatsu, 1978). This species, which accounts for about 14% of the world production of fungi, is now becoming popular in Europe and the USA, where accelerated methods of cultivation, using mixtures of sawdust and cereal brans, are being investigated (Lelley, 1985; Leatham, 1979). *Volvariella volvacea* (Straw mushroom), grown either on rice straw or cotton wastes (Chang, 1978), and *Pleurotus* spp. (Oyster mushrooms), grown on cereal straw or sawdust mixtures (Zadrazil, 1978), each account for about 8% of the total production figure. World production figures for both of these fungi have also increased remarkably in recent years (Tables 2 and 3). The remaining 10-15% of production can be attributed to the following: *Auricularia* spp. (Ear fungi) (Cheng and Tu, 1978) and *Tremella fuciformis* (White jelly fungus), grown mainly on wood logs of broad-leaved trees (Chen and Hou, 1978); *Flammulina velutipes*, commonly called the Winter mushroom (Tonomura, 1978), and *Pholiota nameko* (Viscid mushroom), grown on sterilised sawdust/cereal bran mixtures (Arita, 1978). *Coprinus* spp. (Lelley, 1983), *Stropharia rugosoannulata* (Szudyga, 1978) and *Kuehneromyces mutabilis* (Gramss, 1978) are also cultivated, but on a very small scale.

Table 2. World production (thousands of tonnes) of *V. volvacea* (Delcaire, 1978; Chang, 1987).

	1976	1986
Mainland China	30	100
Thailand	5.8	60
Taiwan	5.7	12
Indonesia	-	4
Philippines	0.2	0.2
Hong Kong	-	1.2
India	-	0.3
World total	41.7	177.7

In the major mushroom-producing countries, lignocellulosic wastes are readily available and likely to remain so for many years. Thus, it is unlikely that a major change in substrate formulations for the cultivation of the main species will occur. Nevertheless, as agricultural and industrial wastes continue to accumulate, especially in developing countries, researchers will be encouraged to concentrate their efforts on upgrading low cost waste materials into high value food products such as mushrooms (Table 4). Although they are not complete foods, mushrooms have high contents of protein and vitamins, and are suitable for supplementing protein- and vitamin-deficient diets (Hayes, 1969; Flegg and Maw, 1975).

Table 3. World production (thousands of tonnes) of *Pleurotus* spp. (Delcaire, 1978; Chang (1987).

	1976	1986	% of EC total
Mainland China	?	100	
South Korea	0.13	36	
Japan	5.5	18.3*	
Italy	1.3	10	69.0
Taiwan	4	8	
Thailand	0.02	5.8	
West Germany	-	2.5	17.2
France	0.6	2.0	13.8
Switzerland	0.4	?	
Hungary	0.05	2.0	
Poland	-	1.0	
Total world	12.0	185.6	

* production in 1983 ("Statistics on Mushrooms" by Forestry Agency of Japan).

Table 4. Traditional substrates and some alternatives for the cultivation of edible fungi.

	Alternatives	
A. bisporus Stable bedding	Wheat straw/N additives	Hayes and Randle (1969)
		Gerrits (1974)
	Corn cobs/hay	Block (1965)
		Schisler (1974)
	Rice straw	Wu (1967); Kim (1976)
	Sugar cane bagasse	Kneebone and Mason (1972)
	Sawdust	Block and Rao (1962)
	Paper/wheat straw	Smith (1983)
	Oil palm pericarp	Castro de Jimenez *et al.* (unpublished)
L. edodes Wood logs	Sawdust/bran	Ando (1976)
		Han *et al.* (1981)
		Leatham (1979); Lelley (1985)
***Pleurotus* spp.** Wood logs	Sawdust mixtures	Lohwag (1951)
		Block *et al.* (1959)
	Wheat straw	Stanek and Rysava (1971)
		Zadrazil and Scheidereit, (1972)
	Rice straw	Bano and Srivastava (1962)
	Rice straw/newspaper	Hashimoto and Takahashi (1976)
	Banana pseudostems/rice straw	Jandalk and Kapoor (1976)
	Cotton seed hulls	Cho *et al.* (1981)
	Cotton waste	Khan and Ali (1981)
		Tan (1981)
	Cocoa shells	Senyah *et al.* (1989)
	Coffee pulp	Martinez-Carrera (1987)
V. volvacea Rice straw	Cotton waste	Yau and Chang (1972)
	Sugar cane	Hu *et al.* (1976)
	Wheat straw/hay/corncobs	San Antonio and Fordyce (1972)
	Oil palm pericarp	Chen and Graham (1973)
***Auricularia* spp.** Wood logs	Sawdust/rice bran	Cheng and Tu (1978)
	Composted sawdust	Vilela and Silverio (1982)
Pholiota nameko Bed logs	Sawdust/rice bran	Hashimoto *et al.* (1966)

LABORATORY-SCALE COMPOSTING TECHNIQUE USED TO EVALUATE A RANGE OF LIGNOCELLULOSIC WASTES

Part of the mushroom programme at the Institute of Horticultural Research, Littlehampton, is to evaluate agricultural, food-processing and forestry waste lignocellulosic materials as substrates for *Agaricus bisporus* and closely related species. We have endeavoured in our studies to look at low-cost substrate production methods. Therefore, we have introduced specificity into the substrates by controlled fermentation methods. It is essential, when testing any new material, to know its content of cellulose, hemicellulose, lignin, nitrogen and soluble carbohydrate. The actual contents will have an important bearing on the eventual compost formulation. The physical characteristics of the compost are also very important since the size and shape of the individual materials can greatly effect the water-holding capacity and porosity of the substrate. Many materials (e.g sugarcane waste) are too fine and consolidate too readily, while some cereal (e.g. barley, rice) straws degrade quickly during composting. Accordingly, adjustments in supplementation or composting time are necessary when using such ingredients. It is also essential during the composting period to have a favourable environment (70-75% water content) in which the naturally-occurring microflora can thrive in an aerobic atmosphere and convert soluble carbon and nitrogen forms into insoluble forms such as microbial biomass. These factors have been shown to be extremely important in compost specificity (Fermor *et al.*, 1985).

Purpose-built fermentor units, similar to those described by Lelley and Flick (1981), were designed in order to assess any new growth material at the laboratory level. These incroporated three 2-L multi-adaptor flask units (Fig. 1). Normally, compost formulations are thoroughly wetted to give a moist mixture but without free draining liquid. Aliquots, generally 800 g, so prepared are placed on a perforated steel platform within each flask. The flasks are then immersed in a water bath at a depth sufficient to bring the water level (outside the flask) above the level of the compost (within the flask). The thermostat setting of the heater/stirrer is initially set at 45-50°C, to allow a gradual build up of the naturally-occurring microflora and each flask is independantly aerated with humidified air throughout the total composting period. When the compost temperature reaches 50°C, the thermostat is reset at 60°C. The compost temperature is maintained at 60°C for 3-6 h to pasteurize the material before being returned to 53-55°C, the ideal temperature for growth of thermophilic bacteria and actinomycetes. Composting is terminated when the ammonia concentration within the compost falls below 20 ppm.

Selective substrates for cultivation of *A. bisporus* were produced from a variety of lignocellulosic wastes in 5-6 days using this procedure provided that the total nitrogen content of the compost ingredients at the commencement of composting did not exceed 1.5% of the total dry matter (Smith and Spencer, 1976) and that there was adequate soluble carbohydrate present to stimulate the growth of naturally-occurring microflora. Most lignocellulosic wastes have a low total nitrogen content, i.e. below 1% of the total dry matter. Consequently, they require additional nitrogen supplementation. Some wastes are also deficient in readily-available soluble carbohydrates and additional supplementation may be necessary to raise the content of such compounds in the compost mixture to 3% of dry matter at the commencement of composting (Smith and Spencer, 1977). Generally most lignocellulosic wastes can be used as part of or, in many cases, as the main constituent in composted substrates provided the compost recipe has been correctly formulated.

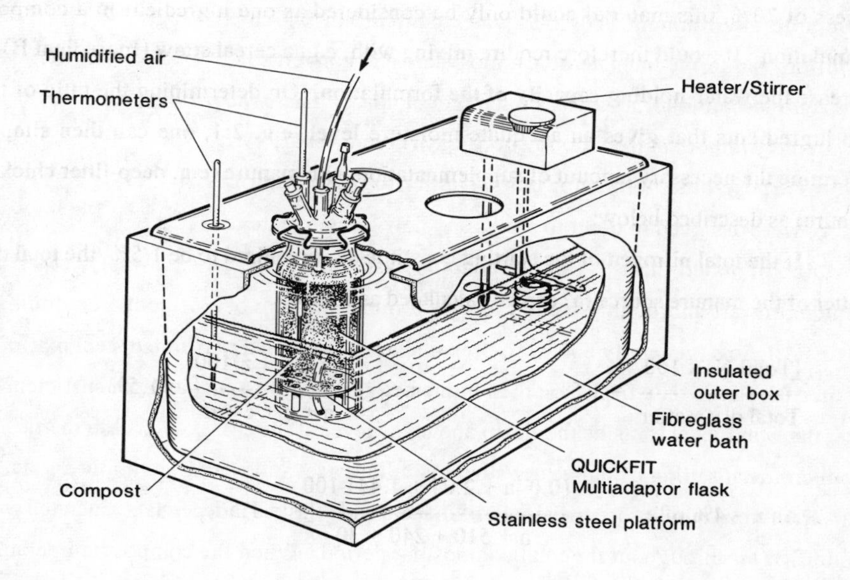

Fig. 1. Multi-adaptor flask fermentation unit.

Table 5. Determination of the required amount of supplementation with manure in a hypothetical compost formulation to be used for short-duration composting.

	Fresh wt. taken (g)	Total N (% DM)	Moisture (% H_2O)	Total dry matter (g)	Total N. (g)
Manure source	?	4.0	20	? (a)	? (x)
Ingredient A	600	0.4	15	510 (b)	2.04 (y)
Ingredient B	300	0.6	20	240 (c)	1.44 (z)
Gypsum	30	-	-	30 (d)	-

Consider a hypothetical waste product from a food-processing plant (Ingredient A) with reasonably high cellulose, hemicellulose and lignin contents, an adequate soluble carbohydrate content, a total nitrogen content of 0.4%, but with a poor water-holding capacity (Table 5). In order to prepare a substrate with a suitable moisture content in excess of 70%, this material could only be considered as one ingredient in a compost formulation. It would therefore require mixing with, e.g. a cereal straw (Ingredient B) to increase the water-holding capacity of the formulation. On determining the ratio of the two ingredients that gives an adequate moisture level, e.g. 2:1, one can then simply determine the necessary amount of supplementation with manure (e.g. deep-litter chicken manure) as described below:

If the total nitrogen of the starting constituents is required to be 1.5%, the total dry matter of the manure source (a) can be calculated as follows.

$$\frac{(\text{Total N}) \times 100}{\text{Total dry matter}} = 1.5\% \quad \text{i.e.} \quad \frac{(x + y + z)100}{a + b + c + d} = 1.5\%$$

$$\text{As } x = 4\% \text{ of a} \quad \frac{(0.04a + 2.04 + 1.44)\,100}{a + 510 + 240 + 30} = 1.5$$

Thus, $4a + 348 = 1.5a + 1170$

$2.5a = 822$

$a = 329$ g (i.e. 411 g fresh weight material)

Recent work using this composting technique has demonstrated that oil palm pericarp waste or coffee bean hulls (Castro de Jiminez, Smith and Love, unpublished observations) can be mixed with wheat straw in ratios 1:1 and 3:1, respectively, and supplemented with deep-litter chicken manure as the main nitrogen source, to produce a suitable substrate for cultivation of *A. bisporus* and *A. bitorquis.* In all of these small-scale trials, the extent of conversion of compost dry matter to mushroom fruitbodies has been such that we consider that economically-acceptable yields could be obtained if the composting process was used on a commercial-scale. This simple method of composting small quantities of materials is also being used in studies to provide a wide range of degraded substrates for screening a number of selected wild *Agaricus* species (Smith *et al.*, 1988b). Successful substrates have also been prepared for a tropical species of *A. bitorquis* shown to have commercial potential (Smith and Love, 1989).

CONCLUSION

In the last 10 years the production of edible fungi on lignocellulosic wastes has increased remarkably. This is likely to continue as solid-substrate fermentation remains the most economical way of upgrading plant wastes to valuable food products. The most likely surge in production will occur in developing countries, especially in those with tropical and subtropical climates, as many such areas have an abundance of lignocellulosic wastes and limited means of disposing of them. The discovery of new strains of mushroom together with the 'genetic engineering' of common strains, so that they can be grown at tropical and sub-tropical temperatures, will also have an important bearing on the rate of cultural development in such countries. The short duration composting technique outlined in this article can be regarded as a laboratory tool for identifying the feasibility of including any new lignocellulosic waste in mushroom compost formulations. However, many materials, by virtue of their physical nature alone, lend themselves to short-duration composting. The information gained from such small-scale studies can, in some cases, be applied directly on a macro-scale. In the case of bulkier materials, a longer, more conventional technique of composting may be necessary. In such cases, additional carbon and nitrogen sources will have to be added so as to introduce a self-heating capacity into the compost fomulations.

REFERENCES

Arita, I. (1978). *Pholioto nameko.* In *The Biology and Cultivation of Edible Mushrooms,* eds. S.T. Chang and W.A. Hayes, Academic Press, New York, pp. 475-496.

Ando, M. (1976). Fruit-body formation of *Lentinus edodes* (Berk.) Sing on the artificial media. Mushroom Sci. 9(1), 415-422

Bano, Z. and Srivastava, H.C. (1962). Studies on cultivation of *Pleurotus* spp. on paddy straw. Food Sci. 12, 363-65.

Block, S.S. (1965). Mushrooms in 19 days. Mushroom Sci. 6, 351-57

Block, S.S. and Rao, S.N. (1962). Sawdust compost for mushroom growing. Mushroom Sci. 5, 134-142

Block, S.S., Tsao, G. and Hau, L. (1959). Experiments in the cultivation of *Pleurotus ostreatus.* Mushroom Sci. 4, 393-399.

Chang, S.T. (1978). *Volvariella volvacea.* In *The Biology and Cultivation of Edible Mushrooms,* eds. S.T. Chang and W.A. Hayes, Academic Press, New York, pp. 573-603.

Chang, S.T. (1987). World production of cultivated edible mushrooms in 1986. Mushroom J. Tropics, 7, 117-120.

Chang, S.T. and Hayes, W.A., eds. (1978). *The Biology and Cultivation of Edible Mushrooms.* Academic Press, New York.

Chen, P.C. and Hou, H.H. (1978). *Tremella fuciformis.* In *The Biology and Cultivation of Edible Mushrooms,* eds. S.T. Chang and W.A. Hayes, Academic Press, New York, pp. 629-643.

Chen, Y.Y. and Graham, K.M. (1973). Studies on the Padi Mushroom (*Volvariella volvacea*). Use of oil palm pericarp waste as an alternative substrate. Mal. Agric. Res. 2, 15-22.

Cheng, S. and Tu, C.C. (1978). *Auricularia* spp. In *The Biology and Cultivation of Edible Mushrooms,* eds. S.T. Chang and W.A. Hayes, Academic Press, New York, pp. 605-625.

Cho, K.Y., Nair, N.G., Bruniges, P.A. and New, P.B. (1981). The use of cotton seed hulls for the cultivation of *Pleurotus sajor-caju* in Australia. Mushroom Sci. 11, (Part 1), 679-690.

Delcaire, J.R. (1978). Economics of Cultivated Mushrooms. In *The Biology and Cultivation of Edible Mushrooms,* eds. S.T. Chang and W.A. Hayes, Academic Press, New York, pp. 728-793.

Fermor, T.R., Randle, P.E. and Smith, J.F. (1985). Compost as a substrate and its preparation. In *The Biology and Technology of the Cultivated Mushroom,* eds. P.B. Flegg, D.M. Spencer and D.A. Wood, Wiley, Chichester, pp. 81-109.

Flegg, P.B. and Maw, G.A. (1975) The mushroom as a source of dietary protein. Annual Report of the Glasshouse Crops Research Institute for 1974, pp. 137-151.

Gerrits, J.P. (1974). Development of a synthetic compost for mushroom growing based on wheat straw and chicken manure. Netherlands J. Agric. Sci. **22**, 175-194.

Goldstein, I.S., ed.(1981). *Organic Chemicals from Biomass.* CRC Press, Boca Raton, Florida.

Gramss, G. (1978). *Kuehneromyces mutabilis.* In *The Biology and Cultivation of Edible Mushrooms,* eds. S.T. Chang and W.A. Hayes, Academic Press, New York, pp. 423-443.

Han, Y.H., Ueng, W.T., Chen, L.C. and Cheng, S. (1981). Physiology and ecology of *Lentinus edodes* (Berk.) Sing. Mushroom Sci. **11**(2), 623-658

Hashimoto, K., Isobe, N. and Takahashi, Z. (1966). Biochemical studies on mushrooms. 11. Nutritional requirements for vegetative growth of *Cortinellus shiitake* and *Pholiota nameko.* Rep. Tokyo. Jr. Coll. Food Technol. **7**, 208-214.

Hashimoto, K. and Takahashi, Z. (1976). Studies on the growth of *Pleurotus ostreatus.* Mushroom Sci. **9**(1), 585-593.

Hayes, W.A. (1969) New techniques with mushroom composts. Span. **12**, 162-166.

Hayes, W.A. and Randle, P.E. (1969). Use of molasses as an ingredient of wheat straw mixtures for the preparation of mushroom composts. Annual Report of the Glasshouse Crops Research Institute for 1968, pp. 142-147.

Hu, K., Song, S., Lui, P. and Peng, J. (1976). Studies on sugarcane rubbish for Chinese Mushroom culture and its growth factor. Mushroom Sci. **9**(1), 691-700.

Jandaik, C.L. and Kapoor, J.N. (1976). Studies on the cultivation of *Pleurotus sajor-caju* (Fr.) Singer. Mushroom Sci. **9**(1), 667-672.

Khan, S.M. and Ali, M.A. (1981). Cultivation of oyster mushroom (*Pleurotus* spp.) on cotton ball locules. Mushroom Sci. **11**, 691-696.

Kim, D.S. (1976). An introduction to mushroom growing in Korea. Mushroom Sci. **9**(1), 121-126.

Kirk, T.K. (1983). Degradation and conversion of lignocelluloses. In *The Filamentous Fungi,* Vol. **4**, *Fungal Technology,* eds. J.E. Smith, D.R. Berry and B. Kristiansen, Edward Arnold, London, pp. 265-295.

Kneebone, L.R. and Mason, E.C. (1972). Sugar cane bagasse as a bulk ingredient in mushroom compost. Mushroom Sci. **8**, 321-330.

Leatham, G.F. (1979). *Selected physiological and biochemical studies of growth and development of Shiitake, the edible Japanese forest mushroom, Lentinus edodes (Berk.) Sing.* Ph.D. thesis, Biochemistry Department, University of Wisconsin, Madison, WI, USA.

Lelley, J. (1983). Investigations on the culture of the Ink Cap, *Coprinus comatus* (Mull. ex Fr.) Gray. Mushroom J. **130**, 364-370.

Lelley, J. (1985). The research institute for mushroom cultivation. Mushroom J. **149**, 171-173.

Lelley, I.J. and Flick, M.E. (1981). Equipment for biological preparation and fungal fermentation of solid substrates under laboratory conditions. Mushroom Sci. **11**, 893-897.

Lohwag, K. (1951). Crops from sawdust. Mushroom Growers' Association Bulletin, **22**, 20-21.

Martinez-Carrera, D. (1987). Design of a mushroom farm for growing *Pleurotus* on coffee pulp. Mushroom J. Tropics, **7**(1), 13-23.

San Antonio, J.P. and Fordyce, C. Jr. (1972). Cultivation of paddy straw mushroom, *Volvariella volvacea* (Bull. Ex. Fr.). Sing. Hort. Sci. **7**, 461-464.

Senyah, J.K., Robinson, R.K. and Smith, J.F. (1989). The cultivation of the Oyster - *Pleurotus ostreatus* (Jacq. Ex. Fr.) Kummer - on cocoa shell waste. Mushroom Sci. **12**(2), 207-218.

Schisler, L.C. (1974). Possible substitutes for corn cobs in mushroom composts. Mushroom News, **22**(5), 6.

Smith, J.F. (1983). Paper bedding from horses and poultry. Mushroom J. **127**, 245-250.

Smith, J.F. and Love, M.E. (1988). A tropical *Agaricus* with commercial potential. Mushroom Sci. **12**(1), 305-315.

Smith, J.F. and Spencer, D.M. (1976). Rapid preparation of composts suitable for the production of the cultivated mushroom. Scientia Horticulturae, **5**, 23-31.

Smith, J.F. and Spencer, D.M. (1977). The use of high energy carbon sources in rapidly prepared mushroom composts. Scientia Horticulturae, **7**, 197-205.

Smith, J.F., Elliott, T.J. and Love, M.E. (1988). Variations on a theme: an assessment of alternative mushroom species. Grower, **109**(15), 20-21.

Smith, J.F., Fermor, T.R. and Zadrazil, F. (1988). Pretreatment of lignocellulosics for edible fungi. In *Treatment of Lignocellulosics with White-rot Fungi,* eds. F. Zadrazil and P. Reiniger, Elsevier Applied Science, London, pp. 3-13.

Stanek, M. and Rysava, J. (1971). Application of thermophilic micro-organisms in the fermentation of a nutrient substrate for the cultivation of *Pleurotus ostreatus* (Jacq. Ex. Fr.) Kummer. Mykol. Sb. **8**, 59-60.

Szudyga, K. (1978). *Stropharia rugoso-annulata.* In *The Biology and Cultivation of Edible Mushrooms,* eds. S.T. Chang and W.A. Hayes, Academic Press, New York, pp. 559-571.

Tan, K.K. (1981). Cotton waste is a good substrate for cultivation of *Pleurotus ostreatus,* the oyster mushrrom. Mushroom Sci. **11**, 705-710.

Tokimoto, K. and Komatsu, M. (1978). Biololgical nature of *Lentinus edodes.* In *The Biology and Cultivation of Edible Mushrooms,* eds. S.T. Chang and W.A. Hayes, Academic Press, New York, pp. 445-459.

Tonomura, H. (1978). *Flammulina velutipes.* In *The Biology and Cultivation of Edible Mushrooms,* eds. S.T. Chang and W.A. Hayes, Academic Press, New York, pp. 410-421.

Tsao, G.T. and Chiang, L.C. (1983). Cellulose and hemicellulose technology. In *The Filamentous Fungi,* Vol. **4**, *Fungal Technology,* eds. J.E. Smith, D.R. Berry and B. Kristiansen, Edward Arnold, London, pp. 296-326.

Vilela, L.C. and Silverio, C.M. (1982). Cultivation of *Auricularia* on composted sawdust in the Philippines. In *Tropical Mushrooms: Biological Nature and Cultivation Methods,* eds. S.T. Chang and T.H. Quimio, Chinese University Press, Hong Kong, pp 427-435.

Wood, D.A. (1984). Microbial processes in mushroom cultivation: a large-scale solid-substrate fermentation. J. Chem. Tech. Biotechnol. **9**: 34B, 232-240.

Wood, D.A. (1985). Useful biodegradation of lignocellulose. Ann. Proc. Phytochem. Soc. Eur. **26**, 295-309.

Wu, K.W. (1967). Study on the preparation of synthetic compost for mushroom growing in Taiwan. Mushroom Sci. **6**, 303-305.

Yau, C.K. and Chang, S.T. (1972). Cotton waste for indoor cultivation of straw mushroom. World Crops, **25**, 302-3.

Zadrazil, F. (1978). Cultivation of *Pleurotus.* In *The Biology and Cultivation of Edible Mushrooms,* eds. S.T. Chang and W.A. Hayes, Academic Press, New York, pp. 521-527.

Zadrazil, F. and Scheidereit, M. (1972). Die Grundlagen für die Inkulturnahme einer bisher nicht kultivierten *Pleurotus*- Art. Champignon, **163**, 7 - 22.

USE OF WHITE-ROT FUNGI FOR THE CLEAN-UP OF CONTAMINATED SITES

D. Loske[1], A. Hüttermann[1], A. Majcherczyk[1], F. Zadrazil[2], H. Lorsen[3] and P. Waldinger[3]

[1]Institüt für Forstbotanik, Büsgenweg 2, D-3400 Göttingen, FRG

[2]Institüt für Bodenbiologie, FAL, Bundesallee 53, 3300 Braunschweig, FRG

[3]NOELL GmbH, Alfred-Nobel-Str. 20, 8700 Würzburg 3, FRG

A major problem in soil contamination is caused by industrial residues of the type known as polycyclic aromatic hydrocarbons (PAHs). Autochtone soil microflora are generally unable to degrade PAHs completely to non-toxic metabolites. Aromatic constituents represent a considerable part of the molecular structure of lignin, one of the most complex of structures in nature and one that is biodegraded effectively only by white-rot fungi. In this communication we describe the use of the enzymic abilities of a white-rot fungus for soil decontamination. The degradation of PAHs in batch cultures and during a field experiment in contaminated soil was monitored.

INTRODUCTION

Urban industrialization during the 19th and 20th centuries has left it's traces in the environment worldwide. Air pollution and contamination of ground-water, drinking-water and soil are subjects of increasing interest, not only to scientists but also to politics and industry. To-date in the FRG, 42,000 sites have been registered as possibly contaminated. About 10% of these have to be cleaned up in the near future. The main contaminants in these soils belong to the following classes (Fig. 1): (a) Polycyclic Aromatic Hydrocarbons (PAHs), viz. residues from the processing of oil, tar, coal and comparable substances; (b) Polychlorinated Biphenyls (PCBs). Because of their flame-hindering qualities and thermostability, these are used as cooling agents for transformers. They are also found in pesticides and as a softener in plastics, paints and many other products; (c) Dioxines. These are possibly the most dangerous of all antropogenic substance classes. They are byproducts of chemical manufacturing and are found in fly-ashes from combustion processes. With respect to biodegradation, dioxines and PCBs are very recalcitrant substances because of their aromatic structures and high degree of chlorination.

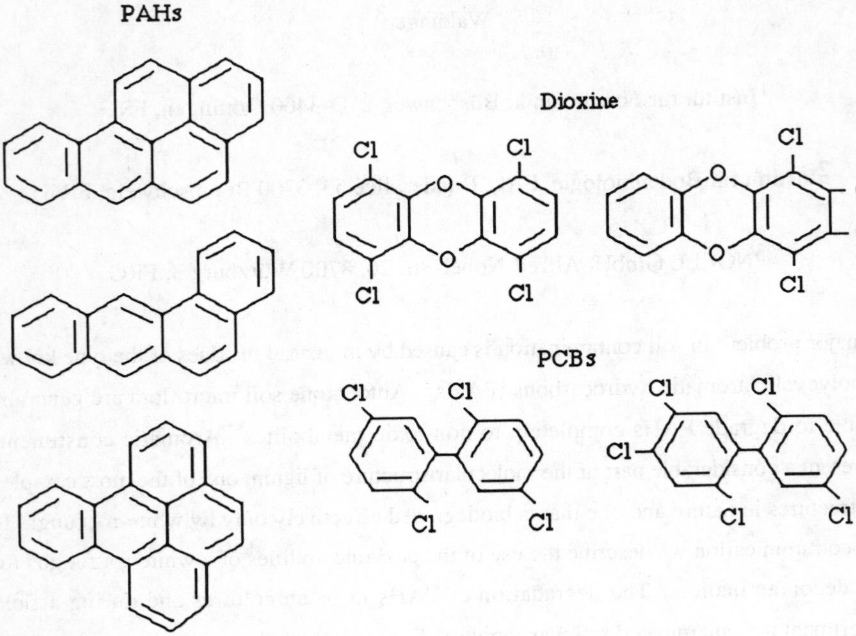

PAHs

Dioxine

PCBs

Fig. 1. Some of the most important of the recalcitrant compounds of anthropogenic origin in contaminated soils.

Current solutions for the clean-up of soils contaminated with these chemicals are less than satisfactory. These include the use of bacterial degradation, a method first developed for wastewater treatment. Some bacteria possess metabolic pathways for the degradation of aromatics but the rate of decomposition is rather slow. Furthermore, according to latest results, the hydroxylated forms of the original substances are assumed to be the endproducts of bacterial degradation.

DEGRADATION OF AROMATIC COMPOUNDS IN NATURE

A common feature of all of the above chemicals, the most problematic compounds to deal with in soil reclamation procedures, is that they are aromatic. The most important of the aromatic compounds in nature is lignin. It is also the second most abundant of organic compounds in the biosphere (Glasser and Kelley, 1987). Two different pathways exist for the biodegradation of lignin and other aromatic compounds.

The degradation of lignin, the most recalcitrant of substrates in nature, takes place within the carbon cycle through the action of white-rot fungi. These fungi secrete certain oxidative enzymes (Kirk and Farrell, 1987) that, via a radical-mediated mechanism, degrade lignin completely to water and carbon dioxide (Crawford, 1981). The enzymes exhibit very broad substrate specificity and act on virtually all aromatics, including PAHs, PCBs and dioxins (Bumpus *et al*., 1985). Litter from both roots and leaves contains in addition to lignin a certain amount of other aromatic substances, including phenolics and tannins. In soils these aromatics are not readily degraded but to a great extent are incorporated into the humic substances. Without this process, humification in soils would not take place. The bulk of the aromatics and chlorinated compounds added to soils containing a normal microflora end up in the high molecular weight humic acid fraction (Bartha, 1981).

In general, it can be stated that the metabolic processes occurring in soil biota are geared to the degradation of aromatic compounds. However, aromatic structures are also conserved to a very high degree. The more condensed an aromatic compound is, and the more it is substituted with chlorine, the more recalcitrant it is to degradation by the endemic soil microflora (Hüttermann *et al*., 1990). For example, the residence time of dioxin in humus-rich garden soils has been estimated to be several hundreds of years (Neidhard and Herrmann, 1987).

CONCEPT

In our opinion the best solution to the problem of the degradation of aromatics in soil is to use the extraordinary enzymic abilities of the white-rot fungi. However, one must note that these fungi, as wood-colonizing organisms, cannot compete with the autochtone microflora present in soils (Hüttermann and Haars, 1987). Therefore, these fungi cannot, without assistance, exploit their lignin-degrading activities even in sterile soils.

In order to circumvent these difficulties and to enable fungal growth and enzyme production in soils, the white-rot species are pregrown on substrates (e.g. straw) containing lignin and are then brought into the contaminated soil (Hüttermann *et al*., 1988, 1989). Under these conditions, the fungi will continue to degrade the straw and retain their lignin-degrading capacity. Since fungal growth is accompanied to a certain extent by the penetration of hyphae into the soil (Fig. 2) and since the enzymes in question are mainly extracellular, the toxic substances present in the soil will be degraded concomitantly with growth.

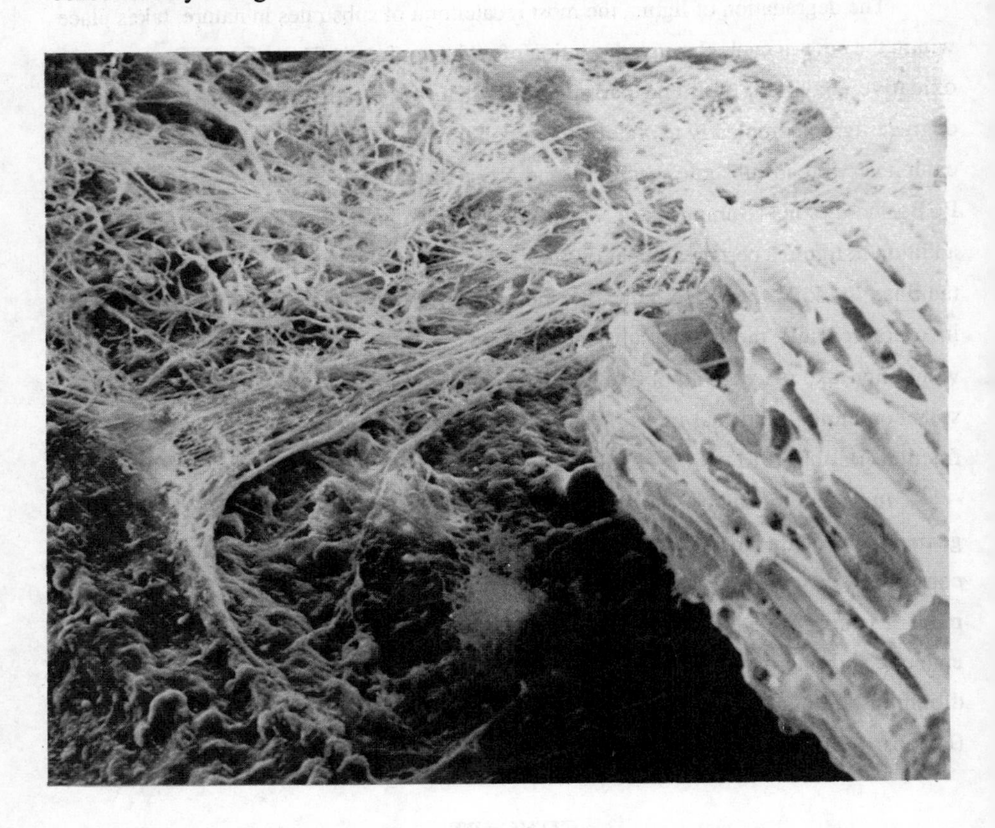

Fig. 2. Scanning electron micrograph of a white-rot fungus growing on straw and seen to be colonizing the surrounding substrate.

The above procedure for treating soils has several advantages: (a) degradation of the xenobiotic substances is not dependent on their concentration, since the real substrate used by the fungi is the lignocellulose fraction of the straw; (b) the procedure works with defined organisms whose properties are known; (c) the organisms used are already present in the environment. Thus, it is not necessary to introduce new strains or species into nature; (d) the whole process is under complete control. No emissions either to air or ground-water occur; (e) during the course of the sanitation the toxic substances are degraded, the straw is converted to humus and the endogenous soil microflora are reactivated; (f) after decomposition of the straw the newly-introduced white-rot fungi cannot compete with the newly-revived soil microflora anymore and die off naturally; (g) soils treated in this way are biologically sound and active and can be used for any purpose; (h) the reclamation process utilizes an agricultural waste product, i.e. straw, and thereby increases the economy of grain production.

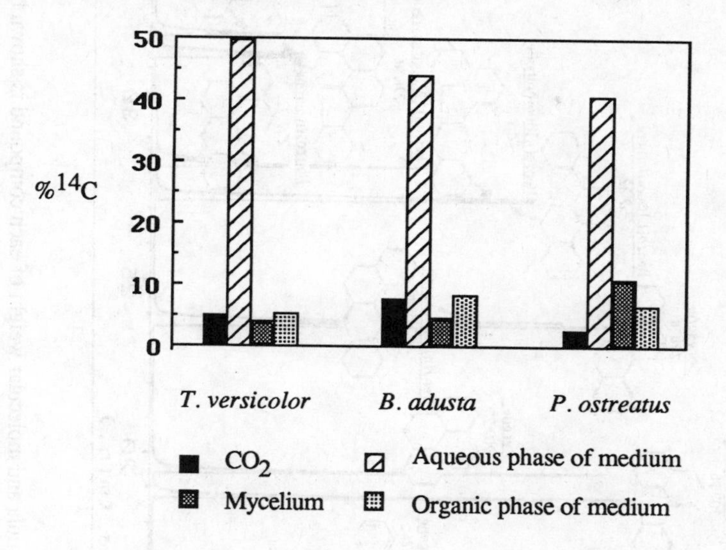

Fig. 3. Distribution of ^{14}C in CO_2, mycelium, aqueous and organic phases of the growth media. The cultures of three different white-rots had been grown in nitrogen-limited media with ^{14}C-labelled anthracene.

RESULTS

Preliminary experiments in batch cultures with ^{14}C-labelled anthracene as model substance for PAHs demonstrated the lack of specificity of the lignin-degrading system. Fig. 3 shows the distribution of ^{14}C in harvested cultures of three different white-rot fungi, viz. *Trametes versicolor*, *Bjerkandera adusta* and *Pleurotus ostreatus*. The bulk of the labelled carbon was found in the water-soluble phase of the media and in the CO_2 during growth. Only about 10% of the ^{14}C remained in the organic phase of the media.

In order to transfer the enzymic abilities of white-rots into the soil habitat we proceeded as described above. Some laboratory experiments conducted with soil samples from contaminated sites gave average degradation rates of about 40% (results not shown). GC/MS methods were developed to measure the 16 PAHs recommended by the Environmental Protection Agency (USA). The separation of the components of the

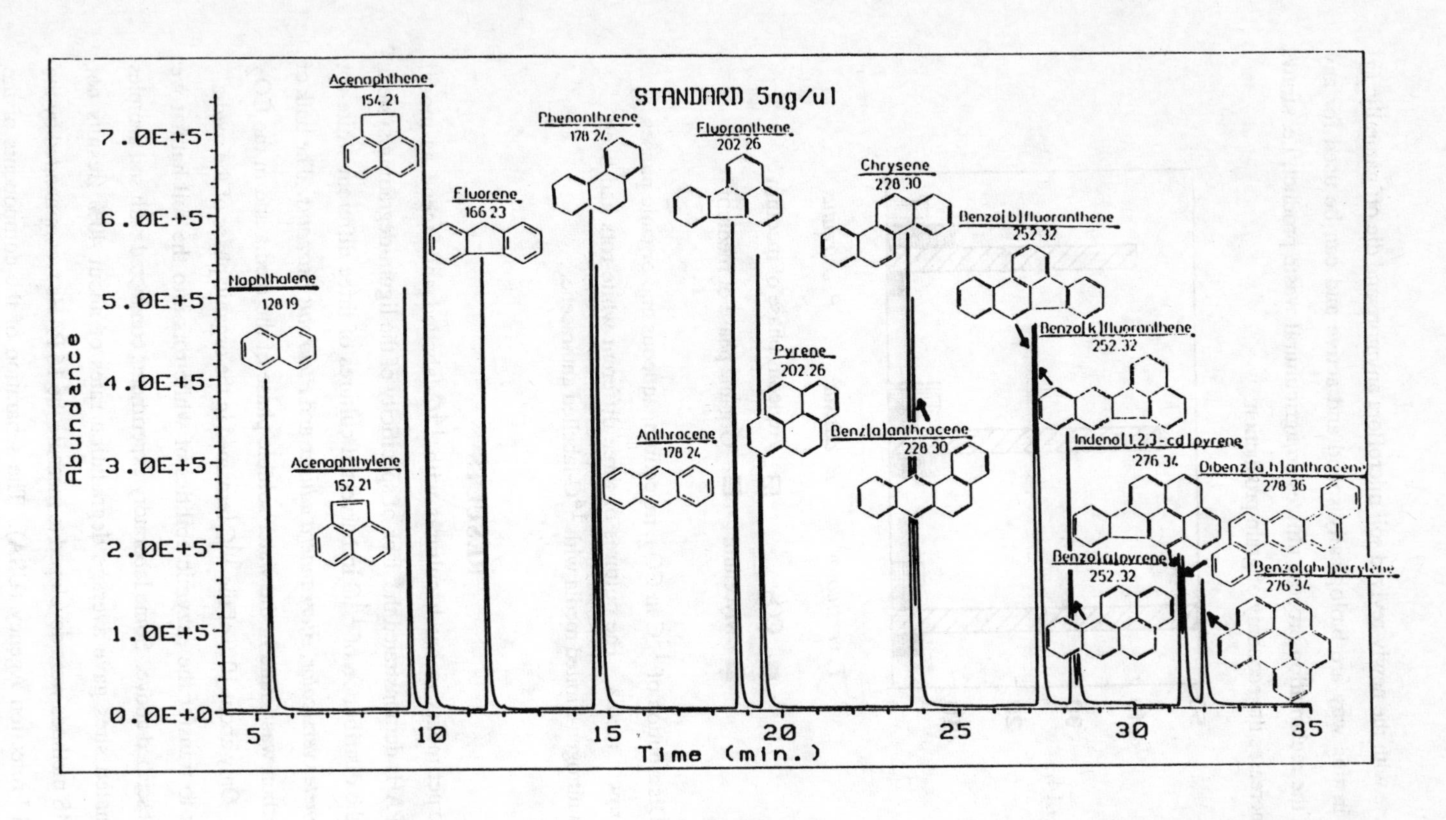

Fig. 4. GC-MS analysis of the PAH standards used. The formula and molecular weight of each compound is shown beside the relevant peak.

standard mixture used is shown in Fig. 4.

During the summer 1988 an experiment was started to test the system under field conditions. The site of a former gas works contained 35 m3 of PAH-contaminated soil. The distribution of the contaminants was not homogenous. Peak values ranged up to 220 mg/kg soil but the average was rather low, c. 20 mg/kg soil (sum of 16 measured PAHs).

Because of the irregular distribution of contaminants on the site, the excavated soil was air-dried and some (12 m^3) was sieved to a size of about 1.5 to 2 cm. Different types of composting techniques were then tested. The remainder of the soil was not sieved but was freed from bigger stones and metal pieces. The white-rot organisms used, viz. *Pleurotus ostreatus* spp., had been grown on straw by a professional mushroom grower from whom they were obtained.

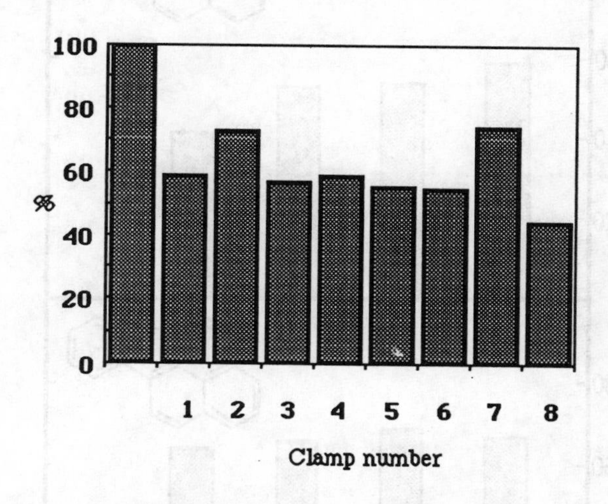

Fig. 5. Extent of decontamination in 8 different clamps after 8 weeks treatment with *Pleurotus ostreatus*. The 16 measured PAHs were added. The bars represent the percentage of contaminants remaining in soil. The 100% bar represents the starting level.

The first samples, taken after 8 weeks, showed the average degradation to be 40% (Fig. 5). The extent of degradation of lowly-condensed and highly-condensed PAHs did not differ significantly (compare Figs. 4 and 6). After 16 weeks in one of the clamps a degradation of 80% was measured (Fig. 7).

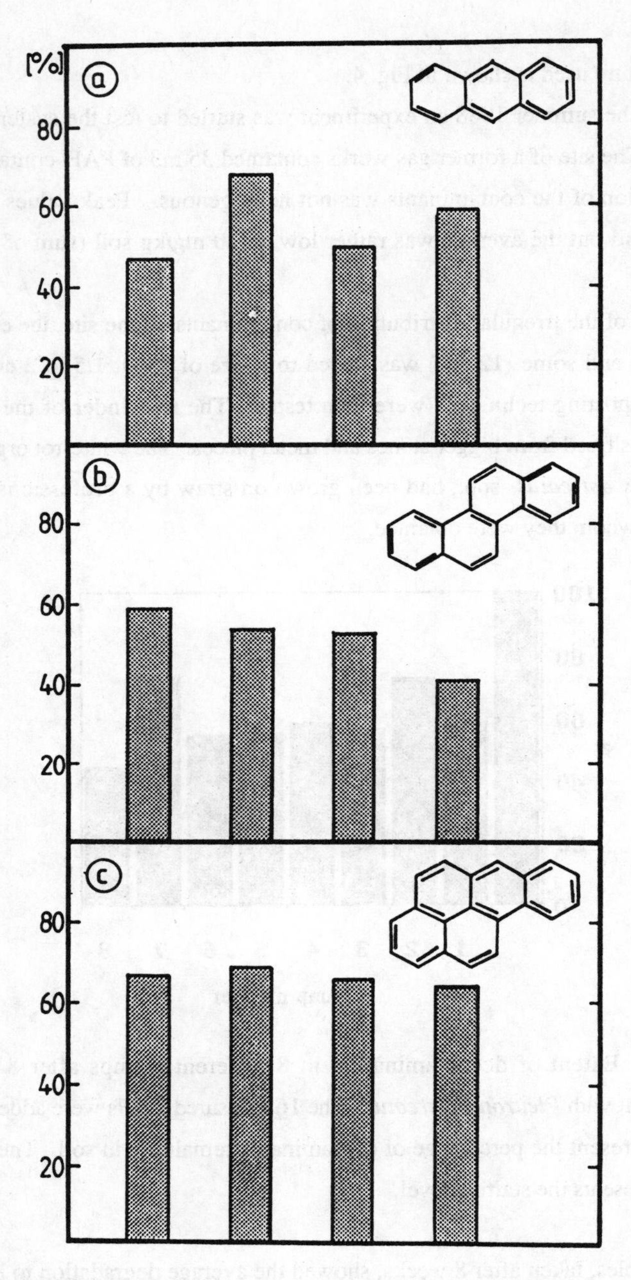

Fig. 6. Degradation of anthracene (a), chrysene (b) and benz(a)pyrene (c) in some parts of the field experiment, 8 weeks after starting the treatment. Each bar represents a different clamp. Degradation is given as a % of the original content.

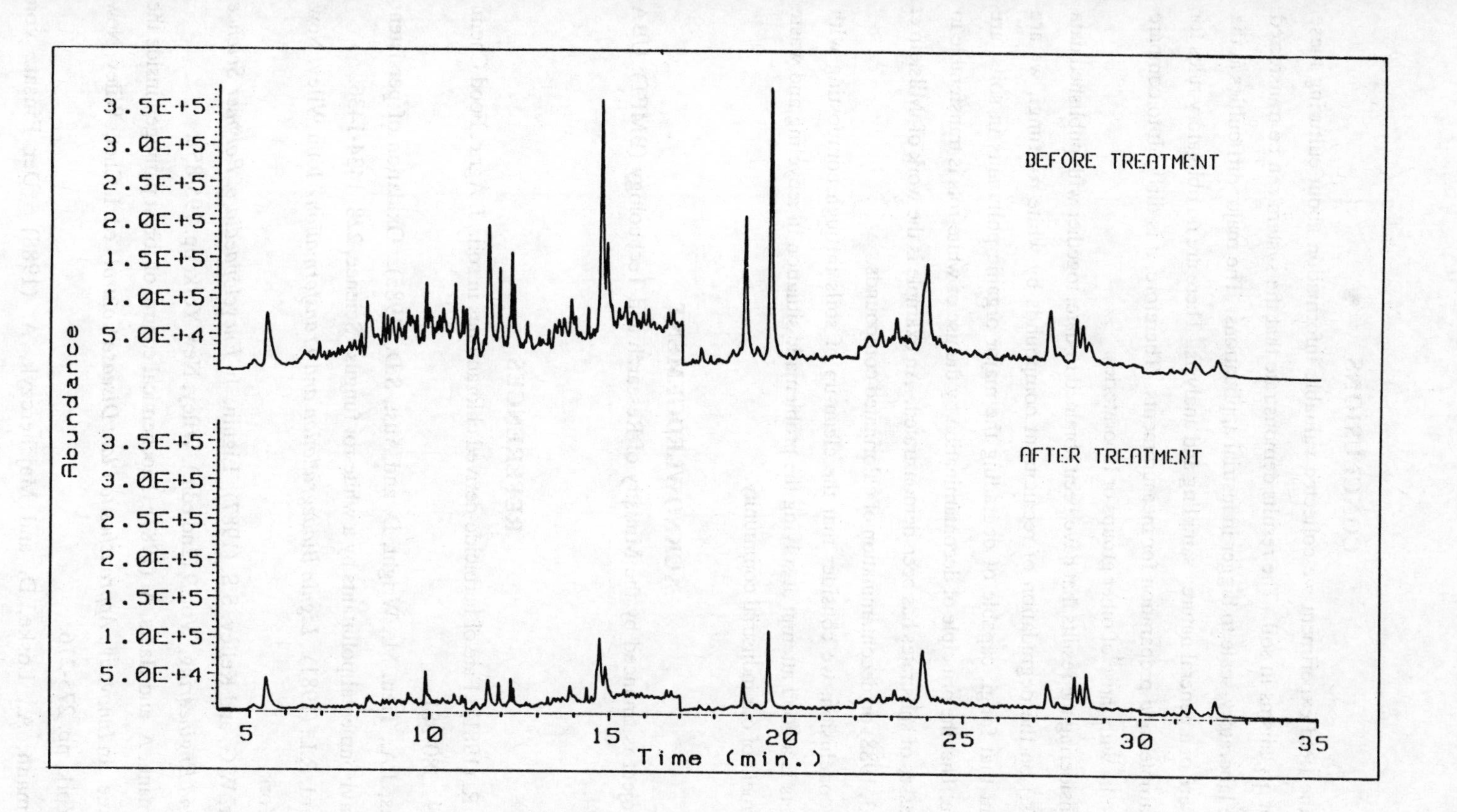

Fig. 7. GC-MS analysis of soil samples taken before and after treatment for 4 months in a clamp.

CONCLUSIONS

During the field experiment we collected valuable information about culturing these white-rot organisms in soil. The results demonstrate that the system can be transferred from the laboratory-scale to bigger industrial applications. The major difficulties at the moment are of a general nature: sampling and analysis. There are no obligatory rules for taking samples and preparation for measurements. Therefore, it is difficult to compare these results with those of other groups or laboratories.

Considering the results that have been obtained to-date, together with published data (see refs.) on the degradation of recalcitrant compounds by white-rot fungi, we are confident that fungi, capable of degrading the major organic pollutants in soils, are available. That the principle of decontamination by the use of white-rots is transferable to other classes of substrates has been demonstrated. An example is the work of Milstein *et al.* (1983, 1988) on decontamination of chlorinated compounds.

In conclusion, we consider that the clean-up of soils through composting with white-rots is a good attempt at solving the problematic situation in recycling and waste management of our industrial community.

ACKNOWLEDGEMENT

This project is financed by the Ministry of Research and Technology (BMFT): UBA 1470494.

REFERENCES

Bartha, R. (1981). Fate of herbicide-derived chloranilines in soil. J. Agric. Food Chem. **19**, 380-381.

Bumpus, J.A., Tien, M., Wright, D. and Aust, S.D. (1985). Oxidation of persistent environmental pollutants by a white-rot fungus. Science, **228**, 1434-1436.

Crawford, R.L. (1981). *Lignin Biodegradation and Transformation*. John Wiley, New York.

Glasser, W.G. and Kelley, S.S. (1987). Lignin. In *Encyclopaedia of Polymer Science and Engineering,* Vol. 9 (2nd edn.), Wiley, New York, pp. 795-852.

Hüttermann, A. and Haars, A. (1987). Biochemical control of forest pathogen inside the tree. In *Innovative Approaches to Plant Disease Control,* ed. I. Chet, Wiley, New York, pp. 275-276.

Hüttermann, A., Loske, D. and Majcherczyk, A. (1988). Der Einsatz von

Weissfäulepilzen bei der Sanierung besonders problematischer Altlasten. In *Altlasten 2,* ed. K.J. Thome-Kozmiensky, EF-Verlag, Berlin, pp. 713-726.

Hüttermann, A., Loske, D., Majcherczyk, A. and Zadrazil, F. (1989) Einsatz von Weissfäulepilzen bei der Sanierung kontaminierter Böden. In *Fortschrittliche Anwendungen der Biotechnologie,* Der Bundesminister für Forschung und Technologie, Bonn, pp. 115-121.

Hüttermann, A., Loske, D. and Majcherczyk, A. (1989). Biologischer Abau von Organochlor-Verbindungen in Boden, im Abwasser und in der Luft. In *Halogenierte Organische Verbindungen in der Umwelt,* VDI-Berichte 745, VDI-Verlag, Düsseldorf, 911-926.

Kirk, T.K. and Farrell, R.L. (1987). Enzymatic "combustion": The microbial degradation of lignin. Ann. Rev. Microbiol. **41**, 465-505.

Milstein, O., Haars, A., Majcherczyk, A., Trojanowski, J., Tautz, D., Zanker, H. and Hüttermann, A. (1988). Removal of chlorophenols and chlorolignins from bleaching effluent by combined chemical and biological treatment. Wat. Sci. Tech. **20**, 161-170.

Milstein, O., Vered, Y., Shragina, L., Gressel, J., Flowers, H.M. and Hüttermann, A. (1983). Metabolism of lignin-related aromatic compounds by *Aspergillus japonicus.* Arch. Microbiol. **135**, 147-154.

Neidhard, H. and Herrmann, M. (1987). Abbau, Persistenz, Transport polychlorierter Dibenzodioxine und Dibenzofurane in der Umwelt. In *Dioxin,* VDI Berichte, VDI-Verlag, Düsseldorf, pp. 303-316.

BIOFILTRATION OF POLLUTED AIR BY A COMPLEX FILTER BASED ON WHITE-ROT FUNGI GROWING ON LIGNOCELLULOSIC SUBSTRATES

A. Majcherczyk, A. Braun-Lüllemann and A. Hüttermann

Forstbotanisches Institut, der Universitat Göttingen
Büsgenweg 2, D-3400 Göttingen-Weende, FRG

In this paper we propose the use of a biofilter based on white-rot fungi growing on straw or other agricultural residues for the development of a possible alternative to current biofilter systems for the decontamination of polluted air. Some data on the operation of the biofilter in question are presented.

INTRODUCTION

Human activities are characterized by the fact that they result in the dissemination of various chemicals into the environment in a dilute form. Air is one of the pathways of distribution. In Central Europe hundreds of thousands of tons of ammonia, hydrogen sulphide and various organic chemicals are emitted each year by factories and farms. Apart from annoying the population by their odour, these substances are hazardous pollutants of the environment. Overall, large amounts of these chemicals are emitted. Judging from measurements of the average emission per unit area, 800,000 tons of ammonia are emitted annualy in the coastal plains of The Netherlands, Germany and Denmark (Krzak *et al.*, 1988). However, owing to the very high mass stream of air used in factories and farms, the actual concentrations of these pollutants are rather low. Concentrations range from a few ppm to a few hundred ppm and, as such, chemical treatments for cleaning of the polluted air are not feasible.

The idea of using microorganisms to clean polluted air streams is not a new one. Reports of filters based on bacteria have been published since 1979 (Steinmüller *et al.*, 1979; Gust *et al.*, 1979). Currently the following are the main systems in use:

1. Bioscrubbers: These are essentially spray columns in which the pollutant flow is countercurrent to that of the aerosol stream. Thus, the pollutant is washed out by the water droplets. The water is then passed on to a regeneration compartment wherein the pollutants are removed by a bacterial sludge.

2. Trickle filters: These consist of columns filled with high-surface area packings on which a biofilm of microbes develops. The column is continuously rinsed with water containing nutrients and the polluted air is passed over the package against the flow of water. The water-soluble components in the air are transferred to the liquid film on the package where they are eliminated by the microorganisms.

3. Biofilters: These consist of a package of compost or peat mixed with heather branches or chips of wood on which a bacterial flora can develop. They are kept wet by supplying a nutrient solution to the filter. The air is then passed through the filter and the pollutants are removed by the microorganisms. Such filters are usually used for odour abatement in waste water treatment or in composting works.

A common feature of all three systems is that the microorganisms that ultimately remove the pollutants from the gas phase live in an aqueous environment. Thus, they may be totally submerged in water or exist in a more-or-less thick film of water on the package of the trickle filter or on the surface of the biofilter material. Thus, the transition of the pollutant from the air to the water may be a crucial factor in determining the macrokinetics of the degradation process. This may be a problem with substances, such as hydrophobic organic compounds, that are poorly soluble in water (Ottengraf, 1986).

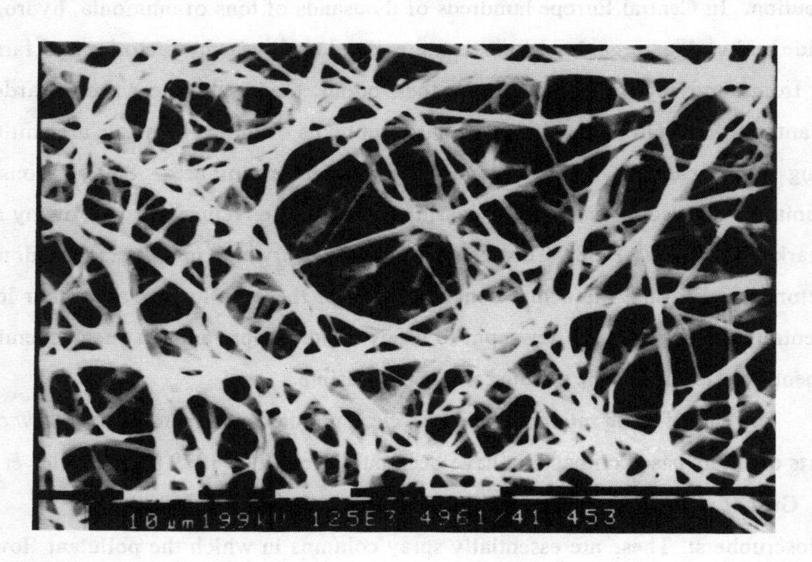

Fig. 1. Scanning electron micrograph of the aerial hyphae of fungi growing in the biofilter.

In this paper we propose the use of a biofilter based on white-rot fungi growing on straw or other agricultural residues for the development of a possible alternative to current biofilter systems.

The rationale underlying this concept has been outlined in broad terms in earlier publications on the use of white-rot fungi for the decontamination of polluted soils (Hüttermann and Trojanowski, 1987; Hüttermann *et al.*, 1989a). It can be summarized as follows: Because these fungi grow as aerial hyphae (Fig. 1), they form a very high surface area that is in direct contact with the air streaming through the filter. Thus, the pollutant comes into direct contact with the cell surfaces without phase-transition problems. When growing on straw, the fungus secretes oxidative enzymes that catalyze the degradation of lignin (Kirk and Farrell, 1987). These extracellular enzymes are very non-specific since they work via oxygen- and hydroxyl-radicals, products of their action. In addition to lignin and phenols, various kinds of organic compounds are "burnt" by this system (Kirk and Farrell, 1987). This fact can be exploited for many environmental applications (see e.g. Bumpus *et al.*, 1985; Chang *et al.*, 1985; and a review by Hüttermann et al., 1989b).

The fungi growing on the lignocellulosic substrate create a highly oxidative milieu around the surface of the hyphae with broad specificity towards different chemical structures. Our idea is to exploit this combination of a large biologically-active surface area combined with oxidative potential for the removal of pollutants from air streams flowing through the filter.

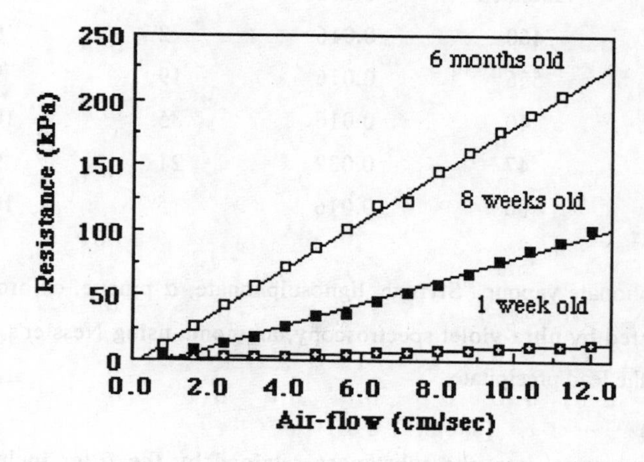

Fig. 2. Resistance of the filter (1 m long) as a function of air-flow velocity.

RESULTS AND DISCUSSION

Flow characteristics

The resistance of columns of biofilters composed of fungi, growing on straw, as a function of flow velocity was measured (Fig. 2). The observed pattern followed the behaviour of laminar air streams with a linear relationship between resistance and velocity. Fig. 2 also shows that old columns, on which the fungi had been allowed to grow for 6 months, still allowed reasonable air velocities with relatively low resistances. This means that within a wide range of air flow velocities the column retained the characteristics of laminar flow and that the filter tolerated rather high air velocities, at least insofar as flow characteristics were concerned. The resistance of the filter packing with respect to height and air stream velocity was in the same range as that reported for the conventional compost filter (Ottengraf, 1987).

Table 1. Retention of air pollutants on the biofilter consisting of white-rot fungi growing on straw.

Compound	Conc. $mg.m^{-3}$	Flow rate $m.sec^{-1}$	Exposure time (days)	Retention %
Styrene	1245	0.016	8	99.9
Styrene	3300	0.016	1	80
Styrene	543	0.016	17	95
LS vapour*	Saturated	0.016	1.5	100
α-Pinene	7460	0.016	2	61
Ammonia	556	0.016	19	99
Ammonia	40	0.016	35	100
Cl-phenols	47	0.032	21	95
H_2S	80	0.016	5	100

* Lignosulphonate vapour. Styrene, lignosulphonate, α-pinene, chlorophenols were measured by ultra-violet spectroscopy, ammonia using Nessler's reagent, and H_2S as the lead precipitate.

In our experiments to date, the substances retained by the filter included styrene, lignosulphonate vapour, α-pinene, ammonia, chlorophenols and H_2S (Table 1).

The data obtained in the current experiments are remarkable in several respects. First, the efficiency of the column was extremely high. Although, relatively large amounts of pollutant were applied to the column for relatively long times, the degree of retention was surprisingly good. It can be expected that the extent of retention of pollutants would still be sufficient even with much greater loading of the filter. Second, it can be expected that the absorbed organic compounds would be completely degraded by the fungi. Even after exposure to a mixture of chlorophenols, such substances were neither detectable nor extractable from the column after termination of the experiment.

Until now no technical means for the removal of ammonia and H_2S from polluted air streams existed. In this context, the data obtained with the fungal biofilter are especially attractive with respect to the future implementation of such devices. After exposure of the column to H_2S, no sulphides could be extracted from it. It must, therefore, be assumed that the sulphide ions were oxidized to sulphate ions by the fungal extracellular enzymes, i.e. the column acted as a catalyst for the oxidation of sulphide to sulphate. Provided that the appropriate base is supplied to neutralize the sulphuric acid eventually formed, this system has a very high capacity for the the removal of H_2S from polluted air streams.

We have, as yet, been unable to determine the fate of the absorbed ammonia. From the results of the longest experiment, it can be calculated that about 4 g of nitrogen had been bound by the column. This means that the C/N ratio of the straw in the column had been lowered from a value of about 100 to a value of about 25.

Special features, problems and perspectives

One of the surprising of the lignocellulose-straw complex was its very high resistance to infection. Although, unsterile air was passed through the column, no infection was observed even after long exposure times. Another striking feature was the stability of the biofilter against acids. In a factory experiment designed to remove lignosulphonate pyrolysis products from the air streams issuing from a furnace, the fungi were exposed not only to these pyrolysis products but also to SO_2 and SO_3. This resulted in a pH of 1.5 in the condensing water after cooling of the air. Nevertheless, the fungi tolerated these extreme conditions and still removed most of the organic load from the air.

The most important factor in determining the longevity of the filter is the relative humidity of the air being passed over it. Extreme care must be taken to bring humidity as close as possible to 100% so as to avoid the transfer of water from the column to the air and so prevent the column from drying out. One may note, however, that feasible technical solutions for avoiding this problem are available on the market.

In conclusion, we are confident that the approach presented here could provide the means for decontamination of polluted air, a problem for which no other solution is available.

ACKNOWLEDGEMENT

This research was funded by a grant from BMFT/Umweltbundesamt. We are indebted to Dr. Frank zadrazil for his cooperation.

REFERENCES

Bumpus, J.A., Tien, M., Wright, D. and Aust, S.D. (1985). Oxidation of persistent environmental pollutants by white-rot fungi. Science, **228**, 1434-1436.

Chang, H.-M., Joyce, T.W., Kirk, T.K. and Huynh, V.B. (1985). Process of degrading chloro-organics by white-rot fungi. US Patent 4,554,075.

Gust, M., Grochowski, H. and Schirz, S. (1979). Grundlagen der biologischen Abluftreinigung. Teil V. Abgasreinigung durch Mikroorganismen mit Hilfe von Biofiltern. Staub Reinhalt. Luft. **39**, 397-438.

Hüttermann, A. and Trojanowski, J. (1987). Ein Knozept für eine in situ Sanierung von mit schwer abbaubaren Aromaten belasteten Böden durch Inkubation mit dafür geeigeten Weißfäulepilzen und Stroh. In *Sanierung kontaminierter Standorte 1986 - Neue Verfahren zur Bodenreinigung "Abfallwirtschaft in Forschung und Fraxis",* Band 18, (V. Franzius, ed.), Erich-schmidt-verlag, berlin, pp. 205-218.

Hüttermann, A., Loske, D. and Majcherczyk, A. (1989a). Biologischer Abau von Organochlor-verbindung in Böden, im Abwasser und in der Luft. VDI Berichte 745: Halogenierte organische Verbindungen in der Umwelt. Band II, VDI-Verlag, Düsseldorf, pp. 911-926.

Hüttermann, A., Loske, D., Majcherczyk, A., Zadrazil, F., Waldinger, P. and Lorson, H. (1989b). Reclamation of PAH-contaminated soils with active fungus-straw-substrata. In *Recycling International,* Vol. 3 (K.J. Thome-Kozmiensky, ed.), EF-Verlag, Berlin, pp. 2191-2199.

Hüttermann, A., Zadrazil, F. and Majcherczyk, A. (1987). Verfahren zum Dekontaminieren von sauerstoffhaltigen gasen. insbesondere von Abgasen. German patent PS 38 07 033.

Kirk, T.K. and Farrell, R.L. (1987). Enzymatic "combustion': the microbial degradation of lignin. Ann. Rev. Microbiol. **41**, 465-505.

Krzak, J., Dong, P.H., Büttner, Hüttermann, A., Kramer, H. and Ulrich, B. (1988). Photosynthesis, nutrient, growth and soil investigations of a declining Norway spruce [*Picea abies* (Karst)] stand in the coastal region of Northern Germany. J. Forest Ecol. Manag. **24**, 263-281.

Ottengraf, S.P.P. (1987). Biological systems for waste gas elimination. Trends Biotechnol. **5**, 132-136.

Steinmüller, W., Claus, G. and Kutzner, H.-J. (1979). Grundlagen der biologischen Abluftreinigung. Teil II. Mikrobiologischer Abbau von luftverunreinigenden Stoffen. Staub Reinhalt. Luft. **39**, 149-152.

UPGRADING OF LIGNOCELLULOSICS FROM AGRICULTURAL AND INDUSTRIAL PRODUCTION PROCESSES INTO FOOD, FEED AND COMPOST-BASED PRODUCTS

K. Grabbe

Institut für Bodenbiologie,
Bundesforschungsanstalt für Landwirtschaft
Bundesallee 50, D-3300 Braunschweig, F.R.G.

The formation of plant biomass for commercial purposes leads to the accumulation of huge amounts of lignocellulosics containing residues more-or-less polluted by the preceding treatments. Attempts to upgrade this material have been successful. Biological- and technolically-orientated procedures are available for the conversion of lignocellulosics into food and feedstuffs and for various other uses. Nevertheless, lignocellulosics containing water causes ewnvironmental problems. Composting might be a solution - if the aims of utilization are well defined. In this paper, I discuss some results on the characterization of well-stabilized compost. The importance of the incorporation of nitrogen into humic acids and lignin is demonstrated. Under defined conditions, compost can be used as a basic material for the formulation of organo-mineral fertilizers. Furthermore, the compaction and disposal of compost might obviate the oxidation of reduced carbon to carbon dioxide.

INTRODUCTION

Efficient production processes in agriculture and forestry, and waste management in technologically-orientated and highly-civilized societies, lead to accumulation of vast amounts of residues. From an ecological viewpoint, these residues have to be treated in such a way as to avoid environmental pollution, the outcome of which could be catastrophic.

Eighty percent of the biogenic organic matter consists of lignocellulosics which are highly stable to microbial degradation. The associated cellulose can be used to a greater or lesser extent by aerobic and anaerobic cellulolytic bacteria depending on the

Table 1. World production of cultivated mushrooms in 1986. (modified from Chang [1987]).

Species (Latin name)		Common name	Production (tons)	Substrate
Agaricus bisporus	(J. Lange) Imbach	Cultivated mushroom or button mushroom	1227000	Stable manure, straw
Lentinus edodes	(Berk.) Singer	Shiitake, oak mushroom	314000	Wood, sawdust*
Volvariella volvacea	(Bull.: Fr.) Singer	Paddy straw mushroom		
Pleurotus ostreatus	(Jacq.: Fr.) Kummer and other spp.	Oyster mushroom	178000	Plant residues, wood, straw
Auricularia auricula-judae	(Fr.) Schreit. and other spp.	Jew's ear mushroom	169000	Wood, sawdust*
Flammulina velutipes	(Curtis: Fr.) Singer	Winter mushroom	100000	Sawdust
Tremella fuciformis	(Berk.)	White jelly fungus, silver ear fungus	40000	Wood, sawdust*
Pholiota nameko	(T. Ito) S. Ito et Imai	Nameko, viscid mushroom	25000	Wood, sawdust*
Kuehneromyces, Hericium, et al.			10000	Wood, sawdust*
		Total	2182000	

*Supplemented with 10-20% wheat bran

degree of encrustation by lignin. Among the fungi with the ability to degrade cellulose, the basidiomycetes are specialists in that they can use cellulose as well as lignin. Many species belonging to the so-called white-rot fungi can mineralize the total complex. Lignocellulose can also be partially metabolized by those micro- organisms that lack the capacity to degrade polymerized aromatics but which can split side chains and methoxylic groups. Induction of synthesis of the relevant enzymes by these species leads to the generation of partially denatured structures.

Lignocellulose may be considered to be a well-defined polymer. In residues that are neither converted chemically nor attacked by microbes, the lignocellulose is often admixed with all kinds of organic substances. This implies a severe loss of quality.

Since lignocellulosics cannot be used directly by animals, the conversion is based on the activity of specialized microorganisms. This selectivity can be considered as an advantage for the development of desired processes in biotechnology.

This paper shall be a contribution to the discussion on the utilization of lignocellulosics from the ecological viewpoint, i.e. in the context of environmental protection.

LIGNOCELLULOSICS FOR FOOD PRODUCTION

There seems to be only one way to produce food by formation of microbial biomass via growth of basidiomycetes as edible fungi. Approximately 2.2 million tonnes of these fungi are produced annually. The process is based upon the preparation of 20 million tonnes of substrate, consisting of all kinds of straw, stable manures, sawdust, less valuable wood and residues from industrial refinements of food. In 1989, 40,000 tonnes of mushrooms were produced in West Germany. The standard recipe used is: 70% horse manure, 20% straw and 10% poultry manure. Nearly all of the horse manure, obtained mainly from racing stables, is used by the mushroom industry. This service is highly important but it makes little impact on the total amount of residues accumulated annually. This situation would not be significantly changed even if the pharmaceutical industry began to produce special metabolites by solid-state cultivation of higher fungi (as mycelia with or without fruitbody formation) on the residues in question.

LIGNOCELLULOSICS FOR FEED PRODUCTION

During the seventies many attempts were made to find a suitable physical or chemical treatment of straw that would increase its digestibility in ruminants (Kamra and Zadrazil, 1988). Cracking by steam or hydrolysis by caustic lye led to acceptable products but the price could not be carried through on the market. As an alternative, the microbial pretreatment under on-farm conditions was taken into consideration. The preparation of fodder is ruled by law insofar as defined substrates have to be incubated with defined microorganisms. Straw inoculated with basidiomycetes is accepted in principle. But, the problem is to make the processing sufficiently safe enough from the standpoint of hygiene. The invasion of moulds has to be strictly avoided and procedures for preservation of the well-grown substrate should be determined.

Much time was spent in screening for suitable species. Among the species that were found to increase digestibility, *Pleurotus* was considered to be one of the best (Kamra and Zadrazil, 1988). Based upon the cultivation in bulk, important parameters were determined for the optimization of the fermentation process. The acceptance of this type of fodder was demonstrated in feeding experiments with cows and sheep.

There is no doubt that upgrading of lignocellulosics by using this biotechnology could be an additional source of feed. Under the present market conditions in Europe, there is little impetus for the introduction of this system into modern, highly-sophisticated agriculture but it might be worthwhile for developing countries. Nevertheless, some problems must still be solved. The technology to handle large amounts of lignocellulose by solid-state fermentation has to be improved in order to make it safe for the practitioner. Knowledge gained in this field will positively influence the creation of similar technologies, e.g. biopulping or the cultivation of those species of edible fungi that are devoid of antagonistic properties (i.e. with which to keep competitors in check).

UPGRADING OF LIGNOCELLULOSICS BY COMPOSTING

The procedures above are based on the use of unpolluted lignocellulose, from freshly-harvested straw or material derived from wood, without any undefined supplementation. During fermentation, the readily available carbon and nitrogen sources are consumed by the microflora initially associated with the raw material (Fig. 1). This activity does not affect the quality of the lignocellulose. Rather, it allows of the growth of the desired basidiomycetes under conditions that precludes the growth of competitors (mainly moulds). The cultivation of primary-rot fungi for the production of food or feed is a

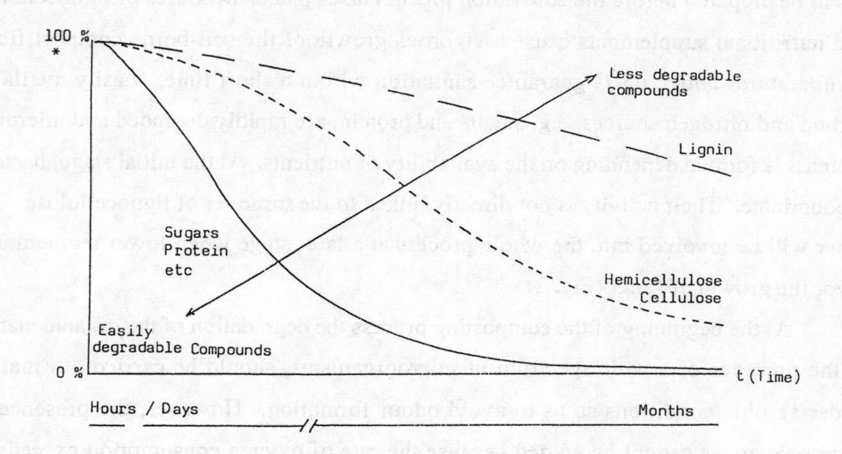

Fig. 1. Schematized rates of degradation of low- and high-molecular weight residues. * The real amounts of the individual components are set at 100%.

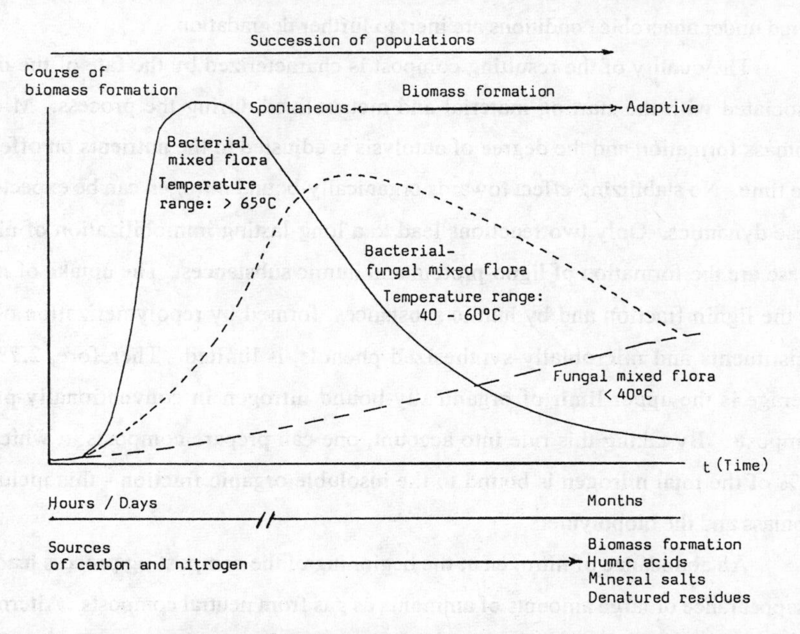

Fig. 2. The composition of the living biomass involved in the degradation of organic matter during composting over the incubation period.

preliminary step in naturally-occurring degradation and is part of an ecologically-orientated recycling process. Secondary-rot fungi prefer composted organic matter. This has to be prepared before the cultivation process takes place. Mixtures of lignocellulose and nutritional supplements cause a vigorous growth of the soil-borne compost flora. Temperatures above 65°C guarantee sanitation within a short time. Easily available carbon and nitrogen sources, e.g. sugars and protein, are rapidly degraded and microbial biomass is formed depending on the availability of nutrients. At the initial stage, bacteria predominate. Their activity is not directly linked to the turnover of lignocellulose. The latter will be involved into the whole process at a later stage when lower temperatures favor the growth of fungi (Fig. 2).

At the beginning of the composting process the degradation of the organic matter, in the presence of a wide spectrum of microorganisms, should be carried out mainly under aerobic conditions so as to avoid odour formation. However, the presence of anaerobic zones cannot be aoided because the rate of oxygen consumption exceeds its supply by convection. Aerobic conditions are strictly required in the latter stages for the growth of those fungi that effect mineralization of the polymers. Otherwise, composts stored under anaerobic conditions are inert to further degradation.

The quality of the resulting compost is characterized by the fate of the nitrogen associated with the starting material and metabolized during the process. Microbial biomass formation and the degree of autolysis is adjusted to the nutrients on offer at any one time. No stabilizing effect towards organically-bound nitrogen can be expected from these dynamics. Only two reactions lead to a long-lasting immobilization of nitrogen. These are the formation of ligno-protein and humic substances. The uptake of nitrogen by the lignin fraction and by humic substances, formed by repolymerization of lignin constituents and microbially-synthesized phenols, is limited. Therefore, 2.7% N on average is the upper limit of organically-bound nitrogen in conventionally-prepared composts. By taking this rule into account, one can prepare composts in which up to 90% of the total nitrogen is bound to the insoluble organic fraction - this includes the biomass and the biopolymers.

An abundance of nitrogen at the beginning of the composting process leads to the disappearance of large amounts of ammonia as gas from neutral composts. Alternatively, the ammonia may be oxidized to nitrite and nitrate and eventually to dinitrogen oxide (N_2O) or elementary nitrogen (N_2) under anaerobic conditions in the middle of the heap.

The cultivation of compost-colonizing edible fungi requires a substrate free of ammonia. However, a nutritious substrate is desirable if one aims to obtain better yields. Therefore, the composting process has been carefully analyzed in the past (Hunte and Grabbe, 1989). The results may be of common interest for composting processes in order to integrate biogenic matter into natural cycles. The incorporation of fresh organic matter into soil leads to an uncontrolled turnover of nutrients at any time. Thus, pretreatment of lignocellulosics on the basis of established biological principles and supported by suitable technical equipment would offer new concepts for environmental protection, both ecologically and economically.

The utilization of lignocellulosics for commercial purposes, as described for the production of food and feed by basidiomycetes, does not lead to the total removal of the substrate. Approximately 50% of the initial organic matter will remain more or less undegraded. The so-called spent composts are well-defined with respect to their behaviour on being returned to soils. Residues from the cultivation of *Pleurotus* spp. or *Lentinus edodes* (Shiitake) are poor in nitrogen. Their degradation will be accompanied by an uptake of nitrogen from the surrounding soil and the consequent nitrogen deficiency will depress plant growth. Residues from the cultivation of *Agaricus bisporus* are rich in nitrogen and highly-stabilized by a close C:N ratio.

Table 2. Incorporation of nitrogen into lignin and humic acids during composting of lignocellulose-containing residues.

Material	Ash %	N %	Lignin %	N %	Humic acid %	N %	C:N ratio	Immobilized N (% of N_t)
SMC*	40	1.5	20	2.6	11	3.8	15:1	>60%
MSSC*	45	1.1	19	2.5	5	4.8	20:1	>60%

* SMC, spent mushroom compost; MSSC, municipal sewage sludge compost.

RESULTS

It could be demonstrated that the microbial treatment of lignocellulosics containing organic matter generally yields materials comparable to spent compost from the mushroom industry. Therefore, composting processes should be carried out under conditions that adhere to sound biological principles.

Composting of spent mushroom compost yields humus in the which the lignin and humic acid components contain more than 60% of the total nitrogen (Table 2). A similar result was achieved in the case of municipal sewage sludge compost.

The utilization of bark instead of peat makes it necessary to take fully-rotted from the bark disposal or to stabilize shredded bark by incubating it with 1 kg of urea.cm^{-3}. Within 3 months a comparable status of stabilization may be achieved (Table 3). Although the C:N ratios varied, the composts were ready for use when tested by the method described by Zöttl (1981). This test is based on the microbial conversion of ammonium to nitrate without incorporation into microbial biomass. This would indicate a competitive uptake of nutrients by microorganisms during plant growth. By comparison with the data obtained with fresh bark, a remarkable incorporation into the high molecular weight fractions took place.

The stability, as a defined property of bark compost, could not be proven by any chemical analysis that destroyed the particle size. Moreover, parameters such as the total nitrogen content, the formation of humic acids or the conversion of lignin, are not suitable for describing the quality of the compost (Meinken, 1985). But, if composts are prepared without high contents of mobile nitrogen - mainly ammonia - they can be considered as a source of humus with a calculable release of nutrients (Fig. 3).

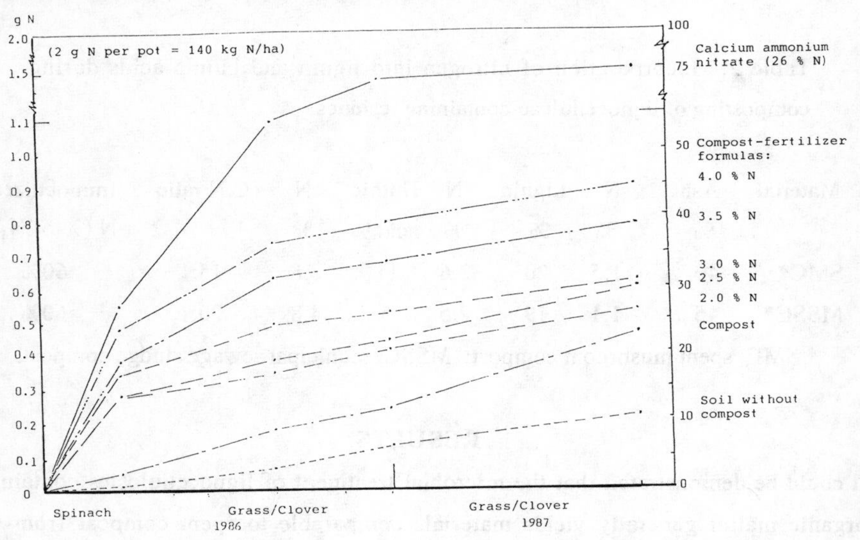

Fig. 3. Experiments with mixtures of defined compost and fertilizers. Withdrawal of nitrogen by plants over a period of 2 years.

Table 3. C:N ratios of bark and the distribution of nitrogen in lignin and humic acids.

Material	Stability	% Total nitrogen	C:N	% Immobilized nitrogen of total nitrogen in	
				Humic acids	Lignin
Fresh spruce bark	Unstable	0.65	78:1	5.9	38
Spruce bark compost (+N) (3 months)	Stable	1.36	29:1	19.0	57
Spruce bark compost Pine bark compost (disposal 1 year)	Stable	1.20	33:1	16.0	76
Oak bark compost (0.5 year)	"Stable"	1.20	34:1	9.6	71
Oak bark compost (1 year)	Stable	1.30	25:1	10.0	57
Oak bark compost (disposal)	Stable	2.09	17:01	20.0	55

Well defined composts were upgraded with mineral and organic fertilizers. The influence on plant growth had been tested with spinach and grass/clover in Mitcherlich pots (Fig. 3). A continuous release of nitrogen could be observed in the case of the pure compost. After 2 years, 25% of the total nitrogen had been withdrawn by the plants. The combination of compost plus fertilizer supported mainly the growth of the spinach, but also that of the vegetation in the following year, on a higher level than that afforded by compost alone. No compost fertilizer combination (forming typical Mitcherlich curves) released more than 50% nitrogen.

DISCUSSION

On the basis of current knowledge, composting under defined conditions may be considered as an upgrading of lignocellulosics suitable for further treatments with the ecological viewpoint in mind. Some highly-industrialized countries are looking for ways to reduce wastes and so to minimize environmental pollution. The consumers in some such countries are used to separating domestic wastes into those relevant to different recycling processes. Throughout such countries, biogenic organic matter is collected and transported to compost plants. However, a simple calculation points out the problems arising from these efforts. On average, 150 kg of degradable organic matter are collected per head per year . Based upon the population in Germany, 10 million tonnes of fresh organic matter could be converted to approximately 5 million tonnes of compost or 8 million cubic metres, respectively. The operation of existing and planned compost plants, for treating bark and residues from agricultural production processes, will lead to the accumulation of a huge amount of compost in the near future. The question will then be how to manage the abundance of compost.

There are 3 possible outlets, each with a specific goal in mind, for the utilization of compost (Fig. 4):

(i) Soil amelioration: The application of stable humus to soil increases the fertility by improving the availability of nutrients, the water-holding capacity and the soil aeration. Furthermore, the stress in soils caused by tillage and environmental pollutants is significantly reduced.

(ii) Upgrading of stable humus with nutrients: The loading of stable humus with mineral fertilizers allows of the use of combinations that release nutrients at a defined rate. Such combinations might allow farmers to include compost applications into modern

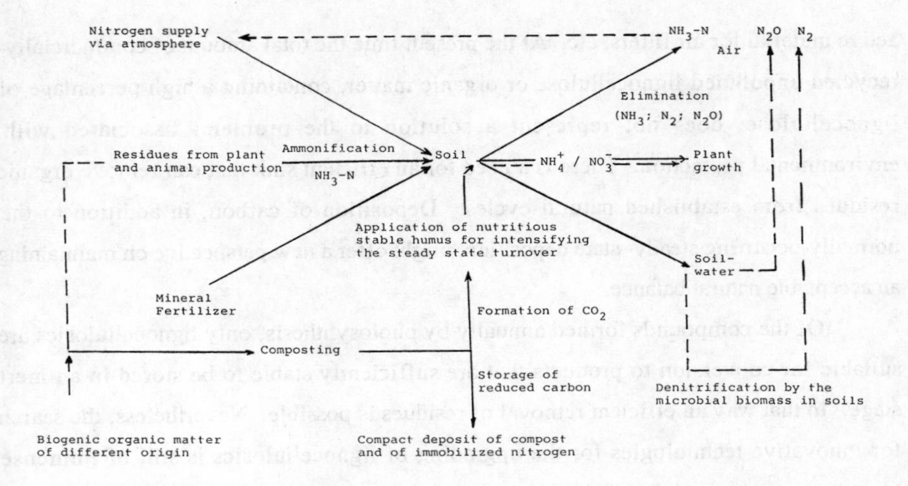

Fig. 4. Schematized concept for the utilization of compost in the context of environmental protection.

agricultural production systems. The consequent reduction in the utilization of mineral fertilizer would be beneficial from an environmental viewpoint.

(iii) Conversion of lignocellulosics to an inert carbon depot: The accumulation of CO_2 in the atmosphere cannot be avoided by technical or chemical means. The removal of CO_2 can only be achieved by keeping carbon in its reduced form. Fresh organic matter consists of constituents that are degradable under anaerobic as well as aerobic conditions. Disposal at the landfill site will cause an unacceptable environmental pollution by the uncontrolled production of nitrite, nitrate and reduced nitrogen forms, soluble organic metabolites, methane and foul-smelling, incompletely-oxidized substances. The conversion of lignocellulosics to stable humus guarantees an inert fraction when being compacted and disposed of.

CONCLUSION

Upgrading of lignocellulosics is a challenge for scientist of various disciplines. Development of a convincing process for handling the huge amounts of residues and wastes that accumulate is essential. The utilization of lignocellulose as a raw material in biotechnology has been investigated and has often been realized commercially: examples include, mushroom cultivation, products for agriculture and horticulture, biologically-active material for air filters, etc. At the present time the total amount of commercially-recycled unpolluted lignocellulose or organic matter, containing a high percentage of lignocellulose, does not represent a solution to the problems associated with environmental protection. There is a need for an efficient sink that can remove organic residues from established natural cycles. Deposition of carbon, in addition to the normally-occurring steady- state deposition, might offer a new perspective on maintaining an acceptable natural balance.

Of the compounds formed annually by photosynthesis, only lignocellulosics are suitable for conversion to products that are sufficiently stable to be stored in an inert stage. In that way an efficient removal of residues is possible. Nevertheless, the search for innovative technologies for the upgrading of lignocellulosics is still of immense importance.

REFERENCES

Chang, S.T. (1987). World production of cultivated mushrooms in 1986. Mushr. f. Tropics, 7, 117-120.

Hunte, W. and Grabbe, K. (1989). *Champignonanbau.* Paul Parey Verlag, Berlin.

Kamra, D.N. and Zadrazil, F. (1988). Microbiological improvement of lignocellulosics in animal feed production. A review. In *Treatment of Lignocellulosics with White-rot Fungi,* F. Zadrazil and P. Reiniger, eds., Elsevier Applied Science, London, pp. 56-63.

Meinken, E. (1985). Verfügbarkeit von Pflanzennährstoffen in Kultursubstraten aus Baumrinde. Dissertation Fachbereich Gartenbau, Universität Hannover, FRG.

Zöttl, H.W. (1981). Bestimmung und Beseitigung der Stickstoffimmobilisierung in Rindehumus. Taspo - SH 1. Rindenprodukte für den Gartenbau, 7-11.

CHAIRMAN'S REPORT ON SESSION V

Use of white-rot fungi for food, feed and industrial processes

White-rot and related fungi are already being used with success in commercial processes to upgrade lignocellulosic wastes, viz. solid-substrate fermentation methods producing edible fungi for human consumption.

The combination of the use of low- or negative-value lignocellulosic wastes, cost-effective microbial technology and application of the product, mushrooms, as a human foodstuff ensures this process can be operated economically. A wide range of lignocellulosic wastes can be treated either by composting for *Agaricus* production or, in some cases directly, to produce substrates for these fungi. Laboratory assays for predicting the suitability of such wastes as ingredients for mushroom cultivation have been developed. Such technology has considerable scope in lignocellulosic waste management in all areas of the EC. Mushroom cultivation is an on-farm process and wastes can be treated directly without significant transport costs. Further progress in this area will come from substrate manipulation and strain improvement aimed at increasing the bioconversion of the lignocellulosic polymers.

Other areas of the potential commercial use of white-rot fungi in lignocellulosic treatment are dependent on satisfactory process costs and the price of the endproduct.

The future use of white-rot fungi in the pulp and paper industry for biopulping and biobleaching is linked to the costs of alternative and currently more reliable chemical processes and the extent to which EC legislation will require the application of more environmentally-sensitive technology to pulp and paper production. The current biological processes are not yet competitive with chemical processes for large-scale pulping and bleaching operations. The pulp and paper industry produces large volumes of lignocellulosic wastes and it would be prudent to continue research programmes aimed at the development of satisfactory and reliable biological processes.

The ability of white-rot fungi to detoxify recalcitrant aromatic chemical molecules is now being examined. For example, contaminated soils have been treated with solid-substrate grown cultures of white-rot fungi and the rates of degradation of various pollutants were monitored. Considerable decontamination of such soils has been achieved and this technology shows promise for reclaiming heavily-polluted soils. The economics of the process remain to be studied as do the environmental factors affecting rate and quantity of decontamination. Also reported was related work aimed at using solid-substrate cultures to decontaminate various volatile pollutants. Once again these efforts met with considerable success but examination of the economic criteria have yet to be carried out.

Consideration of the quantity of waste lignocellulosic residues produced within the EC indicates that current agricultural and industrial technologies are incapable of processing these materials at a rate sufficient to ensure their efficient disposal. Additionally, recycling of agricultural surpluses by biological treatment or burning contributes to global CO_2 concentration. One route to avoid increasing the CO_2 load would be to remove stabilized wastes, after composting pretreatment, to sites where only low biological activity could occur. Such sequestration of carbon would diminish the CO_2 overloading problem. This particular strategy for utilization and removal of biologically-stabilized lignocellulosic wastes requires investigation at the scientific, technical and economic levels.

David A. Wood

PARTICIPANTS

Amaral Collaço, M.T.
DTIA-LNETI
Rua Vale Formoso 1
1900 Lisboa, Portugal
FAX: 351-1-8582312
TX: 42486 LNETI P
TEL: 8 58 43 01

Ander, P.
STFI, Box 5604
S-114 86 Stockholm
Sweden
FAX: 46-8-108340
TX: 10880 WOODRES S
TEL: 08-22 43 40

Avelino, H.T.
DTIA-LNETI
Rua Vale Formoso 1
1900 Lisboa
Portugal
FAX: 351-1-8582312
TX: 42486 LNETI P
TEL: 8 58 43 01

Bento, H.
DTIA-LNETI
Rua Vale Formoso 1
1900 Lisboa, Portugal
FAX: 351-1-8582312
TX: 42486 LNETI P
TEL: 8 58 43 01

Caetano, M.
INDAL
Rua de Santo António, 68-3° Esq°.
8001 Faro - Codex, Portugal
TX: 56582
TEL: 089-23208

Coughlan, M.P.
Dept. Biochemistry
University College
Galway, Ireland
FAX: 353-91-25700
TX: 28823 EI
TEL: 353-91-24411

Fernandez, E.
IFI, Ap. 56
28500 Arganda del Rey
Madrid, España
FAX: 34-1-4113077
TX: CSIC E
TEL: 8713328

Fonty, G.
Lab. de Microbiologie
Centre de Recherches de
Clermont-Ferrand-Theix
63122 St. Genès Champagne,
France
FAX: 00-33-73624450
TX: INRATEX 990 227 F
TEL: 33-73624000

France, J.
IGAP, Hurley
Maidenhead, Surrey, UK
FAX: 44-062-882-3630
TX: ARS002 265451 MONREF G
TEL: 0628-82-3630

Galletti, G.C.
CSCF,
Univ. di Bologna
Via Filippo Re, 8
I-40126 Bologna, Italia
FAX: 39-51-242136
TX: 511650 UNIVBO I
TEL: 051-24 44 45

Garrido, D.
IFI, Ap. 56
28500 Arganda del Rey
Madrid, España
FAX: 34-1-4113077
TX: 42182 CSIC E
TEL: 8713328

Gerrits, J.P.G.
Mushroom Exptl. Stn.
P.O. Box 6042
5960 AA Horst
The Netherlands
FAX: 31-4764-1567
TEL: 04764-1944

Giovannozzi-Sermanni, G.
Univ. Degli Studi della Tuscia
Dipt. Agrobiol. Agrochim.
Via S. Canullo de Lellis
01100 Viterbo, Italy
FAX: 39-761-35 25 41
TEL: 0761-25 03 15

Grabbe, K.
Inst. für Bodenbiologie
FAL
D-3300 Braunschweig
FRG
FAX: 0531-596814
TEL: 0531-596-1

Hovell, D.
Rowett Res. Inst.
Bucksburn
Aberdeen AB2 9SB
Scotland
FAX: 44-224-715349
TX: 739988 ROWETT G
TEL: 00-44-224-712751

Hüttermann, A.
Forstbotanisches Inst.
Univ. Göttingen
Büsgenweg 2
D-3400 Göttingen, FRG
FAX: 49-551/399629
TX: 96703 UNIGOE D
TEL: 0551-39 34 82

Jansen, H.
Tech. Handels-en Adviesbureau
Napoleonsweg 82
6086 AH Neer
Postbus 7226
5995 ZG Zessel
The Netherlands
FAX: 31-4759-4436
TX: 58087 JAKES NL
TEL: 4759-4242

Marques, M. J.
IDAL
2130 Benavente
Portugal
TX: 13209
TEL: 063-54473/54424

Martinez, A.T.
CIB CSIC
Velasquez 144
28006 Madrid, España
FAX: 34-1-2627518
TX: 42182 CSIC E
TEL: 1-2611800

Mascarenhas, A.F.
Inst. Univ. de Trás-oa-Montes
e Alto Douro
5000 Vila Real, Portugal
FAX: 059-23688
TX: 24436 UTAD

Medeiros, J.
INOVA
Ponta Delgada
9500 Açores, Portugal
FAX: 096-22481
TEL: 096-22481

Mulder, M.
Inst. Atomic Molec. Phys.
Kevislan 407
1098 S.J. Amsterdam
The Netherlands
FAX: 020-6684106
TEL: 020-946711

Ørskov, E.R.
Rowett Res. Inst.
Bucksburn
Aberdeen AB2 9SB
Scotland
FAX: 44-224-715349
TX: 739988 ROWETT G
TEL: 44-224-712751

Peito, A.
DTIA-LNETI
Rua Vale Formoso 1
1900 Lisboa, Portugal
FAX: 351-1-382312
TX: 42486 LNETI P
TEL: 8 58 43 01

Pereira, M.
ESTG
Inst. Politécnico de Faro
Quinta da Penha
8000 Faro, Portugal
TX: 56168 IP FARO P
TEL: 089-29357/29367

Puls, J.
Inst. für Holzchemie
Leuschnerstraße 91
D-2050 Hamburg
FRG
TEL: 040 73962-1
FAX: 040 73962-480

Reiniger, P.
CEC
Rue de la Loi 200
B-1049 Brussels
Belgium
FAX: 23-63024
TX: COMEU B 21877
TEL: 23-59586

Reeves, J.
USDA-ARS
US Dept. Agriculture
Bldg. 200
Beltsville, MD, USA
FAX: 301-344-1553
TEL: 301-344-2294

Rodeia, N.
Fac. Ciencias de Lisboa
Dept. Biol. Vegetal
Rua Ernesto Vasconcelos
Bloco C2, Campo Grande
1700 Lisboa, Portugal

Ruel, K.
Centre de Recherche sur
les Macromol. Végétales
Univ. de Grenoble, B.P. 53 X
38041 Grenoble Cedex
France
FAX: 33-76 54 22 03
TX: CETEPAP 980642
TEL: 76 54 11 45

Severo, A.
DTIA-LNETI
Rua Vale Formoso 1
1900 Lisboa, Portugal
FAX: 351-1-382312
TX: 42486 LNETI P
TEL: 8 58 43 01

Smith, J.F.
AFRC
Inst. Horticultural Res.
Worthing Rd., Littlehampton
West Sussex BN17 6LP
UK
FAX: 44-903-726780
TEL: 0903-716123

TEIFKE, J.
Inst. Verfahr. Kerntech.
Langer Kamp 7
3300 Braunschweig, FRG
FAX: 05-31-391-4577
TX: 952526 TUSW D
TEL: 9531-391-2781

Thonart, P.
Fac. Sci. Agron. de l'Etat
Passage des Déportés 2
B-5800 Gembloux
Belgique
FAX: 32-081-615965
TX: FSAGX 59482
TEL: 081-622309

Tuohy, M.G.
Dept. Biochemistry
University College
Galway, Ireland
FAX: 353-91-25700
TX: 28823 EI
TEL: 353-91-24411

Vasconcelos, L.
Central de Cerveças
Av. Almirante Reis 115
1197 Lisboa Codex
Portugal
TX: SLX-P 13749
TEL: 53 68 41/53 50 71

Wood, D.A.
AFRC
Inst. Horticultural Res.
Worthing Rd., Littlehampton
West Sussex BN17 6LP
UK
FAX: 44-903-726780
TEL: 0903-716123

Zadrazil, F.
Inst. für Bodenbiologie
FAL
Bundesallee 50
D-3300 Braunschweig
FRG
FAX: 0531-596814
TEL: 0531-596-1

INDEX